Lecture Notes in Computer Science 12741

Fabio Gadducci · Timo Kehrer (Eds.)

Graph Transformation

14th International Conference, ICGT 2021
Held as Part of STAF 2021
Virtual Event, June 24–25, 2021
Proceedings

 Springer

Editors
Fabio Gadducci ⓘ
Università di Pisa
Pisa, Italy

Timo Kehrer ⓘ
Humboldt-Universität zu Berlin
Berlin, Germany

ISSN 0302-9743 ISSN 1611-3349 (electronic)
Lecture Notes in Computer Science
ISBN 978-3-030-78945-9 ISBN 978-3-030-78946-6 (eBook)
https://doi.org/10.1007/978-3-030-78946-6

LNCS Sublibrary: SL1 – Theoretical Computer Science and General Issues

This Springer imprint is published by the registered company Springer Nature Switzerland AG
The registered company address is: Gewerbestrasse 11, 6330 Cham, Switzerland

Preface

This volume contains the proceedings of ICGT 2021, the 14th International Conference on Graph Transformation held during June 24–25, 2021. Due to the pandemic situation leading to COVID-19 countermeasures and travel restrictions, for the second time in a row the conference was held online. ICGT 2021 was affiliated with STAF (Software Technologies: Applications and Foundations), a federation of leading conferences on software technologies, and it took place under the auspices of the European Association of Theoretical Computer Science (EATCS), the European Association of Software Science and Technology (EASST), and the IFIP Working Group 1.3 on Foundations of Systems Specification.

The ICGT series aims at fostering exchange and the collaboration of researchers from different backgrounds working with graphs and graph transformation, either by contributing to their theoretical foundations or by highlighting their relevance in different application domains. Indeed, the use of graphs and graph-like structures as a formalism for specification and modeling is widespread in all areas of computer science as well as in many fields of computational research and engineering. Relevant examples include software architectures, pointer structures, state space graphs, control/data flow graphs, UML and other domain-specific models, network layouts, topologies of cyber-physical environments, and molecular structures. Often, these graphs undergo dynamic change, ranging from reconfiguration and evolution to various kinds of behavior, which can be captured by rule-based graph manipulation. Thus, graphs and graph transformation form a fundamental universal modeling paradigm that serves as a means for formal reasoning and analysis, ranging from the verification of certain properties of interest to the discovery of new computational insights.

ICGT 2021 continued the series of conferences previously held in Barcelona (Spain) in 2002, Rome (Italy) in 2004, Natal (Brazil) in 2006, Leicester (UK) in 2008, Enschede (the Netherlands) in 2010, Bremen (Germany) in 2012, York (UK) in 2014, L'Aquila (Italy) in 2015, Vienna (Austria) in 2016, Marburg (Germany) in 2017, Toulouse (France) in 2018, Eindhoven (the Netherlands) in 2019 and online in 2020, following a series of six International Workshops on Graph Grammars and Their Application to Computer Science from 1978 to 1998 in Europe and in the USA.

This year, the conference solicited research papers describing original contributions in the theory and applications of graph transformation as well as tool presentation papers that demonstrate new features and functionalities of graph-based tools. The Program Committee selected 16 out of 26 submissions for inclusion in the conference program. All submissions went through a thorough peer-review process and were discussed online. There was no preset number of papers to accept, and each paper was evaluated and assessed based on its own strengths and weaknesses. The topics of the accepted papers cover a wide spectrum, from theoretical approaches to graph transformation to their application in specific domains.

The papers presented new results on the DPO/SPO dichotomy and their rule application conditions, and introduced novel rewriting formalisms. Furthermore, model checking issues were explored along with the use of graph transformation in application domains such as chemical reaction modeling or AI-supported computer games. In addition to the submitted papers and tool presentations, the conference program included an invited talk, given by Joost-Pieter Katoen (RWTH Aachen University, Germany), on the reliability and criticality analysis of dynamic reliability models, including various fault tree dialects. In particular, Katoen discussed how to simplify such fault trees prior to their expensive analysis by using graph transformation.

We would like to thank all the people who contributed to the success of ICGT 2021, the invited speaker Joost-Pieter Katoen, the authors of the submitted papers, the members of the Program Committee, and all the reviewers for their valuable contributions to the selection process. We are grateful to Reiko Heckel, the chair of the Steering Committee of ICGT, for his fruitful suggestions, and to Adrian Rutle, the STAF 2021 general chair, for the organisation and the close collaboration during the difficult pandemic situation.

May 2021   Fabio Gadducci
                                                            Timo Kehrer

Organization

Steering Committee

Paolo Bottoni	Sapienza University of Rome, Italy
Andrea Corradini	University of Pisa, Italy
Gregor Engels	University of Paderborn, Germany
Holger Giese	Hasso Plattner Institute, University of Potsdam, Germany
Reiko Heckel (Chair)	University of Leicester, UK
Dirk Janssens	University of Antwerp, Belgium
Barbara König	University of Duisburg-Essen, Germany
Hans-Jörg Kreowski	University of Bremen, Germany
Leen Lambers	Hasso Plattner Institute, University of Potsdam, Germany
Ugo Montanari	University of Pisa, Italy
Mohamed Mosbah	University of Bordeaux, France
Manfred Nagl	RWTH Aachen University, Germany
Fernando Orejas	Polytechnic University of Catalonia, Spain
Francesco Parisi-Presicce	Sapienza University of Rome, Italy
John Pfaltz	University of Virginia, USA
Detlef Plump	University of York, UK
Arend Rensink	University of Twente, the Netherlands
Leila Ribeiro	Federal University of Rio Grande do Sul, Brazil
Grzegorz Rozenberg	University of Leiden, the Netherlands
Andy Schürr	Technical University of Darmstadt, Germany
Gabriele Taentzer	Philipps University of Marburg, Germany
Jens Weber	University of Victoria, Canada
Bernhard Westfechtel	University of Bayreuth, Germany

Program Committee

Paolo Baldan	University of Padua, Italy
Paolo Bottoni	Sapienza University of Rome, Italy
Andrea Corradini	University of Pisa, Italy
Juan De Lara	Autonomous University of Madrid, Spain
Juergen Dingel	Queen's University, Canada
Reiko Heckel	University of Leicester, UK
Thomas Hildebrandt	University of Copenhagen, Denmark
Wolfram Kahl	McMaster University, Canada
Aleks Kissinger	University of Oxford, UK
Jean Krivine	CNRS, France
Barbara König	University of Duisburg-Essen, Germany

Leen Lambers	Hasso Plattner Institute, University of Potsdam, Germany
Yngve Lamo	Western Norway University of Applied Sciences, Norway
Koko Muroya	Kyoto University, Japan
Fernando Orejas	Polytechnic University of Catalonia, Spain
Detlef Plump	University of York, UK
Arend Rensink	University of Twente, the Netherlands
Leila Ribeiro	Federal University of Rio Grande do Sul, Brazil
Andy Schürr	Technical University of Darmstadt, Germany
Gabriele Taentzer	Philipps University of Marburg, Germany
Matthias Tichy	University of Ulm, Germany
Uwe Wolter	University of Bergen, Norway
Steffen Zschaler	King's College London, UK

Additional Reviewers

Campbell, Graham	Jalali, Arash
Courtehoute, Brian	Kosiol, Jens
Ehmes, Sebastian	Minas, Mark
Fritsche, Lars	Schneider, Sven
Heindel, Tobias	Zambon, Eduardo

Verification Conquers Reliability Engineering
(Abstract of Invited Talk)

Joost-Pieter Katoen[1,2]

[1] RWTH Aachen University, Germany
[2] University of Twente, the Netherlands

Reliability engineering is "a sub-discipline of systems engineering that emphasizes the ability of equipment to function without failure"[1]. Prominent objectives in reliability engineering include assessing the likely reliability of complex system designs—reliability analysis—and identifying critical system components that are major causes of system failures—so-called criticality analysis. These analyses have a enormous broad spectrum of applications. This includes automotive, aerospace engineering, avionics, electricity networks, nuclear power plants, and so forth.

Fault trees (in fact: directed acyclic graphs) are pivotal in reliability engineering. They have been introduced in 1962 by Watson and are still widely used. They model the different component failures that can occur in a system and prescribe how such failures can propagate through the system. Fault trees are widely applied at industrial scale, and have been subject to international standards in many application areas. Dedicated state-of-the-art techniques for classical static fault trees include e.g., symbolic analysis with binary decision diagrams.

Static fault trees have however restricted expressive power. Modern fault-tree dialects can model redundancies, functional dependencies, repairs, spare elements, activation mechanisms, failure restrictions, and so forth. This includes dynamic fault trees, state-event fault trees, component fault trees, and Boolean-driven Markov processes. For recent surveys see [6, 9]. The reliability and criticality analysis of these *dynamic* reliability models is a serious bottleneck: state-of-the-art analysis techniques either do not scale, or only provide statistical guarantees, or require manual effort.

We will show that various formal methods can effectively be used to:

a) give a formal semantics to fault-tree dialects using Petri nets [5, 7]
b) simplify fault trees prior to their expensive analysis using graph rewriting [4]
c) prove such rewriting correct with theorem proving [2]
d) analyse the simplified fault trees by probabilistic model checking [10], and
e) treat gigantic models by an iterative "generate partial state-space and verify" paradigm that provides sound bounds [8, 10].

We will treat the key algorithmic principles and showcase their usage on some industrial cases: the safety for autonomous vehicle guidance [3], the criticality of components in railway station areas [11], and the power supply of a nuclear power plant [1].

---

[1] https://en.wikipedia.org/wiki/Reliability_engineering.

Acknowledgements. This is based on joint work with Dennis Guck (TWT), Arend Rensink, Mariëlle Stoelinga and Matthias Volk (Univ. of Twente), Shahid Khan, Nils Nießen (RWTH Aachen), Norman Weik (DLR), Sebastian Junges (Univ. of Berkeley), Marc Bouissou (EDF), Majdi Ghadhab and Matthias Kuntz (BMW), Yasmeen Elderkalli and Osman Hasan (NUST), and Sofiène Tahar (Concordia Univ.).

References

1. Bouissou, M., Khan, S., Katoen, J.-P., Krcál, P.: Various ways to quantify BDMPS. In: MARS@ETAPS, EPTCS, vol. 316, pp. 1–14 (2020)
2. Elderhalli, Y., Volk, M., Hasan, O., Katoen, J.-P., Tahar, S.: Formal verification of rewriting rules for dynamic fault trees. In: Ölveczky, P., Salaün, G. (eds.) SEFM. LNCS, vol. 11724, pp. 513–531. Springer, Cham (2019). https://doi.org/10.1007/978-3-030-30446-1_27
3. Ghadhab, M., Junges, S., Katoen, J.-P., Kuntz, M., Volk, M.: Safety analysis for vehicle guidance systems with dynamic fault trees. Reliab. Eng. Syst. Saf. **186**, 37–50 (2019)
4. Junges, S., Guck, D., Katoen, J.-P., Rensink, A., Stoelinga, M.: Fault trees on a diet: automated reduction by graph rewriting. Formal Aspects Comput. **29**(4), 651–703 (2017)
5. Junges, S., Katoen, J.-P., Stoelinga, M., Volk, M.: One net fits all - a unifying semantics of dynamic fault trees using GSPNs. In:Petri Nets, LNCS, vol. 10877 , pp. 272–293. Springer, Cham (2018). https://doi.org/10.1007/978-3-319-91268-4_14
6. Kabir, S.: An overview of fault tree analysis and its application in model based dependability analysis. Expert Syst. Appl. **77**, 114–135 (2017)
7. Khan, S., Katoen, J.-P., Bouissou, M.: Explaining boolean-logic driven markov processes using GSPNs. In: EDCC, pp. 119–126. IEEE (2020)
8. Khan, S., Katoen, J.-P., Vold, M., Bouissou, M.: Scalable reliability analysis by lazy verification. In: Dutle, A., Moscato, M.M., Titolo, L., Muñoz, C.A., Perez, I. (eds.) NFM, LNCS. Springer, Cham (2021). https://doi.org/10.1007/978-3-030-76384-8_12
9. Ruijters, E., Stoelinga, M.: Fault tree analysis: a survey of the state-of-the-art in modeling, analysis and tools. Comput. Sci. Rev. .**15**, 29–62 (2015)
10. Volk, M., Junges, S., Katoen, J.-P.: Fast dynamic fault tree analysis by model checking techniques. IEEE Trans. Ind. Inform. **14**(1), 370–379 (2018)
11. Volk, M., Weik, N., Katoen, J.-P., Nießen, N.: DFT modeling approach for infrastructure reliability analysis of railway station areas. In:Larsen, K., Willemse, T. (eds.) FMICS, LNCS, vol. 11687, pp. 40–58. Springer, Cham (2019). https://doi.org/10.1007/978-3-030-27008-7_3

Contents

Tool Presentations

Theoretical Advances

Concurrency Theorems for Non-linear Rewriting Theories

Nicolas Behr[1]([⊠])[iD], Russ Harmer[2][iD], and Jean Krivine[1][iD]

[1] Université de Paris, CNRS, IRIF, 8 Place Aurélie Nemours,
75205 Paris Cedex 13, France
{nicolas.behr,jean.krivine}@irif.fr
[2] Université de Lyon, ENS de Lyon, UCBL, CNRS, LIP,
46 allée d'Italie, 69364 Lyon Cedex 07, France
russell.harmer@ens-lyon.fr

Abstract. Sesqui-pushout (SqPO) rewriting along non-linear rules and for monic matches is well-known to permit the modeling of fusing and cloning of vertices and edges, yet to date, no construction of a suitable concurrency theorem was available. The lack of such a theorem, in turn, rendered compositional reasoning for such rewriting systems largely infeasible. We develop in this paper a suitable concurrency theorem for non-linear SqPO-rewriting in categories that are quasi-topoi (subsuming the example of adhesive categories) and with matches required to be regular monomorphisms of the given category. Our construction reveals an interesting "backpropagation effect" in computing rule compositions. We derive in addition a concurrency theorem for non-linear double pushout (DPO) rewriting in rm-adhesive categories. Our results open non-linear SqPO and DPO semantics to the rich static analysis techniques available from concurrency, rule algebra and tracelet theory.

1   Introduction

Sesqui-pushout (SqPO) graph transformation was introduced [17] as an extension of single-pushout rewriting that accommodates the possibility of non-input-linear[1] rules. The result of such a rewrite is specified abstractly by the notion of *final pullback complement (FPC)* [21], a categorical generalization of the notion of set difference: the FPC of two composable arrows, $f : A \to B$ and $g : B \to D$ is the largest, *i.e.* least general, C together with arrows $g' : A \to C$ and $f' : C \to D$ for which the resulting square is a pullback (PB). The extension of graph transformation to input-non-linear rules allows for the expression of the natural operation of the *cloning* of a node, or an edge (when the latter is meaningful), as

---

[1] In this paper, we follow the conventions of compositional rewriting theory [9], i.e., we speak of "input"/"output" motifs of rules, as opposed to "left"/"right" motifs in the traditional literature [22].

An extended version of this paper containing additional technical appendices is available online [7].

© Springer Nature Switzerland AG 2021
F. Gadducci and T. Kehrer (Eds.): ICGT 2021, LNCS 12741, pp. 3–21, 2021.
https://doi.org/10.1007/978-3-030-78946-6_1

explained in [14,17,18]. More recently, such rules have also been used to express operations such as concept refinement in schemata for graph databases [11] and, more generally, in graph-based knowledge representation [29]. In combination with output-non-linear rules, as for (non-linear) double- or single-pushout rewriting, SqPO thus allows the expression of all the natural primitive operations on graphs: addition and deletion of nodes and edges; and cloning and merging of nodes and edges.

In this paper, we study the categorical structure required in order to support SqPO rewriting and establish that *quasi-topoi* [1,15,16,27,33] naturally possess all the necessary structure to express the effect of SqPO rewriting and to prove the concurrency theorem for fully general non-linear rules. This significantly generalizes previous results on concurrency theorems for linear SqPO-rewriting over adhesive categories [2] and for linear SqPO-rewriting for linear rules with conditions in $\mathcal{M}$-adhesive categories [8,9]. In terms of SqPO-rewriting for generic rules, previous results were rather sparse and include work on polymorphic SqPO-rewriting [36] and on reversible SqPO rewriting [19,30], where [30] in particular introduced a synthesis (but not an analysis) construction for reversible non-linear SqPO rules without application conditions which motivated the present paper.

An interesting technical aspect of basing our constructions on quasi-topoi concerns the rewriting of simple directed graphs, which constitutes one of the running examples in this paper: unlike the category of directed multigraphs (which constitutes one of the prototypical examples of an adhesive category [34]), the category of simple graphs is neither adhesive nor quasi-adhesive [33], but it is in fact only a quasi-topos [1,33], and as such also an example of an rm-quasi-adhesive [27] and of an $\mathcal{M}$-adhesive category [23–25,31].

Our proof of the concurrency theorem relies on the existence of certain structures in quasi-topoi that, to the best of our knowledge, have not been previously noted in the literature (cf. Sect. 2.2): restricted notions of *multi-sum* and *multi-pushout complement (mPOC)*, along the lines of the general theory of multi-(co-)limits due to Diers [20], and a notion of *FPC-pushout-augmentation (FPA)*. The notion of multi-sum provides a generalization of the property of effective unions (in adhesive categories) that guarantees that all necessary monos are regular. The notions of mPOC and FPA handle the "backward non-determinism" introduced by non-linear rules: given a rule and a matching from its output motif, we cannot—unlike with linear or reversible non-linear rules—uniquely determine a matching from the input motif of the rule.

Related Work. Conditions under which FPCs are guaranteed to exist have been studied in [21], and more concretely and of particular relevance to our approach in [18], which provides a direct construction assuming the existence of appropriate partial map classifiers [16,31]. We make additional use of these partial map classifiers in order to construct mPOCs in a quasi-topos (Sect. 2.2). Our construction is a mild, but necessary for our purposes, generalization of the notion of minimal pushout complement defined in [14] that requires the universal property with respect to a larger class of encompassing pushouts (POs)—precisely analogous to the definition of FPC. However, there is the additional complexity that, for our purposes, PO complements are not uniquely determined, and we must therefore specify a family of solutions that collectively satisfy this universal property (à la Diers [20]). We also exploit the epi-regular mono factorization [1] in quasi-topoi in order to construct multi-sums—with respect to co-spans of regular monos—and FPAs. Our overall approach relates closely to the work of Garner and Lack on rm-quasi-adhesive categories [27], which provide an abstract setting for graph transformation that accommodates the technical particularities of simple graphs—notably the fact that the 'exactness' direction of the van Kampen condition fails in general for cubes where the vertical arrows, between the two PO faces, are not regular.

2 Quasi-topoi

In this section, we will demonstrate that quasi-topoi form a natural setting within which non-linear sesqui-pushout (SqPO) rewriting is well-posed. Quasi-topoi have been considered in the context of rewriting theories as a natural generalization of adhesive categories in [35]. While several adhesive categories of interest to rewriting are topoi, including in particular the category **Graph** of directed multigraphs (cf. Definition 4), it is not difficult to find examples of categories equally relevant to rewriting theory that fail to be topoi. A notable such example is the category **SGraph** of directed simple graphs (cf. Definition 5).

We will demonstrate that quasi-topoi combine all technical properties necessary such as to admit the construction of non-linear sesqui-pushout semantics over them. We will first list these abstract properties, and illustrate them via the two aforementioned paradigmatic examples of topoi and quasi-topoi.

Let us first recall a number of results from the work of Cockett and Lack [15,16] on restriction categories. We will only need a very small fragment of their theory, namely the definition and existence guarantees for $\mathcal{M}$-partial map classifiers, so we will follow mostly [18]. We will in particular not be concerned with the notion of $\mathcal{M}$-partial maps itself.

Definition 1 ([15], **Sec. 3.1**). *For a category* **C**, *a stable system of monics* $\mathcal{M}$ *is a class of monomorphisms of* **C** *that (i) includes all isomorphisms, (ii) is stable under composition, and (iii) is stable under pullbacks (i.e., if* (f', m') *is a pullback of* (m, f) *with* $m \in \mathcal{M}$, *then* $m' \in \mathcal{M}$). *Throughout this paper, we will reserve the notation* $\rightarrowtail$ *for monics in* $\mathcal{M}$, *and* $\hookrightarrow$ *for generic monics.*

Definition 2 ([18], **Sec. 2.1; compare** [16], **Sec. 2.1**). *For a stable system of monics* $\mathcal{M}$ *in a category* **C**, *an* $\mathcal{M}$-*partial map classifier* (T, η) *is a functor* $T : \mathbf{C} \to \mathbf{C}$ *and a natural transformation* $\eta : ID_{\mathbf{C}} \dot{\to} T$ *such that*

1. *for all* $X \in \mathsf{obj}(\mathbf{C})$, $\eta_X : X \to T(X)$ *is in* $\mathcal{M}$
2. *for each span* $(A \xleftarrow{m} X \xrightarrow{f} B)$ *with* $m \in \mathcal{M}$, *there exists a unique morphism* $A \xrightarrow{\varphi(m,f)} T(B)$ *such that* (m, f) *is a pullback of* $(\varphi(m, f), \eta_B)$.

Proposition 1 ([18], **Prop. 6**). *For every $\mathcal{M}$-partial map classifier (T, η), T preserves pullbacks, and η is Cartesian, i.e., for each $X \xrightarrow{f} Y$, (η_x, f) is a pullback of $(T(f), \eta_Y)$.*

Definition 3 ([33], **Def. 9**). *A category $\mathbf{C}$ is a quasi-topos iff*

1. *it has finite limits and colimits*
2. *it is locally Cartesian closed*
3. *it has a regular-subobject-classifier.*

Based upon a variety of different results from the rich literature on quasi-topoi, we will now exhibit that quasi-topoi indeed possess all technical properties required in order for non-linear SqPO-rewriting to be well-posed:

Corollary 1. *Every quasi-topos $\mathbf{C}$ enjoys the following properties:*

- *It has (by definition) a stable system of monics $\mathcal{M} = \mathsf{rm}(\mathbf{C})$ (the class of regular monos), which coincides with the class of extremal monomorphisms [1, Cor. 28.6], i.e., if $m = f \circ e$ for $m \in \mathsf{rm}(\mathbf{C})$ and $e \in \mathsf{epi}(\mathbf{C})$, then $e \in \mathsf{iso}(\mathbf{C})$.*
- *It has (by definition) a $\mathcal{M}$-partial map classifier (T, η).*
- *It is rm-quasi-adhesive, i.e., it has pushouts along regular monomorphisms, these are stable under pullbacks, and pushouts along regular monos are pullbacks [27].*
- *It is $\mathcal{M}$-adhesive [31, Lem. 13].*
- *For all pairs of composable morphisms $A \xrightarrow{f} B$ and $B \xrightarrow{m} C$ with $m \in \mathcal{M}$, there exists a final pullback-complement (FPC) $A \xrightarrow{n} F \xrightarrow{g} C$, and with $n \in \mathcal{M}$ ([18, Thm. 1]; cf. Theorem 2).*
- *It possesses an epi-$\mathcal{M}$-factorization [1, Prob. 28.10]: each morphism $A \xrightarrow{f} B$ factors as $f = m \circ e$, with morphisms $A \xrightarrow{e} \bar{B}$ in $\mathsf{epi}(\mathbf{C})$ and $\bar{B} \xrightarrow{m} A$ in $\mathcal{M}$ (uniquely up to isomorphism in $\bar{B}$).*
- *It possesses a strict initial object $\varnothing \in \mathsf{obj}(\mathbf{C})$ [32, A1.4], i.e., for every object $X \in \mathsf{obj}(\mathbf{C})$, there exists a morphism $i_X : \varnothing \to X$, and if there exists a morphism $X \to \varnothing$, then $X \cong \varnothing$.*

If in addition the strict initial object $\varnothing$ is $\mathcal{M}$-initial, i.e., if for all objects $X \in \mathsf{obj}(\mathbf{C})$ the unique morphism $i_X : \varnothing \to X$ is in $\mathcal{M}$, then $\mathbf{C}$ has disjoint coproducts, i.e., for all $X, Y \in \mathsf{obj}(\mathbf{C})$, the pushout of the $\mathcal{M}$-span $X \hookleftarrow \varnothing \hookrightarrow Y$ is $X \rightarrowtail X + Y \hookleftarrow Y$ (cf. [37, Thm. 3.2], which also states that this condition is equivalent to requiring $\mathbf{C}$ to be a solid quasi-topos), and the coproduct injections are $\mathcal{M}$-morphisms as well. Finally, if pushouts along regular monos of $\mathbf{C}$ are van Kampen, $\mathbf{C}$ is a rm-adhesive category [27, Def. 1.1].

2.1   The Categories of Directed Multi- and Simple Graphs

Throughout this paper, we will illustrate our constructions with two prototypical examples of (quasi-)topoi, namely categories of two types of directed graphs.

Definition 4. *The* category **Graph** *of* directed multigraphs *is defined as the presheaf category* **Graph** $:= (\mathbb{G}^{op} \to \mathbf{Set})$, *where* $\mathbb{G} := (\cdot \rightrightarrows \star)$ *is a category with two objects and two morphisms [34]. Objects* $G = (V_G, E_G, s_G, t_G)$ *of* **Graph** *are given by a set of vertices* V_G, *a set of directed edges* E_G *and the source and target functions* $s_G, t_G : E_G \to V_G$. *Morphisms of* **Graph** *between* $G, H \in \mathrm{obj}(\mathbf{Graph})$ *are of the form* $\varphi = (\varphi_V, \varphi_E)$, *with* $\varphi_V : V_G \to V_H$ *and* $\varphi_E : E_G \to E_H$ *such that* $\varphi_V \circ s_G = s_H \circ \varphi_E$ *and* $\varphi_V \circ t_G = t_H \circ \varphi_E$.

Definition 5. *The* category **SGraph** *of* directed simple graphs[2] *is defined as the category of binary relations* **BRel** $\cong$ **Set** $/\!/ \Delta$ *[33]. Here,* $\Delta : \mathbf{Set} \to \mathbf{Set}$ *is the pullback-preserving diagonal functor defined via* $\Delta X := X \times X$, *and* **Set** $/\!/ \Delta$ *denotes the full subcategory of the slice category* **Set**$/\Delta$ *defined via restriction to objects* $m : X \to \Delta X$ *that are monomorphisms. More explicitly, an object of* **Set** $/\!/ \Delta$ *is given by* $S = (V, E, \iota)$, *where* V *is a set of vertices,* E *is a set of directed edges, and where* $\iota : E \to V \times V$ *is an injective function. A morphism* $f = (f_V, f_E)$ *between objects* S *and* S' *is a pair of functions* $f_V : V \to V'$ *and* $f_E : E \to E'$ *such that* $\iota' \circ f_E = (f_V \times f_V) \circ \iota$ *(see (2)).*

These two categories satisfy the following well-known properties:

Theorem 1. *The category* **Graph** *is an* adhesive category *and (by definition) a* presheaf topos *[34] (and thus in particular a quasi-topos), with strict-initial object* $\varnothing = (\emptyset, \emptyset, \emptyset \to \emptyset, \emptyset \to \emptyset)$ *the empty graph, and with the following additional properties:*

- *Morphisms are in the classes* $\mathsf{mono}(\mathbf{Graph})/\mathsf{epi}(\mathbf{Graph})/\mathsf{iso}(\mathbf{Graph})$ *if they are component-wise injective/surjective/bijective functions, respectively. All monos in* **Graph** *are regular, and* **Graph** *therefore possesses an epi-mono-factorization.*
- *For each* $G \in \mathrm{obj}(\mathbf{Graph})$ *[18, Sec. 2.1],* $\eta_G : G \to T(G)$ *is defined as the embedding of* G *into* $T(G)$, *where* $T(G)$ *is defined as the graph with vertex set* $V'_G := V_G \uplus \{\star\}$ *and edge set* $E_G \uplus E'_G$. *Here,* E'_G *contains one directed edge* $e_{n,p} : v_n \to v_p$ *for each pair of vertices* $(v_n, v_p) \in V'_G \times V'_G$.

The category **SGraph** *is not adhesive, but it is a quasi-topos [33], and with the following additional properties:*

- *In* **SGraph** *[33] (compare [14, Prop. 9]), morphisms* $f = (f_V, f_E)$ *are monic (epic) if* f_V *is monic (epic), while isomorphisms satisfy that both* f_V *and* f_E *are bijective. Regular monomorphisms in* **SGraph** *are those for which* (ι, f_E) *is a pullback of* $(\Delta(f_V), \iota')$ *[33, Lem. 14(ii)], i.e., a monomorphism is regular iff it is* edge-reflecting. *As is the case for any quasi-topos,* **SGraph** *possesses an epi-regular mono-factorization.*

---

[2] Some authors prefer to not consider directly the category **BRel**, but rather define **SGraph** as some category equivalent to **BRel**, where simple graphs are of the form $\langle V, E \rangle$ with $E \subseteq V \times V$. This is evidently equivalent to directly considering **BRel**, whence we chose to not make this distinction in this paper.

– *The regular mono-partial map classifier (T, η) of* **SGraph** *is defined as follows [1, Ex. 28.2(3)]: for every object $S = (V, E, \iota) \in \mathrm{obj}(\mathbf{SGraph})$,*

$$T(S) := (V_\star = V \uplus \{\star\}, E_\star = E \uplus (V \times \{\star\}) \uplus (\{\star\} \times V) \uplus \{(\star, \star)\}, \iota_\star), \quad (1)$$

where $\iota_\star$ is the evident inclusion map, and moreover $\eta_S : S \rightarrowtail T(S)$ is the (by definition edge-reflecting) inclusion of S into $T(S)$.
– **SGraph** *possesses a regular mono-initial object $\varnothing = (\emptyset, \emptyset, \emptyset \to \emptyset)$.*

Proof. While most of these results are standard, we briefly demonstrate that the epi-regular mono-factorization of **SGraph** [33] is "inherited" from the epi-mono-factorization of the adhesive category **Set**. To this end, given an arbitrary morphism $f = (f_V, f_E)$ in **SGraph** as on the left of (2), the epi-mono-factorization $f_V = m_V \circ e_V$ lifts via application of the diagonal functor Δ to a decomposition of the morphism $f_V \times f_V$. Pulling back $(\Delta(m_v), \iota')$ results in a span $(\tilde{\iota}, f''_E)$ and (by the *universal property of pullbacks*) an induced morphism f'_E that makes the diagram commute. By stability of monomorphisms under pullbacks, $\tilde{\iota}$ is a monomorphism, thus the square marked $(*)$ precisely constitutes the data of a regular monomorphism in **SGraph**, while the square marked $(\dagger)$ is an epimorphism in **SGraph** (since $e_V \in \mathrm{epi}(\mathbf{Set})$).

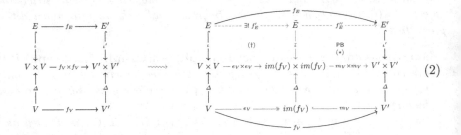

$$(2)$$

2.2   FPCs, $\mathcal{M}$-Multi-POCs, $\mathcal{M}$-Multi-sums and FPAs

Compared to compositional SqPO-type rewriting for $\mathcal{M}$-linear rules [2], in the generic SqPO-type setting we require both a generalization of the concept of pushout complements that forgoes uniqueness, as well as a certain form of FPC-augmentation. To this end, it will prove useful to recall from [18] the following constructive result:

Theorem 2 ([18], **Thm. 1**). *For a category* **C** *with $\mathcal{M}$-partial map classifier (T, η), the final pullback complement (FPC) of a composable sequence of arrows $A \xrightarrow{f} B$ and $B \xrightarrow{m} C$ with $m \in \mathcal{M}$ is guaranteed to exist, and is constructed via the following algorithm:*

1. *Let $\bar{m} := \varphi(m, id_B)$ (i.e., the morphism that exists by the universal property of (T, η), cf. square (1) below).*

2. *Construct* $T(A) \xleftarrow{\tilde{n}} F \xrightarrow{g} C$ *as the pullback of* $T(A) \xrightarrow{T(f)} T(B) \xleftarrow{\tilde{m}} C$ *(cf. square (2) below); by the universal property of pullbacks, this in addition entails the existence of a morphism* $A \xrightarrow{n} F$.

Then (n, g) *is the FPC of* (f, m), *and* n *is in* $\mathcal{M}$.

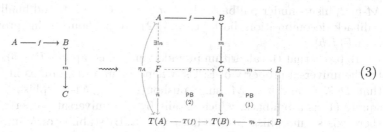

$$(3)$$

This guarantee for the existence of FPCs will prove quintessential for constructing $\mathcal{M}$-multi-pushout complements, which are defined as follows:

Definition 6. *For a category* **C** *with an* $\mathcal{M}$-*partial map classifier, the* $\mathcal{M}$-*multi-pushout complement (*$\mathcal{M}$-*multi-POC)* $\mathcal{P}(f, b)$ *of a composable sequence of morphisms* $A \xrightarrow{f} B$ *and* $B \xrightarrow{b} D$ *with* $b \in \mathcal{M}$ *is defined as*

$$\mathcal{P}(f, b) := \{(A \xrightarrow{a} P, P \xrightarrow{d} D) \in \mathrm{mor}(\mathbf{C})^2 \mid a \in \mathcal{M} \wedge (d, b) = \mathrm{PO}(a, f)\}. \quad (4)$$

Proposition 2. *In a quasi-topos* **C** *and for* $\mathcal{M} = \mathrm{rm}(\mathbf{C})$ *the class of regular monomorphisms, let* $\mathcal{P}(f, b)$ *be an* $\mathcal{M}$-*multi-POC.*

- **Universal property of** $\mathcal{P}(f, b)$: *for every diagram such as in* (5)(i) *where* (1) + (2) *is a pushout along an* $\mathcal{M}$-*morphism* n, *and where* $m = m' \circ b$ *for some* $m', b \in \mathcal{M}$, *there exists an element* (a, d) *of* $\mathcal{P}(f, b)$ *and an* $\mathcal{M}$-*morphism* $p \in \mathcal{M}$ *such that the diagram commutes and* (2) *is a pushout. Moreover, for any* $p' \in \mathcal{M}$ *and for any other element* (a', d') *of* $\mathcal{P}(f, b)$ *with the same property, there exists an isomorphism* $\delta \in \mathrm{iso}(\mathbf{C})$ *such that* $\delta \circ a = a'$ *and* $d' \circ \delta = d$.
- **Algorithm to compute** $\mathcal{P}(f, b)$:
  1. *Construct* (n, g) *in diagram* (5)(ii) *by taking the FPC of* (f, b).
  2. *For every pair of morphisms* (a, p) *such that* $a \in \mathcal{M}$ *and* $a \circ p = n$, *take the pushout* (1), *which by universal property of pushouts induces an arrow* $D \xrightarrow{e} C$; *if* $e \in \mathrm{iso}(\mathbf{C})$, (a, d) *is a contribution to the* $\mathcal{M}$-*multi-POC of* (f, b).

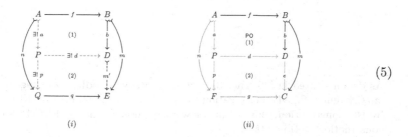

$$(5)$$

Proof. The universal property of $\mathcal{P}(f,b)$ follows from pushout-pullback decomposition: pushouts along $\mathcal{M}$-morphisms are pullbacks, so (1) + (2) is a pullback; taking the pullback (p,d) of (q,m') yields by the universal property of pullbacks a morphism a (which is unique up to isomorphism), and thus by pullback-pullback decomposition that (1) and (2) are pullbacks. By stability of $\mathcal{M}$-morphisms under pullbacks, both a and p are in $\mathcal{M}$, and finally by pushout-pullback decomposition, both (1) and (2) are pushouts. This proves that (a,d) is in $\mathcal{P}(f,b)$.

To prove that the algorithm provided indeed computes $\mathcal{P}(f,b)$, note first that by the universal property of FPCs, whenever in a diagram as in (5)(ii) we have that $D \cong C$ and $b \in \mathcal{M}$, since pushouts along $\mathcal{M}$-morphisms are pullbacks, square (1) is a pullback, which entails by the universal property of FPCs that there exists a morphism p such that $p \circ a = n$. By stability of $\mathcal{M}$-morphisms under pullbacks, we find that a must be in $\mathcal{M}$, so indeed every possible contribution to $\mathcal{P}(f,b)$ must give rise to a diagram as in (5)(ii), which proves the claim.

An example of an $\mathcal{M}$-multi-POC construction both in **SGraph** and in **Graph** is given in the diagram below. Note that in **Graph**, the $\mathcal{M}$-multi-POC does not contain the FPC contribution (since in **Graph** the pushout of the relevant span would yield to a graph with a multi-edge).

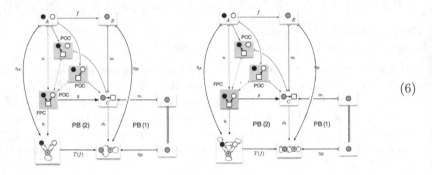

$$(6)$$

Definition 7 ($\mathcal{M}$-FPC-augmentations). *In a quasi-topos[3] $\mathbf{C}$ with $\mathcal{M} = \mathrm{rm}(\mathbf{C})$, consider a pushout square along an $\mathcal{M}$-morphism such as square (1) in the diagram below (where $\alpha, \bar{\alpha} \in \mathcal{M}$):*

$$(7)$$

---

[3] As demonstrated in [26, Fact 3.4], every finitary $\mathcal{M}$-adhesive category $\mathbf{C}$ possesses an (extremal $\mathcal{E}$, $\mathcal{M}$)-factorization, so if $\mathbf{C}$ is known to possess FPCs as required by the construction, this might allow to generalize the $\mathcal{M}$-FPC-PO-augmentation construction to this setting.

We define an $\mathcal{M}$-FPC augmentation (FPA) of the pushout square (1) as a diagram formed from an epimorphism $e \in \mathsf{epi}(\mathbf{C})$ and that satisfies the following properties:

- The morphism $e \circ \bar{\alpha}$ is an $\mathcal{M}$-morphism.
- $(\bar{\alpha}, id_B)$ is a pullback of $(e, e \circ \bar{\alpha})$.
- Square (1) + (2) is an FPC, and the induced morphism n that exists[4] by the universal property of FPCs, here w.r.t. the FPC $(n \circ \alpha, f)$ of $(a, e \circ \bar{\alpha})$, is an $\mathcal{M}$-morphism.

For a pushout as in (1), we denote by $\mathsf{FPA}(\alpha, a)$ its class of FPAs:

$$\mathsf{FPA}(\alpha, a) := \{(n, f, e) \mid e \in \mathsf{epi}(\mathbf{C}) \wedge e \circ \bar{\alpha}, n \in \mathcal{M} \wedge (f, n \circ \alpha) = FPC(a, e \circ \bar{\alpha})\} \quad (8)$$

As induced by the properties of pushouts and of FPCs, FPAs are defined up to universal isomorphisms (in D, E and F), and for a given pushout square there will in general exist multiple non-isomorphic such augmentations.

The final technical ingredient for our rewriting theoretic constructions is a notion of *multi-sum* adapted to the setting of quasi-topoi, a variation on the general theory of multi-(co-)limits due to Diers [20].

Definition 8. In a quasi-topos $\mathbf{C}$, the multi-sum $\sum_{\mathcal{M}}(A, B)$ of two objects $A, B \in \mathsf{obj}(\mathbf{C})$ is defined as a family of cospans of regular monomorphisms $A \xrightarrow{f} Y \xleftarrow{g} B$ with the following universal property: for every cospan $A \xrightarrow{a} Z \xleftarrow{b} B$ with $a, b \in \mathsf{rm}(\mathbf{C})$, there exists an element $A \xrightarrow{f} Y \xleftarrow{g} B$ in $\sum_{\mathcal{M}}(A, B)$ and a regular monomorphism $Y \xrightarrow{y} Z$ such that $a = y \circ f$ and $b = y \circ g$, and moreover (f, g) as well as y are unique up to universal isomorphisms.

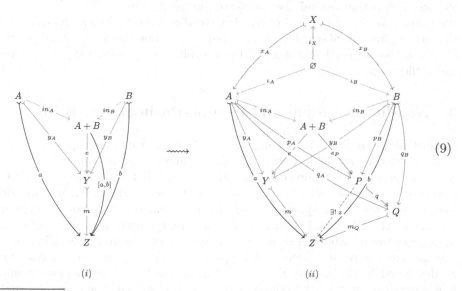

$$(9)$$

(i)   (ii)

---

[4] Note that square (1) pasted with the pullback square formed by the morphisms $\bar{\alpha}, id_B, e, e \circ \bar{\alpha}$ yields a pullback square that is indeed of the right form to warrant the existence of a morphism n into the FPC square (1) + (2).

Lemma 1. *If* **C** *is a quasi-topos, the multi-sum* $\sum_{\mathcal{M}}(A, B)$ *arises from the epi-*$\mathcal{M}$*-factorization of* **C** *(for* $\mathcal{M} = \mathsf{rm}(\mathbf{C})$*; compare [29]).*

- **Existence:** *Let* $A \xrightarrow{in_A} A + B \xleftarrow{in_B} B$ *be the disjoint union of* A *and* B. *Then for any cospan* $A \xrightarrow{a} Z \xleftarrow{b} B$ *with* $a, b \in \mathcal{M}$, *the epi-*$\mathcal{M}$*-factorization of the induced arrow* $A + B \xrightarrow{[a,b]} Z$ *into an epimorphism* $A + B \xrightarrow{e} Y$ *and an* $\mathcal{M}$*-morphism* $Y \xrightarrow{m} Z$ *yields a cospan* $(y_A = e \circ in_A, y_B = e \circ in_B)$, *which by the decomposition property of* $\mathcal{M}$*-morphisms is a cospan of* $\mathcal{M}$*-morphisms (cf. (9)(i)).*
- **Construction:** *For objects* $A, B \in \mathsf{obj}(\mathbf{C})$, *every element* $A \xrightarrow{q_A} Q \xleftarrow{q_B} B$ *in* $\sum_{\mathcal{M}}(A, B)$ *is obtained from a pushout of some span* $A \xleftarrow{x_A} X \xrightarrow{x_B} B$ *with* $x_A, x_B \in \mathcal{M}$ *and a morphism* $P \xrightarrow{q} Q$ *in* $\mathsf{mono}(\mathbf{C}) \cap \mathsf{epi}(\mathbf{C})$ *(cf. (9)(ii)).*

Proof. See [7, Appendix B].

Since in an adhesive category all monos are regular [34], in this case the multi-sum construction simplifies to the statement that every monic cospan can be uniquely factorized as a cospan obtained as the pushout of a monic span composed with a monomorphism. It is however worthwhile emphasizing that for generic quasi-topoi **C** one may have $\mathcal{M} \neq \mathsf{mono}(\mathbf{C})$, as is the case in particular for the quasi-topos **SGraph** of simple graphs. We illustrate this phenomenon in the diagram on the right via presenting the multi-sum construction for $A = B = \bullet$. Note in particular the monic-epis that extend the two-vertex graph S_0 into

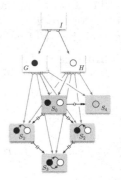

the graphs S_1, S_2 and S_3, all of which have the same vertices as S_0 (recalling that a morphism in **SGraph** is monic/epic if it is so on vertices), yet additional edges, so that in particular none of the morphisms $S_0 \to S_j$ for $j = 1, 2, 3$ is edge-reflecting.

3   Non-linear Sesqui- and Double-Pushout Rewriting

In much of the traditional work on graph- and categorical rewriting theories [22], while it was appreciated early in its development that in particular SqPO-rewriting permits the cloning of subgraphs [17], and that both SqPO- and DPO-semantics permit the fusion of subgraphs (i.e. via input-linear, but output-non-linear rules), the non-uniqueness of pushout complements along non-monic morphisms for the DPO- and the lack of a concurrency theorem in the SqPO-case in general has prohibited a detailed development of non-linear rewriting theories to date. Interestingly, the SqPO-type concurrency theorem for linear rules as developed in [2] exhibits the same obstacle for the generalization to non-linear rewriting as the DPO-type concurrency theorem, i.e., the non-uniqueness of certain pushout complements. Our proof for non-linear rules identifies in addition a new and highly non-trivial "backpropagation effect", which will be highlighted in Sect. 4. It may be worthwhile emphasizing that there exists previous work that aimed at circumventing some of the technical obstacles of non-linear rewriting either via specializing the semantics e.g. from double pushout to a version based upon so-called *minimal* pushout complements [14], or from sesqui-pushout to *reversible* SqPO-semantics [19,30] or other variants such as AGREE-rewriting [18]. In contrast, we will in the following introduce the "true" extensions of both SqPO- and DPO-rewriting to the non-linear setting, with our constructions based upon multi-sums, multi-POCs and FPAs.

Definition 9. *General SqPO-rewriting semantics over a quasi-topos* **C**:

- *The set of SqPO-admissible matches of a rule* rule $r = (O \leftarrow K \rightarrow I) \in$ span(**C**) *into an object* $X \in$ obj(**C**) *is defined as*

$$\mathsf{M}_r^{SqPO}(X) := \{ I \xrightarrow{m} X \mid m \in \mathrm{rm}(\mathbf{C}) \} . \tag{10}$$

A SqPO-type direct derivation[5] of $X \in$ obj(**C**) *with rule* r *along* $m \in$ $\mathsf{M}_r^{SqPO}(X)$ *is defined as a diagram in* (11), *where* (1) *is formed as an FPC, while* (2) *is formed as a pushout.*

$$
\begin{array}{ccccc}
O & \xleftarrow{\ o\ } & K & \xrightarrow{\ i\ } & I \\
{\scriptstyle m^*}\downarrow & {\scriptstyle (2)} & {\scriptstyle \bar{m}}\downarrow & {\scriptstyle (1)} & \downarrow{\scriptstyle m} \\
r_m(X) & \xleftarrow{\ \bar{o}\ } & \bar{X} & \xrightarrow{\ \bar{i}\ } & X
\end{array}
\tag{11}
$$

- *The set of SqPO-type admissible matches of rules* $r_2, r_1 \in$ span(**C**) *(also referred to in the literature as* dependency relations*) is defined as*

$$\mathcal{M}_{r_2}^{SqPO}(r_1) := \{ (j_2, j_1, \bar{j}_1, \bar{o}_1, \bar{\bar{j}}_1, \bar{\bar{i}}_1, \iota_{21}) \mid$$

$$(j_2, j_1) \in \sum\nolimits_{\mathcal{M}} (I_2, O_1) \wedge (\bar{j}_1, \bar{o}_1) \in \mathcal{P}(o_1, j_1) \tag{12}$$

$$\wedge (\bar{\bar{j}}_1, \bar{\bar{i}}_1, \iota_{21}) \in \mathsf{FPA}(\bar{j}_1, i_1) \} / \sim ,$$

where equivalence is defined up to the compatible universal isomorphisms of multi-sums, multi-POCs and FPAs (see below).
- *An SqPO-type rule composition of two general rules* $r_1, r_2 \in$ span(**C**) *along an admissible match* $\mu \in \mathcal{M}_{r_2}^{SqPO}(r_1)$ *is defined via a diagram as in* (13) *below, where (going column-wise from the left) squares* (2_2), (6), *and* (4) *are pushouts,* (1_1) *is the multi-POC element specified as part of the data of the match,* (2_1) *and* (3) *form an FPA-diagram as per the data of the match, and finally* (1_2) *and* (5) *are FPCs:*

---

[5] Note that this part of the definition of general SqPO-semantics coincides precisely with the original definition of [17].

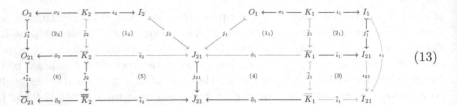

We then define the composite rule via span composition:

$$r_2 \stackrel{\mu}{\triangleleft} r_1 := (\overline{O}_{21} \leftarrow \overline{\overline{K}}_2 \rightarrow \overline{J}_{21}) \circ (\overline{J}_{21} \leftarrow \overline{\overline{K}}_1 \rightarrow \overline{I}_{21}) \tag{14}$$

Definition 10. *General DPO-rewriting semantics over an rm-adhesive category* **C**:

– *The set of DPO-admissible matches of a rule* rule $r = (O \leftarrow K \rightarrow I) \in$ span(**C**) *into an object* $X \in$ obj(**C**) *is defined as*

$$\mathsf{M}_r^{DPO}(X) := \{(m, \bar{m}, \bar{i}) \mid m \in \mathsf{rm}(\mathbf{C}) \land (\bar{m}, \bar{i}) \in \mathcal{P}(i, m)\}. \tag{15}$$

A DPO-type direct derivation of $X \in$ obj(**C**) *with rule* r *along* $m \in \mathsf{M}_r^{DPO}(X)$ *is defined as a diagram in* (11), *where* (1) *is the multi-POC element chosen as part of the data of the match, while* (2) *is formed as a pushout.*
– *The set of DPO-type admissible matches of rules* $r_2, r_1 \in$ span(**C**) *(also referred to as* dependency relations*) is defined as*

$$\mathcal{M}_{r_2}^{DPO}(r_1) := \{(j_2, j_1, \bar{j}_2, \bar{i}_2, \bar{j}_1, \bar{o}_1) \mid$$
$$(j_2, j_1) \in \sum_{\mathcal{M}} (I_2, O_1) \tag{16}$$
$$\land (\bar{j}_2, \bar{i}_2) \in \mathcal{P}(i_2, j_2) \land (\bar{j}_1, \bar{o}_1) \in \mathcal{P}(o_1, j_1)\}/\sim,$$

where equivalence is defined up to the compatible universal isomorphisms of multi-sums and multi-POCs (see below).
– *A DPO-type rule composition of two general rules* $r_1, r_2 \in$ span(**C**) *along an admissible match* $\mu \in \mathcal{M}_{r_2}^{DPO}(r_1)$ *is defined via a diagram as in* (17) *below, where* (1_2) *and* (1_1) *are the multi-POC elements chosen as part of the data of the match, while* (2_2) *and* (2_1) *are pushouts:*

$$
\begin{array}{c}
O_2 \xleftarrow{\ o_2\ } K_2 \xrightarrow{\ i_2\ } I_2 \qquad\qquad O_1 \xleftarrow{\ o_1\ } K_1 \xrightarrow{\ i_1\ } I_1 \\
j_2^* \downarrow \quad (2_2) \quad \bar{j}_2 \downarrow \quad (1_2) \qquad j_2 \searrow \quad \swarrow j_1 \quad (1_1) \quad \bar{j}_1 \downarrow \quad (2_1) \quad j_1^* \downarrow \\
O_{21} \xleftarrow{\ \bar{o}_2\ } \overline{K}_2 \xrightarrow{\ \bar{i}_2\ } J_{21} \xleftarrow{\qquad \bar{o}_1 \qquad} \overline{K}_1 \xrightarrow{\ \bar{i}_1\ } I_{21}
\end{array}
\tag{17}
$$

We then define the composite rule via span composition:

$$r_2 \stackrel{\mu}{\blacktriangleleft} r_1 := (O_{21} \leftarrow \overline{K}_2 \rightarrow J_{21}) \circ (J_{21} \leftarrow \overline{K}_1 \rightarrow I_{21}) \tag{18}$$

The precise reasons for the definitions of SqPO- and DPO-semantics for generic rules and regular monos as matches will only become evident via the concurrency theorems that will be developed in the following sections.

Let us illustrate the notion of SqPO-type rule composition, as given in Definition 9, with the following example in the setting of directed multi-graphs. Note that, since this is an adhesive category, all monos are automatically regular and we therefore have no need to restrict matches to being edge-reflecting monomorphisms.

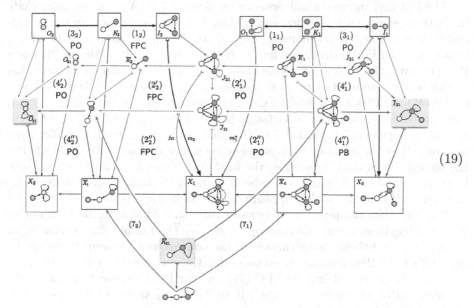

$$(19)$$

In this example, we have two rules. The first clones a node[6], but not its incident edge, then adds a new edge between the original node and its clone and merges the blue node with the original node. The second rule deletes a node and then merges two nodes. The given applications to the graphs X_0 and X_1 illustrate some of the idiosyncrasies of SqPO-rewriting:

- Since the node of X_0 that is being cloned possesses a self-loop, the result of cloning is two nodes, each with a self-loop, with one edge going each way between them.
- In the application of the second rule to X_1, we see the side-effect whereby all edges incident to the deleted node are themselves deleted (as also occurs in SPO-, but not in DPO-rewriting).

The overall effect of the two rewrites can be seen in X_2; as usual, this depends on the overlap between the images of O_1 and I_2 in X_1. This overlap is precisely the multi-sum element J_{21}. Since our example is set in an adhesive category, this can be most easily computed by taking the PB of m_1^* and m_2 and then the

---

[6] Note that we have drawn the rule from right to left so that the *input*, sometimes called the left-hand side, of the rule is the topmost rightmost graph. Note also that the structure of the homomorphisms may be inferred from the node positions, with the exception of the vertex clonings that are explicitly mentioned in the text.

PO of the resulting span. The PO that defines the rewrite from $\overline{X}_0$ to X_1 can now be factorized by computing the PB of j_{21} and the arrow from $\overline{X}_0$ to X_1; this determines $\overline{K}_1$ and its universal arrow from K_1 with consequence that (1_1) and (2_1) are both POs. Let us note that $\overline{K}_1$ is the appropriate member of the multi-POC, as determined by the particular structure of $\overline{X}_0$.

The PO (3_1) induces a universal arrow from I_{21} to X_0; but an immediate inspection reveals that this homomorphism is not a mono (nor an epi in this case). As such, we cannot hope to use I_{21} as the input/left hand side of the composite rule. Furthermore, we find that the square (4_1) is neither a PB nor a PO. However, the FPA $\overline{I}_{21}$ resolves these problems by enabling a factorization of this square, giving rise to a monomorphism m_{21} into X_0, where $(4_1'')$ and $(3_1) + (4_1')$ are PBs and indeed FPCs. This factorization, as determined by e_{21}, can now be *back-propagated* to factorize (2_1) into POs $(2_1')$ and $(2_1'')$ which gives rise to an augmented version $\overline{J}_{21}$ of the multi-sum object in the middle. Note moreover that the effect of back-propagation concerns also the contribution of the second rule in the composition: the final output motif contains an extra self-loop (compared to the motif O_{21} defined by the PO (3_2)), which is induced by the extra self-loop of $\overline{J}_{21}$ that appears due to back-propagation.

We may then compute the composite rule via taking a pullback to obtain $\overline{K}_{21}$, yielding in summary the rule $\overline{O}_{21} \leftarrow \overline{K}_{21} \rightarrow \overline{I}_{21}$. Performing the remaining steps of the "synthesis" construction of the concurrency theorem (compare [7, Appendix C.1]) then amounts to constructing the commutative cube in the middle of the diagram, yielding the FPC (7_1) and the PO (7_2), and thus finally the one-step SqPO-type direct derivation from X_0 to X_2 along the composite rule $\overline{O}_{21} \leftarrow \overline{K}_{21} \rightarrow \overline{I}_{21}$.

Let us finally note, as a general remark, that if the first rule in an SqPO-type rule composition is output- (or right-) linear then the POC is uniquely determined; and if it is input- (or left-) linear then the PO (3_1) is also an FPC and (4_1) is a PB, by Lemma 2(h) of [2]. In this case, the FPA is trivial, and consequently so is the back-propagation process. Our rule composition can thus be seen as a conservative extension of that defined for linear rules in [2].

4   Concurrency Theorem for Non-linear SqPO Rewriting

Part of the reason that a concurrency theorem for generic SqPO-rewriting had remained elusive in previous work concerns the intricate nature of the interplay between multi-sums, multi-POCs and FPAs as seen from the definition of rule compositions according to Definition 9, which is justified via the following theorem, constituting the first main result of the present paper:

Theorem 3. *Let* $\mathbf{C}$ *be a quasi-topos, let* $X_0 \in \mathrm{obj}(\mathbf{C})$ *be an object, and let* $r_2, r_1 \in span(\mathbf{C})$ *be two (generic) rewriting rules.*

1. **Synthesis:** *For every pair of admissible matches* $m_1 \in \mathsf{M}_{r_1}^{SqPO}(X_0)$ *and* $m_2 \in \mathsf{M}_{r_2}^{SqPO}(r_{1_{m_1}}(X_0))$, *there exist an admissible match* $\mu \in \mathcal{M}_{r_2}^{SqPO}(r_1)$ *and an admissible match* $m_{21} \in \mathsf{M}_{r_{21}}^{SqPO}(X_0)$ *(for* r_{21} *the composite of* r_2 *with* r_1 *along* μ*) such that* $r_{21_{m_{21}}}(X_0) \cong r_{2_{m_2}}(r_{1_{m_1}}(X_0))$.

2. **Analysis:** *For every pair of admissible matches $\mu \in \mathcal{M}^{SqPO}_{r_2}(r_1)$ and $m_{21} \in \mathsf{M}^{SqPO}_{r_{21}}(X_0)$ (for r_{21} the composite of r_2 with r_1 along μ), there exists a pair of admissible matches $m_1 \in \mathsf{M}^{SqPO}_{r_1}(X_0)$ and $m_2 \in \mathsf{M}^{SqPO}_{r_2}(r_{1_{m_1}}(X_0))$ such that $r_{2_{m_2}}(r_{1_{m_1}}(X_0)) \cong r_{21_{m_{21}}}(X_0)$.*

3. **Compatibility:** *If in addition $\mathbf{C}$ is finitary [26, Def. 2.8], i.e., if for every object of $\mathbf{C}$ there exist only finitely many regular subobjects up to isomorphisms, the sets of pairs of matches (m_1, m_2) and (μ, m_{21}) are isomorphic if they are suitably quotiented by universal isomorphisms, i.e., by universal isomorphisms of $X_1 = r_{1_{m_1}}(X_0)$ and $X_2 = r_{2_{m_2}}(X_1)$ for the set of pairs (m_1, m_2), and by the universal isomorphisms of multi-sums, multi-POCs and FPAs for the set of pairs (μ, m_{21}), respectively.*

Proof. See [7, Appendix C].

5   Concurrency Theorem for Non-linear DPO-Rewriting

The well-known and by now traditional results on concurrency in DPO-type semantics by Ehrig et al. were formulated for $\mathcal{M}$-linear rules in $\mathcal{M}$-adhesive categories (albeit possibly for non-monic matches; cf. [22, Sec. 5] for the precise details), and notably the non-uniqueness of pushout complements along non-linear morphisms posed the main obstacle for extending this line of results to non-linear DPO rewriting. As we will demonstrate in this section, taking advantage of multi-sums and multi-POCs, and if the underlying category $\mathbf{C}$ is an *rm-adhesive category* [27, Def. 1.1], one may lift this restriction and obtain a fully well-posed semantics for DPO-rewriting along generic rules, and for regular monic matches:

Theorem 4. *Let $\mathbf{C}$ be an rm-adhesive category, let $X_0 \in \mathsf{obj}(\mathbf{C})$ be an object, and let $r_2, r_1 \in span(\mathbf{C})$ be (generic) spans in $\mathbf{C}$.*

- **Synthesis:** *For every pair of admissible matches $m_1 \in \mathsf{M}^{DPO}_{r_1}(X_0)$ and $m_2 \in \mathsf{M}^{DPO}_{r_2}(r_{1_{m_1}}(X_0))$, there exist an admissible match $\mu \in \mathcal{M}^{DPO}_{r_2}(r_1)$ and an admissible match $m_{21} \in \mathsf{M}^{DPO}_{r_{21}}(X_0)$ (for r_{21} the composite of r_2 with r_1 along μ) such that $r_{21_{m_{21}}}(X_0) \cong r_{2_{m_2}}(r_{1_{m_1}}(X_0))$.*

- **Analysis:** *For every pair of admissible matches $\mu \in \mathcal{M}^{DPO}_{r_2}(r_1)$ and $m_{21} \in \mathsf{M}^{DPO}_{r_{21}}(X_0)$ (for r_{21} the composite of r_2 with r_1 along μ), there exists a pair of admissible matches $m_1 \in \mathsf{M}^{DPO}_{r_1}(X_0)$ and $m_2 \in \mathsf{M}^{SqPO}_{r_2}(r_{1_{m_1}}(X_0))$ such that $r_{2_{m_2}}(r_{1_{m_1}}(X_0)) \cong r_{21_{m_{21}}}(X_0)$.*

- **Compatibility:** *If in addition $\mathbf{C}$ is finitary, the sets of pairs of matches (m_1, m_2) and (μ, m_{21}) are isomorphic if they are suitably quotiented by universal isomorphisms, i.e., by universal isomorphisms of $X_1 = r_{1_{m_1}}(X_0)$ and $X_2 = r_{2_{m_2}}(X_1)$ for the set of pairs of matches (m_1, m_2), and by the universal isomorphisms of multi-sums and multi-POCs for the set of pairs of matches (μ, m_{21}), respectively.*

Proof. See [7, Appendix D].

It is worthwhile noting that for an adhesive category **C** (in which every monomorphism is regular) and if we consider *linear* rules (i.e., spans of monomorphisms), the characterization of multi-sums according to Lemma 1 permits to verify that DPO-type rule compositions as in Theorem 4 specialize in this setting precisely to the notion of DPO-type *D-concurrent compositions* [35, Sec. 7.2]. This is because, in this case, each multi-sum element is precisely characterized as the pushout of a monic span (referred to as a D-dependency relation between rules in [35]), so one finds indeed that Theorem 4 conservatively generalizes the traditional DPO-type concurrency theorem to the non-linear setting. Unlike for the generic SqPO-type setting however, quasi-topoi are *not* sufficient for generic DPO-rewriting, since in the "analysis" part of the proof of the DPO-type concurrency theorem the van Kampen property of pushouts along regular monomorphisms is explicitly required (cf. [7, Appendix D]).

6   Conclusion and Outlook

We have defined an abstract setting for SqPO graph transformation in quasi-topoi that captures the important concrete cases of (directed) multi-graphs and simple graphs. In particular, we have established the existence of appropriate notions of $\mathcal{M}$-multi-sums, $\mathcal{M}$-multi-POCs and $\mathcal{M}$-FPC-PO-augmentations in this setting that permit a proof of the concurrency theorem for general non-linear rules.

Our immediate next goal is to prove associativity of our notion of rule composition in order to enable the use of rule algebra constructions [5,8,10] and tracelets [3] for static analysis [4,6] of systems generated by non-linear SqPO or DPO transformations. Intuitively, associativity is necessary in order to guarantee that one may consistently analyze and classify derivation traces based upon nested applications of the concurrency theorem, in the sense that recursive rule composition operations should yield a "catalogue" of all possible ways in which rules can interact in derivation sequences. The latter is formalized as the so-called tracelet characterization theorem in [3], whereby any derivation trace is characterized as an underlying tracelet and a match of the tracelet into the initial state of the trace. As illustrated in the worked example presented in (19), which highlighted the intriguing effect that comparatively complicated intermediate state in derivation traces involving cloning and fusing of graph structures are consistently abstracted away via performing rule compositions, one might hope that this type of effect persists also in n-step derivation traces for arbitrary n, for which however associativity is a prerequisite. Concretely, without the associativity property, the tentative "summaries" of the overall effects of derivation traces via their underlying tracelets would not be mathematically consistent, as they would only encode the causality of the nesting order in which they were calculated via pairwise rule composition operations. Preliminary results indicate however that indeed our generalized SqPO- and DPO-type semantics both satisfy the requisite associativity property, which will be presented in future work.

Beyond known applications to rule-based descriptions of complex systems, such as in Kappa [13] and related formalisms, we hope to exploit this framework in graph combinatorics and structural graph theory [12]—which frequently employ operations such as edge contraction, which requires input-linear but output-non-linear rules, and node expansion, which further requires input-non-linear rules—to provide stronger tools for reasoning about graph reconfigurations as used, for example, in the study of coloring problems. We moreover expect this framework to be useful in strengthening existing approaches to graph-based knowledge representation [28], particularly for the extraction and manipulation of audit trails [30] that provide a semantic notion of version control in these settings.

References

1. Adamek, J., Herrlich, H., Strecker, G.: Abstract and concrete categories: the joy of cats. Reprints Theory Appl. Categ. (17), 1–507 (2006). http://www.tac.mta.ca/ tac/reprints/articles/17/tr17.pdf
2. Behr, N.: Sesqui-pushout rewriting: concurrency, associativity and rule algebra framework. In: Proceedings of GCM 2019. EPTCS, vol. 309, pp. 23–52 (2019). https://doi.org/10.4204/eptcs.309.2
3. Behr, N.: Tracelets and tracelet analysis of compositional rewriting systems. In: Proceedings of ACT 2019. EPTCS, vol. 323, pp. 44–71 (2020). https://doi.org/10. 4204/EPTCS.323.4
4. Behr, N.: On stochastic rewriting and combinatorics via rule-algebraic methods. In: Proceedings of TERMGRAPH 2020, vol. 334, pp. 11–28 (2021). https://doi. org/10.4204/eptcs.334.2
5. Behr, N., Danos, V., Garnier, I.: Stochastic mechanics of graph rewriting. In: Proceedings of LiCS 2016. ACM Press (2016). https://doi.org/10.1145/2933575. 2934537
6. Behr, N., Danos, V., Garnier, I.: Combinatorial conversion and moment bisimulation for stochastic rewriting systems. LMCS **16**(3), 3:1–3:45 (2020). https://lmcs. episciences.org/6628
7. Behr, N., Harmer, R., Krivine, J.: Concurrency theorems for non-linear rewriting theories (long version). CoRR (2021). https://arxiv.org/abs/2105.02842
8. Behr, N., Krivine, J.: Rewriting theory for the life sciences: a unifying theory of CTMC semantics. In: Gadducci, F., Kehrer, T. (eds.) ICGT 2020. LNCS, vol. 12150, pp. 185–202. Springer, Cham (2020). https://doi.org/10.1007/978-3-030- 51372-6_11
9. Behr, N., Krivine, J.: Compositionality of rewriting rules with conditions. Compositionality **3**, 2 (2021). https://doi.org/10.32408/compositionality-3-2
10. Behr, N., Sobocinski, P.: Rule algebras for adhesive categories (extended journal version). LMCS **16**(3), 2:1–2:38 (2020). https://lmcs.episciences.org/6615
11. Bonifati, A., Furniss, P., Green, A., Harmer, R., Oshurko, E., Voigt, H.: Schema validation and evolution for graph databases. In: Laender, A.H.F., Pernici, B., Lim, E.-P., de Oliveira, J.P.M. (eds.) ER 2019. LNCS, vol. 11788, pp. 448–456. Springer, Cham (2019). https://doi.org/10.1007/978-3-030-33223-5_37
12. Bousquet-Mélou, M.: Counting planar maps, coloured or uncoloured. London Mathematical Society Lecture Note Series. Cambridge University Press, pp. 1–50 (2011). https://doi.org/10.1017/CBO9781139004114.002

13. Boutillier, P., et al.: The Kappa platform for rule-based modeling. Bioinformatics **34**(13), i583–i592 (2018). https://doi.org/10.1093/bioinformatics/bty272
14. Braatz, B., Golas, U., Soboll, T.: How to delete categorically—two pushout complement constructions. J. Symb. Comput. **46**(3), 246–271 (2011). https://doi.org/10.1016/j.jsc.2010.09.007
15. Cockett, J., Lack, S.: Restriction categories I: categories of partial maps. Theor. Comput. Sci. **270**(1), 223–259 (2002). https://doi.org/10.1016/S0304-3975(00)00382-0
16. Cockett, J., Lack, S.: Restriction categories II: partial map classification. Theor. Comput. Sci. **294**(1), 61–102 (2003). https://doi.org/10.1016/S0304-3975(01)00245-6
17. Corradini, A., Heindel, T., Hermann, F., König, B.: Sesqui-pushout rewriting. In: Corradini, A., Ehrig, H., Montanari, U., Ribeiro, L., Rozenberg, G. (eds.) ICGT 2006. LNCS, vol. 4178, pp. 30–45. Springer, Heidelberg (2006). https://doi.org/10.1007/11841883_4
18. Corradini, A., Duval, D., Echahed, R., Prost, F., Ribeiro, L.: AGREE – algebraic graph rewriting with controlled embedding. In: Parisi-Presicce, F., Westfechtel, B. (eds.) ICGT 2015. LNCS, vol. 9151, pp. 35–51. Springer, Cham (2015). https://doi.org/10.1007/978-3-319-21145-9_3
19. Danos, V., Heindel, T., Honorato-Zimmer, R., Stucki, S.: Reversible Sesqui-pushout rewriting. In: Giese, H., König, B. (eds.) ICGT 2014. LNCS, vol. 8571, pp. 161–176. Springer, Cham (2014). https://doi.org/10.1007/978-3-319-09108-2_11
20. Diers, Y.: Familles universelles de morphismes, Publications de l'U.E.R. mathématiques pures et appliquées, vol. 145. Université des sciences et techniques de Lille I (1978)
21. Dyckhoff, R., Tholen, W.: Exponentiable morphisms, partial products and pullback complements. J. Pure Appl. Algebra **49**(1–2), 103–116 (1987)
22. Ehrig, H., et al.: Fundamentals of Algebraic Graph Transformation. Monographs in Theoretical Computer Science. Springer, Heidelberg (2006). https://doi.org/10.1007/3-540-31188-2
23. Ehrig, H., Golas, U., Hermann, F.: Categorical frameworks for graph transformation and HLR systems based on the DPO approach. Bull. EATCS **102**, 111–121 (2010)
24. Ehrig, H., Habel, A., Padberg, J., Prange, U.: Adhesive high-level replacement categories and systems. In: Ehrig, H., Engels, G., Parisi-Presicce, F., Rozenberg, G. (eds.) ICGT 2004. LNCS, vol. 3256, pp. 144–160. Springer, Heidelberg (2004). https://doi.org/10.1007/978-3-540-30203-2_12
25. Ehrig, H., et al.: $\mathcal{M}$-adhesive transformation systems with nested application conditions. Part 1: parallelism, concurrency and amalgamation. MSCS **24**(04), 1–48 (2014). https://doi.org/10.1017/s0960129512000357
26. Braatz, B., Ehrig, H., Gabriel, K., Golas, U.: Finitary $\mathcal{M}$-adhesive categories. In: Ehrig, H., Rensink, A., Rozenberg, G., Schürr, A. (eds.) ICGT 2010. LNCS, vol. 6372, pp. 234–249. Springer, Heidelberg (2010). https://doi.org/10.1007/978-3-642-15928-2_16
27. Garner, R., Lack, S.: On the axioms for adhesive and quasiadhesive categories. TAC **27**(3), 27–46 (2012)
28. Harmer, R., Le Cornec, Y.S., Légaré, S., Oshurko, E.: Bio-curation for cellular signalling: the kami project. IEEE/ACM Trans. Comput. Biol. Bioinform. **16**(5), 1562–1573 (2019). https://doi.org/10.1109/TCBB.2019.2906164
29. Harmer, R., Oshurko, E.: Knowledge representation and update in hierarchies of graphs. JLAMP **114**, 100559 (2020)

30. Harmer, R., Oshurko, E.: Reversibility and composition of rewriting in hierarchies. EPTCS **330**, 145–162 (2020). https://doi.org/10.4204/eptcs.330.9
31. Heindel, T.: Hereditary pushouts reconsidered. In: Ehrig, H., Rensink, A., Rozenberg, G., Schürr, A. (eds.) ICGT 2010. LNCS, vol. 6372, pp. 250–265. Springer, Heidelberg (2010). https://doi.org/10.1007/978-3-642-15928-2_17
32. Johnstone, P.T.: Sketches of an Elephant - A Topos Theory Compendium, vol. 1. Oxford University Press, Oxford (2002)
33. Johnstone, P.T., Lack, S., Sobociński, P.: Quasitoposes, quasiadhesive categories and artin glueing. In: Mossakowski, T., Montanari, U., Haveraaen, M. (eds.) CALCO 2007. LNCS, vol. 4624, pp. 312–326. Springer, Heidelberg (2007). https://doi.org/10.1007/978-3-540-73859-6_21
34. Lack, S., Sobociński, P.: Adhesive categories. In: Walukiewicz, I. (ed.) FoSSaCS 2004. LNCS, vol. 2987, pp. 273–288. Springer, Heidelberg (2004). https://doi.org/10.1007/978-3-540-24727-2_20
35. Lack, S., Sobociński, P.: Adhesive and quasiadhesive categories. RAIRO - Theor. Inform. Appl. **39**(3), 511–545 (2005). https://doi.org/10.1051/ita:2005028
36. Löwe, M.: Polymorphic Sesqui-pushout graph rewriting. In: Parisi-Presicce, F., Westfechtel, B. (eds.) ICGT 2015. LNCS, vol. 9151, pp. 3–18. Springer, Cham (2015). https://doi.org/10.1007/978-3-319-21145-9_1
37. Monro, G.: Quasitopoi, logic and heyting-valued models. J. Pure Appl. Algebra **42**(2), 141–164 (1986). https://doi.org/10.1016/0022-4049(86)90077-0

A Generalized Concurrent Rule Construction for Double-Pushout Rewriting

Jens Kosiol[✉] and Gabriele Taentzer

Philipps-Universität Marburg, Marburg, Germany
{kosiolje,taentzer}@mathematik.uni-marburg.de

Abstract. Double-pushout rewriting is an established categorical approach to the rule-based transformation of graphs and graph-like objects. One of its standard results is the construction of concurrent rules and the Concurrency Theorem pertaining to it: The sequential application of two rules can equivalently be replaced by the application of a concurrent rule and vice versa. We extend and generalize this result by introducing *generalized concurrent rules* (GCRs). Their distinguishing property is that they allow identifying and preserving elements that are deleted by their first underlying rule and created by the second one. We position this new kind of composition of rules among the existing ones and obtain a Generalized Concurrency Theorem for it. We conduct our work in the same generic framework in which the Concurrency Theorem has been presented, namely double-pushout rewriting in $\mathcal{M}$-adhesive categories via rules equipped with application conditions.

Keywords: Graph transformation · Double-pushout rewriting · $\mathcal{M}$-adhesive categories · Concurrency Theorem · Model editing

1 Introduction

The composition of transformation rules has long been a topic of interest for (theoretical) research in graph transformation. Classical kinds of rule composition are the ones of *parallel* and *concurrent* [11] as well as of *amalgamated rules* [4]. Considering the *double-pushout approach* to graph transformation, these rule constructions have been lifted from ordinary graphs to the general framework of $\mathcal{M}$-adhesive categories and from plain rules to such with application conditions [5,7,9,17]. These central forms of rule composition have also been developed for other variants of transformation, like *single-* or *sesqui-pushout rewriting* [1,22,23].

In this work, we are concerned with simultaneously generalizing two variants of sequential rule composition in the context of double-pushout rewriting. We develop *generalized concurrent rules* (GCRs), which comprise concurrent as well as so-called *short-cut rules* [14]. The concurrent rule construction, on the one hand, is optimized concerning *transient* model elements: An element that is

© Springer Nature Switzerland AG 2021
F. Gadducci and T. Kehrer (Eds.): ICGT 2021, LNCS 12741, pp. 22–39, 2021.
https://doi.org/10.1007/978-3-030-78946-6_2

created by the first rule and deleted by the second does not occur in a concurrent rule. A model element that is deleted by the first rule, however, cannot be reused in the second one. A short-cut rule, on the other hand, takes a rule that only deletes elements and a monotonic rule (i.e., a rule that only creates elements) and combines them into a single rule, where elements that are deleted and recreated may be preserved throughout the process. GCRs fuse both effects, the omission of transient elements and the reuse of elements, into a single construction.

The reuse of elements that is enabled by short-cut rules has two distinct advantages. First, information can be preserved. In addition, a rule that reuses model elements instead of deleting and recreating them is often applicable more frequently since necessary context does not get lost: Considering the double-pushout approach to graph transformation, a rule with higher reuse satisfies the dangling edge condition more often in general. These properties allowed us to employ short-cut rules to improve model synchronization processes [13, 15, 16]. Our construction of GCRs provides the possibility of reusing elements when sequentially composing arbitrary rules. Hence, it generalizes the restricted setting in which we defined short-cut rules. Thereby, we envision new possibilities for application, for example, the automated construction of complex (language-preserving) editing operations from simpler ones (which are not monotonic in general). This work, however, is confined to developing the formal basis. We present our new theory in the general and abstract framework of double-pushout rewriting in $\mathcal{M}$-adhesive categories [5, 21]. We restrict ourselves to the case of $\mathcal{M}$-matching of rules, though. While results similar to the ones we present here also hold in the general setting, their presentation and proof are much more technical.

In Sect. 2, we introduce our running example and motivate the construction of GCRs by contrasting it to the one of concurrent rules. Section 3 recalls preliminaries. In Sect. 4, we develop the construction of GCRs. We characterize under which conditions the GCR construction results in a rule and prove that it generalizes indeed both, the concurrent as well as the short-cut rule constructions. Section 5 contains our main result: The Generalized Concurrency Theorem states that subsequent rule applications can be synthesized into the application of a GCR. It also characterizes the conditions under which the application of a GCR can be equivalently split up into the subsequent application of its two underlying rules. Finally, we consider related work in Sect. 6 and conclude in Sect. 7. A long version of this paper contains additional preliminaries and all proofs [20].

2   Running Example

In this section, we provide a short practical motivation for our new rule construction. It is situated in the context of model editing, more precisely class refactoring [12]. Refactoring is a technique to improve the design of a software system without changing its behavior. Transformation rules can be used to specify suitable refactorings of class models. For the sake of simplicity, we focus on the class structure here, where classes are just blobs. Two kinds of class relations are specified using typed edges, namely class references and generalizations; they

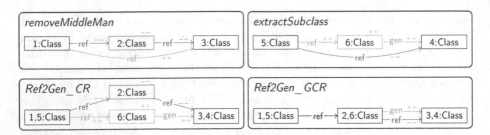

Fig. 1. Two refactoring rules for class diagrams (first line) and sequentially composed rules derived from them (second line). (Color figure online)

are typed with ref and gen, respectively. All rules are depicted in an integrated fashion, i.e., as a single graph where annotations determine the roles of the elements. Black elements (without further annotations) need to exist to match a rule and are not changed by its application. Elements in red (additionally annotated with −−) need to exist and are deleted upon application; green elements (annotated with ++) get newly created.

The refactoring rules for our example are depicted in the first line of Fig. 1. The rule *removeMiddleMan* removes a Class that merely delegates the work to the real Class and directs the reference immediately to this, instead. The rule *extractSubclass* creates a new Class that is generalized by an already existing one; to not introduce unnecessary abstraction, the rule also redirects an existing reference to the newly introduced subclass.

Sequentially combining these two refactorings results in further ones. For example, this allows us to replace the second reference of a chain of two references with a generalization. The according concurrent rule is depicted as *Ref2Gen_CR* in Fig. 1. It arises with *removeMiddleMan* as its first underlying rule and *extractSubclass* as its second, where Classes 1 and 5 and 3 and 4 are identified, respectively. The new reference-edge created by *removeMiddleMan* is deleted by *extractSubclass* and, thus, becomes transient. But the Class 2 that originally delegated the reference is deleted and cannot be reused. Instead, Class 6 has to be newly created to put Classes 1 and 3 into this new context.

In many situations, however, it would be preferable to just reuse Class 2 and only replace the reference with a generalization. In this way, information (such as references, values of possible attributes, and layout information) is preserved. And, maybe even more importantly, when adopting the double-pushout approach, such a rule is typically more often applicable, namely also when Class 2 has adjacent edges. In our construction of generalized concurrent rules, we may identify elements deleted by the first rule and recreated by the second and decide to preserve them. In this example, our new construction allows one to also construct *Ref2Gen_GCR* (Fig. 1) from *removeMiddleMan* and *extractSubclass*. In contrast to *Ref2Gen_CR*, it identifies Class 2 deleted by *removeMiddleMan* and Class 6 created by *extractSubclass* and the respective incoming references and preserves them. This rule specifies a suitable variant of the refactoring *Remove Middle Man*, where the middle man is turned into a subclass of the real class.

3   Preliminaries

In this section, we recall the main preliminaries for our work. We introduce $\mathcal{M}$-adhesive categories, double-pushout rewriting, initial pushouts, $\mathcal{M}$-effective unions, and concurrent rules.

Adhesive categories can be understood as categories where pushouts along monomorphisms behave like pushouts along injective functions in **Set**. They have been introduced by Lack and Sobociński [21] and offer a unifying formal framework for double-pushout rewriting. Later, more general variants, which cover practically relevant examples that are not adhesive, have been suggested [5,7]. In this work, we address the framework of $\mathcal{M}$-*adhesive categories*.

Definition 1 ($\mathcal{M}$-adhesive category). *A category $\mathcal{C}$ is $\mathcal{M}$-adhesive with respect to a class of monomorphisms $\mathcal{M}$ if*

- $\mathcal{M}$ *contains all isomorphisms and is* closed under composition and decomposition, *i.e.,* $f : A \hookrightarrow B, g : B \hookrightarrow C \in \mathcal{M}$ *implies* $g \circ f \in \mathcal{M}$ *and* $g \circ f, g \in \mathcal{M}$ *implies* $f \in \mathcal{M}$.
- $\mathcal{C}$ *has pushouts and pullbacks along $\mathcal{M}$-morphisms and $\mathcal{M}$-morphisms are* closed under pushouts and pullbacks *such that if Fig. 2 depicts a pushout square with $m \in \mathcal{M}$, then also $n \in \mathcal{M}$, and analogously, if it depicts a pullback square with $n \in \mathcal{M}$, then also $m \in \mathcal{M}$.*
- *Pushouts in $\mathcal{C}$ along $\mathcal{M}$-morphisms are* vertical weak van Kampen squares: *For any commutative cube as depicted in Fig. 3 where the bottom square is a pushout along an $\mathcal{M}$-morphism, $b, c, d \in \mathcal{M}$, and the backfaces are pullbacks, then the top square is a pushout if and only if both front faces are pullbacks.*

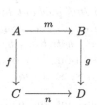

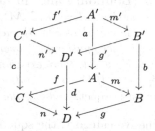

Fig. 2. A pushout square.   **Fig. 3.** Commutative cube over pushout square.

We write that $(\mathcal{C}, \mathcal{M})$ is an $\mathcal{M}$-adhesive category to express that a category $\mathcal{C}$ is $\mathcal{M}$-adhesive with respect to the class of monomorphisms $\mathcal{M}$ and denote morphisms belonging to $\mathcal{M}$ via a hooked arrow. Typical examples of $\mathcal{M}$-adhesive categories are **Set** and **Graph** (for $\mathcal{M}$ being the class of all injective functions or homomorphisms, respectively).

Rules are used to declaratively describe the *transformation* of objects. We use application conditions without introducing *nested conditions* as their formal basis; they are presented in [18,20]. Moreover, we restrict ourselves to the case of $\mathcal{M}$-matching.

Definition 2 (Rules and transformations). *A rule $\rho = (p, ac)$ consists of a plain rule p and an* application condition *ac. The plain rule is a span of $\mathcal{M}$-morphisms $p = (L \xleftarrow{l} K \xrightarrow{r} R)$; the objects are called* left-hand side *(LHS), interface, and* right-hand side *(RHS), respectively. The* application condition *ac is a nested condition over L. A* monotonic rule *is a rule, where l is an isomorphism; it is just denoted as $\rho = (r : L \hookrightarrow R, ac)$. Given a rule $\rho = (L \xleftarrow{l} K \xrightarrow{r} R, ac)$ and a morphism $m : L \hookrightarrow G \in \mathcal{M}$, a (direct) transformation $G \Rightarrow_{\rho,m} H$ from G to H is given by the diagram in Fig. 4 where both squares are pushouts and $m \models ac$. If such a transformation exists, the morphism m is called a* match *and rule ρ is* applicable *at match m.*

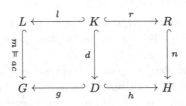

Fig. 4. Definition of a direct transformation via two pushouts.

For some of the following results to hold, we will need $\mathcal{M}$-adhesive categories with further properties. *Initial pushouts* are a way to generalize the set-theoretic complement operator categorically.

Definition 3 (Boundary and initial pushout). *Given a morphism $m : L \to G$ in an $\mathcal{M}$-adhesive category $(\mathcal{C}, \mathcal{M})$, an* initial pushout *over m is a pushout (1) over m (as depicted in Fig. 5) such that $b_m \in \mathcal{M}$ and this pushout factors uniquely through every pushout (3) over m where $b'_m \in \mathcal{M}$. I.e., for every pushout (3) over m with $b'_m \in \mathcal{M}$, there exist unique morphisms b^*_m, c^*_m with $b_m = b'_m \circ b^*_m$ and $c_m = c'_m \circ c^*_m$. If (1) is an initial pushout, b_m is called* boundary *over m, B_m the* boundary object, *and C_m the* context object *with respect to m.*

In an $\mathcal{M}$-adhesive category, the square (2) in Fig. 5 is a pushout and $b^*_m, c^*_m \in \mathcal{M}$. In **Graph**, if m is injective, C_m is the minimal completion of $G \backslash m(L)$ (the componentwise set-theoretic difference on nodes and edges) to a subgraph of G, and B_m contains the boundary nodes that have to be added for this completion [5, Example 6.2].

The existence of $\mathcal{M}$-*effective unions* ensures that the $\mathcal{M}$-subobjects of a given object constitute a lattice.

Definition 4 ($\mathcal{M}$-effective unions). *An $\mathcal{M}$-adhesive category $(\mathcal{C}, \mathcal{M})$ has $\mathcal{M}$-effective unions if, for each pushout of a pullback of a pair of $\mathcal{M}$-morphisms, the induced mediating morphism belongs to $\mathcal{M}$ as well.*

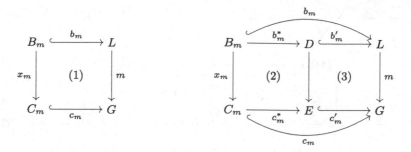

Fig. 5. Initial pushout (1) over the morphism m and its factorization property.

Finally, we recall *E-concurrent rules*, which combine the actions of two rules into a single one. Their definition assumes a given class $\mathcal{E}'$ of pairs of morphisms with the same codomain. For the computation of the application condition of a concurrent rule, we refer to [7, 20].

Definition 5 (*E*-concurrent rule). *Given two rules* $\rho_i = (L_i \overset{l_i}{\hookleftarrow} K_i \overset{r_i}{\hookrightarrow} R_i, ac_i)$, *where* $i = 1, 2$, *an object* E *with morphisms* $e_1 : R_1 \to E$ *and* $e_2 : L_2 \to E$ *is an* E-dependency relation *for* ρ_1 *and* ρ_2 *if* $(e_1, e_2) \in \mathcal{E}'$ *and the pushout complements* (1a) *and* (1b) *for* $e_1 \circ r_1$ *and* $e_2 \circ l_2$ *(as depicted in Fig. 6) exist.*

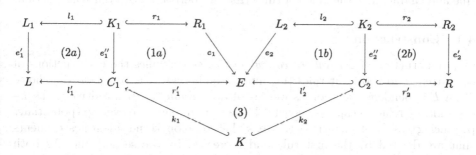

Fig. 6. *E*-dependency relation and *E*-concurrent rule.

Given an E-dependency relation $E = (e_1, e_2) \in \mathcal{E}'$ for rules ρ_1, ρ_2, their E-concurrent rule *is defined as* $\rho_1 *_E \rho_2 := (L \overset{l}{\hookleftarrow} K \overset{r}{\hookrightarrow} R, ac)$, *where* $l := l'_1 \circ k_1$, $r := r'_2 \circ k_2$, (1a), (1b), (2a), *and* (2b) *are pushouts,* (3) *is a pullback (also shown in Fig. 6), and* ac *is computed in a way that suitably combines the semantics of* ac_1 *and* ac_2.

A transformation sequence $G \Rightarrow_{\rho_1, m_2} H \Rightarrow_{\rho_2, m_2} G'$ *is called E-related for the E-dependency relation* $(e_1, e_2) \in \mathcal{E}'$ *if there exists* $h : E \to H$ *with* $h \circ e_1 = n_1$ *and* $h \circ e_2 = m_2$ *and morphisms* $d_i : C_i \to D_i$, *where* $i = 1, 2$, *such that* (4a) *and* (4b) *commute and* (5a) *and* (5b) *are pushouts (see Fig. 7).*

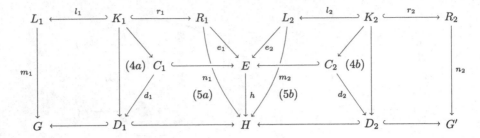

Fig. 7. E-related transformation.

The Concurrency Theorem [5, Theorem 5.23] states that two E-related rule applications may be synthesized into the application of their E-concurrent rule and that an application of an E-concurrent rule may be analyzed into a sequence of two E-related rule applications.

4   Constructing Generalized Concurrent Rules

In this section, we develop our construction of *generalized concurrent rules* (GCRs) in the context of an $\mathcal{M}$-adhesive category $(\mathcal{C}, \mathcal{M})$. We first define GCRs and relate them to the construction of concurrent and short-cut rules. Subsequently, we elaborate the conditions under which our construction results in a rule and characterize the kinds of rules that are derivable with our construction.

4.1   Construction

Our construction of *generalized concurrent rules* combines the constructions of concurrent and short-cut rules [14] into a single one. It is based on the choice of an E-dependency relation as well as of a *common kernel*. Intuitively, the E-dependency relation captures how both rules are intended to overlap (potentially producing transient elements) whereas the common kernel identifies elements that are deleted by the first rule and recreated by the second one. As both concepts identify parts of the interfaces of the involved rules, the construction of a GCR assumes an E-dependency relation and a common kernel that are *compatible*.

Definition 6 ((Compatible) Common kernel). *Given two rules* $\rho_i = (L_i \overset{l_i}{\hookleftarrow} K_i \overset{r_i}{\hookrightarrow} R_i, ac_i)$, *where* $i = 1, 2$, *a common kernel for them is an $\mathcal{M}$-morphism* $k : K_\cap \hookrightarrow V$ *with $\mathcal{M}$-morphisms* $u_i : K_\cap \hookrightarrow K_i$, $v_1 : V \hookrightarrow L_1$, $v_2 : V \hookrightarrow R_2$ *such that both induced squares* (1a) *and* (1b) *in Fig. 8 are pullbacks.*

Given additionally an E-dependency relation $E = (e_1 : R_1 \hookrightarrow E, e_2 : L_2 \hookrightarrow E) \in \mathcal{E}'$ *for* ρ_1 *and* ρ_2, E *and* k *are* compatible *if square* (2) *is a pullback.*

In the following, we will often suppress the morphisms u_i, v_i from our notation and just speak of a common kernel $k : K_\cap \hookrightarrow V$. As an $\mathcal{M}$-morphism k might

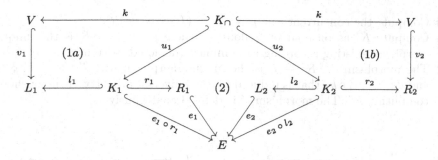

Fig. 8. Compatibility of common kernel with E-dependency relation.

constitute a common kernel for a pair of rules in different ways, we implicitly assume the embedding to be given.

Example 1. Figure 9 shows an E-dependency relation and a common kernel compatible with it for rules *removeMiddleMan* and *extractSubclass* (Fig. 1). The names of the nodes also indicate how the morphisms are defined. The concurrent rule for this E-dependency relation is the rule *Ref2Gen_CR* in Fig. 1.

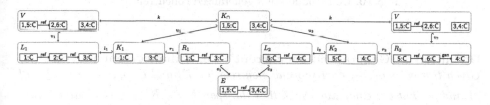

Fig. 9. Common kernel for rules *removeMiddleMan* and *extractSubclass*.

The following lemma is the basis for the construction of generalized concurrent rules; it directly follows from the definition of the interface K of a concurrent rule as pullback.

Lemma 1. *Given two rules ρ_1, ρ_2, an E-dependency relation E, and a common kernel k for ρ_1, ρ_2 that is compatible with E, there exists a unique $\mathcal{M}$-morphism $p : K_\cap \hookrightarrow K$, where K is the interface of the concurrent rule $\rho_1 *_E \rho_2$, such that $k_i \circ p = e_i'' \circ u_i$ for $i = 1, 2$ (compare the diagrams in Definitions 5 and 6).*

A GCR extends a concurrent rule by enhancing its interface K with the additional elements in V of a given common kernel. Formally, this means to compute a pushout along the just introduced morphism p.

Construction 1. *Given two plain rules $\rho_i = (L_i \xleftarrow{l_i} K_i \xrightarrow{r_i} R_i)$, where $i = 1, 2$, an E-dependency relation $E = (e_1 : R_1 \hookrightarrow E, e_2 : L_2 \hookrightarrow E) \in \mathcal{E}'$, and a common kernel $k : K_\cap \hookrightarrow V$ of ρ_1 and ρ_2 that is compatible with E, we construct the span $L \xleftarrow{l'} K' \xrightarrow{r'} R$ as follows (compare Fig. 10):*

(1) Compute the concurrent rule $\rho_1 *_E \rho_2 = (L \xleftarrow{l} K \xrightarrow{r} R)$.
(2) Compute K' as pushout of k along p, where $p : K_\cap \hookrightarrow K$ is the unique morphism existing according to Lemma 1 (depicted twice in Fig. 10).
(3) The morphism $l' : K' \to L$ is the unique morphism with $l' \circ p' = e'_1 \circ v_1$ and $l' \circ k' = l'_1 \circ k_1$ that is induced by the universal property of the pushout computing K'. The morphism r' is defined analogously.

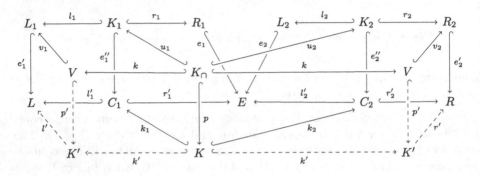

Fig. 10. Construction of a generalized concurrent rule.

Definition 7 (Generalized concurrent rule. Enhancement morphism).
Given two rules ρ_1, ρ_2, an E-dependency relation E, and a common kernel k of ρ_1 and ρ_2 that are compatible such that the span $L \xleftarrow{l'} K' \xrightarrow{r'} R$ obtained from Construction 1 consists of M-morphisms, the generalized concurrent rule of ρ_1 and ρ_2, given E and k, is defined as $\rho_1 *_{E,k} \rho_2 := (L \xleftarrow{l'} K' \xrightarrow{r'} R, ac)$ with ac being the application condition of the concurrent rule $\rho_1 *_E \rho_2 = (L \xleftarrow{l} K \xrightarrow{r} R, ac)$.

The unique M-morphism $k' : K \hookrightarrow K'$ with $l = l'_1 \circ k_1 = l' \circ k'$ and $r = r'_2 \circ k_2 = r' \circ k'$, which is obtained directly from the construction, is called enhancement morphism. We also say that $\rho_1 *_{E,k} \rho_2$ is a GCR enhancing $\rho_1 *_E \rho_2$.

Example 2. Ref2Gen_GCR is a GCR that enhances Ref2Gen_CR (Fig. 1); it is constructed using the common kernel presented in Example 1. Figure 11 illustrates the computation of its interface K' and left-hand morphism l'. The pushout of k and p extends the interface of Ref2Gen_CR by the Class 2,6 and its incoming reference.

Note 1 (Assumptions and notation). For the rest of the paper, we fix the following assumptions: We work in an M-adhesive category $(\mathcal{C}, \mathcal{M})$ with a given class $\mathcal{E}'$ of pairs of morphisms with the same codomain such that $\mathcal{C}$ possesses an

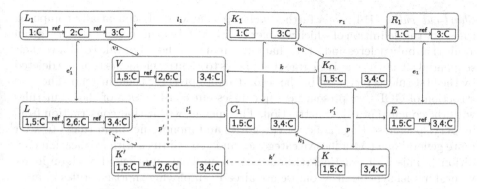

Fig. 11. Computing the interface and the left-hand morphism of *Ref2Gen_GCR*.

$\mathcal{E}'$-$\mathcal{M}$ pair factorization.[1] Further categorical assumptions are mentioned as needed. Furthermore, we always assume two rules ρ_1, ρ_2, an E-dependency relation E, and a common kernel k for them to be given such that E and k are compatible. We consistently use the notations and names of morphisms as introduced above (and in Fig. 10).

4.2 Relating Generalized Concurrent Rules to Other Kinds of Rules

In this section, we relate generalized concurrent rules to other variants of rule composition.

Concurrent Rules are the established technique of sequential rule composition in double-pushout rewriting. By definition, the left- and right-hand sides of a GCR coincide with the ones of the concurrent rule it enhances. One also directly obtains that a GCR coincides with its underlying concurrent rule if and only if its common kernel k is chosen to be an isomorphism.

Proposition 1 (A concurrent rule is a GCR). *Given a concurrent rule* $\rho_1 *_E \rho_2$ *and a GCR* $\rho_1 *_{E,k} \rho_2$ *enhancing it, the enhancement morphism* $k' : K \hookrightarrow K'$ *is an isomorphism if and only if k is one. In particular,* $\rho_1 *_E \rho_2$ *coincides with* $\rho_1 *_{E,k} \rho_2$ *(up to isomorphism) for* $k = id_{K_\cap}$, *where* $K_\cap$ *is obtained by pulling back* $(e_1 \circ r_1, e_2 \circ l_2)$.

---

[1] This means, every pair of morphisms with the same codomain can be factored as a pair of morphisms belonging to $\mathcal{E}'$ followed by an $\mathcal{M}$-morphism. We do not directly need this property in any of our proofs but it is assumed for the computation of application conditions of concurrent and, hence, also generalized concurrent rules. Moreover, it guarantees the existence of E-related transformations [6, Fact 5.29].

Since we restrict ourselves to the case of $\mathcal{M}$-matching, decomposition of $\mathcal{M}$-morphisms then ensures that all occurring pairs $(e_1, e_2) \in \mathcal{E}'$ are in fact even pairs of $\mathcal{M}$-morphisms. This in turn (by closedness of $\mathcal{M}$ under pullbacks) implies that in any common kernel k compatible to a given E-dependency relation, the embedding morphisms u_1, u_2 are necessarily $\mathcal{M}$-morphisms.

Short-cut rules [14] are a further, very specific kind of sequentially composed rules (for the definition of which we refer to [14,20]). In an adhesive category, given a rule that only deletes and a rule that only creates, a short-cut rule combines their sequential effects into a single rule that allows to identify elements that are deleted by the first rule as recreated by the second and to preserve them instead. The construction of GCRs we present here now fuses our construction of short-cut rules with the concurrent rule construction. This means, we lift that construction from its very specific setting (adhesive categories and monotonic, plain rules) to a far more general one ($\mathcal{M}$-adhesive categories and general rules with application conditions). This is of practical relevance as, in application-oriented work on incremental model synchronization, we are already employing short-cut rules in more general settings (namely, we compute short-cut rules from monotonic rules with application conditions rewriting typed attributed triple graphs, which constitute an adhesive HLR category that is not adhesive) [13,15,16].

Proposition 2 (A short-cut rule is a GCR). *Let $\mathcal{C}$ be an adhesive category and the class $\mathcal{E}'$ be such that it contains all pairs of jointly epic $\mathcal{M}$-morphisms. Let $r_i = (r_i : L_i \hookrightarrow R_i)$, where $i = 1, 2$, be two monotonic rules and $k : K_\cap \hookrightarrow V$ a common kernel for them. Then the short-cut rule $r_1^{-1} \ltimes_k r_2$ coincides with the generalized concurrent rule $r_1^{-1} *_{E,k} r_2$, where $E = (e_1, e_2)$ is given via pushout of (u_1, u_2).*

Parallel and amalgamated rules are further kinds of rules arising by composition. Whereas concurrent rules combine the sequential application of two rules, an amalgamated rule combines the application of two (or more) rules to the same object into the application of a single rule [4,17]. In categories with coproducts, the parallel rule is just the sum of two rules; for plain rules (i.e., without application conditions) it is a special case of the concurrent as well as of the amalgamated rule construction. A thorough presentation of all three forms of rule composition in the context of $\mathcal{M}$-adhesive categories, rules with application conditions, and general matching can be found in [7]. When introducing short-cut rules [14], we showed that their effect cannot be achieved by concurrent or amalgamated rules. Thus, by the above proposition, the same holds for GCRs; they indeed constitute a new form of rule composition. The relations between the different kinds of rule composition are summarized in Fig. 12. The lines from parallel rule are dashed as the indicated relations only hold in the absence of application conditions and, for short-cut rules, in the specific setting only in which these are defined.

4.3   Characterizing Derivable Generalized Concurrent Rules

Next, we characterize the GCRs derivable from a given pair of rules. We do so in two different ways, namely (i) characterizing possible choices for the morphisms v_1, v_2 in a common kernel and (ii) characterizing the possible choices for enhancement morphisms k'.

Proposition 3 (Embedding characterization of GCRs). *Let $(\mathcal{C}, \mathcal{M})$ be an $\mathcal{M}$-adhesive category with $\mathcal{M}$-effective unions.*

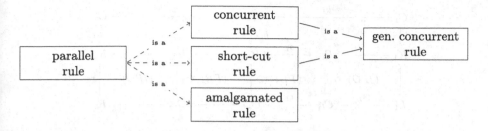

Fig. 12. Relations between the different kinds of rule composition.

(1) *The application of Construction 1 results in a GCR $\rho_1 *_{E,k} \rho_2$ if and only if $v_1, v_2 \in \mathcal{M}$.*

(2) *The assumption that $\mathcal{M}$-effective unions exist is necessary for this result to hold.*

Next, we consider enhancement morphisms in more detail and answer the following question: Given a concurrent rule $\rho_1 *_E \rho_2 = L \hookleftarrow K \hookrightarrow R$, which common $\mathcal{M}$-subobjects K' of L and R that enhance K constitute the interface of a GCR? The following example shows that not all of them do.

Example 3. In the category **Graph** (or **Set**), if p_1 is the trivial rule $\emptyset \hookleftarrow \emptyset \hookrightarrow \emptyset$ and $p_2 = (\bullet \hookleftarrow \emptyset \hookrightarrow \bullet)$, it is straightforward to verify that p_2 can be derived as concurrent rule again (for the E-dependency object $E = \bullet$). However, $\bullet \hookleftarrow \bullet \hookrightarrow \bullet$ cannot be derived as GCR from these two rules since p_1 does not delete a node.

It turns out that only elements that are deleted by the first rule and created by the second can be identified and incorporated into K'. The next proposition clarifies this connection using the language of initial pushouts.

Definition 8 (Appropriately enhancing). *In a category $\mathcal{C}$ with initial pushouts, let $\rho_1 *_E \rho_2 = L \overset{l}{\hookleftarrow} K \overset{r}{\hookrightarrow} R$ be an E-concurrent rule and $k' : K \hookrightarrow K'$ be an $\mathcal{M}$-morphism such that there exist $\mathcal{M}$-morphisms $l' : K' \hookrightarrow L$ and $r' : K' \hookrightarrow R$ with $l' \circ k' = l$ and $r' \circ k' = r$. Then k' is called* appropriately enhancing *if the following holds (compare Fig. 13): The boundary and context objects $B_{k'}$ and $C_{k'}$ of the initial pushout over k' factorize via $\mathcal{M}$-morphisms $s_L, s_R : B_{k'} \hookrightarrow B_{l_1}, B_{r_2}$ and $t_L, t_R : C_{k'} \hookrightarrow C_{l_1}, C_{r_2}$ as pullback through the initial pushouts over $l_1 : K_1 \hookrightarrow L_1$ and $r_2 : K_2 \hookrightarrow R_2$ in such a way that $k_1 \circ b_{k'} = e''_1 \circ b_{l_1} \circ s_L$ and $k_2 \circ b_{k'} = e''_2 \circ b_{r_2} \circ s_R$.*

On the level of graph elements (in fact: arbitrary categories of presheaves over **Set**), this means the following: When considering K' as a subobject of L via l', the elements of $K' \backslash K$ have to be mapped to elements of $L_1 \backslash K_1$. When considering K' as a subobject of R via r', the elements of $K' \backslash K$ have to be mapped to elements of $R_2 \backslash K_2$.

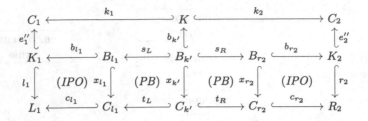

Fig. 13. Definition of *appropriate enhancement.*

Example 4. Figure 14 illustrates the notion of appropriate enhancement using our running example. The two inner squares constitute pullbacks, which means that the additional elements of K', namely Class 2,6 and its incoming reference, are mapped to elements deleted by *removeMiddleMan* via t_L and to elements created by *extractSubclass* via t_R. Moreover, for both C_1 and C_2 the two possible ways to map Class 1,5 from $B_{k'}$ to it coincide.

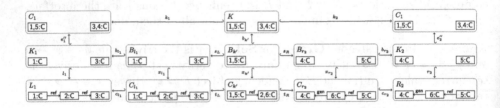

Fig. 14. Illustrating the property of appropriate enhancement.

Proposition 4 (Enhancement characterization of GCRs). *Let $(\mathcal{C}, \mathcal{M})$ be an $\mathcal{M}$-adhesive category with initial pushouts. Given an E-concurrent rule $\rho_1 *_E \rho_2 = L \xleftarrow{l} K \xrightarrow{r} R$ and an $\mathcal{M}$-morphism $k' : K \hookrightarrow K'$ such that there exist $\mathcal{M}$-morphisms $l' : K' \hookrightarrow L$ and $r' : K' \hookrightarrow R$ with $l' \circ k' = l$ and $r' \circ k' = r$, the span $L \xleftarrow{l'} K' \xrightarrow{r'} R$ is derivable as a GCR $\rho_1 *_{E,k} \rho_2$ if and only if k' is appropriately enhancing.*

In **Graph**, the result above characterizes a GCR as a rule whose interface K' enhances the interface K of the enhanced concurrent rule by identifying elements of $L_1 \backslash K_1$ and $R_2 \backslash K_2$ with each other and including them in K'.

Corollary 1 (Enhancement characterization in the category Graph). *In the category of (typed/attributed) graphs, given an E-dependency relation E for two rules, every GCR $\rho_1 *_{E,k} \rho_2$ enhancing $\rho_1 *_E \rho_2$ is obtained in the following way: K' arises by adding new graph elements to K and l' and r' extend the morphisms l and r in such a way that they (i) remain injective graph morphisms and (ii) under l' the image of every newly added element is in $L_1 \backslash K_1$ and in $R_2 \backslash K_2$ under r'.*

Assuming finite graphs only, the number of GCRs enhancing a concurrent rule $\rho_1 *_E \rho_2$ may grow factorially in $\min(|L_1 \backslash K_1|, |R_2 \backslash K_2|)$.

5   A Generalized Concurrency Theorem

In this section, we present our Generalized Concurrency Theorem that clarifies how sequential applications of two rules relate to an application of a GCR derived from them. As a prerequisite, we present a proposition that relates applications of concurrent rules with those of enhancing GCRs. It states that an application of a GCR leads to the same result as one of the enhanced concurrent rule; however, by its application, more elements are preserved (instead of being deleted and recreated).

Proposition 5 (Preservation property of GCRs). *Let $G_0 \Rightarrow_{\rho_1 *_E \rho_2, m} G_2$ be a transformation via concurrent rule $\rho_1 *_E \rho_2$, given by the span $G_0 \xleftarrow{g_0} D \xrightarrow{g_2} G_2$. For any GCR $\rho_1 *_{E,k} \rho_2$ enhancing $\rho_1 *_E \rho_2$, there is a transformation $G_0 \Rightarrow_{\rho_1 *_{E,k} \rho_2, m} G_2$, given by a span $G_0 \xleftarrow{g_0'} D' \xrightarrow{g_2'} G_2$, and a unique $\mathcal{M}$-morphism $k'' : D \hookrightarrow D'$ such that $g_i' \circ k'' = g_i$ for $i = 0, 2$. Moreover, k'' is an isomorphism if and only if the enhancement morphism k' is one.*

Example 5. Every match for the concurrent rule *Ref2Gen_CR* is also one for the GCR *Ref2Gen_GCR*. Moreover, the two results of the according applications will be isomorphic. The above proposition, however, formally captures that the application of *Ref2Gen_CR* will delete more elements than the one of *Ref2Gen_GCR*. The graph intermediately arising during the application of the former (not containing the class to which Class 2 had been matched) properly embeds into the one arising during the application of the latter.

The classical Concurrency Theorem states that a sequence of two rule applications can be replaced by an application of a concurrent rule (synthesis) and, vice versa (analysis). The synthesis is still possible in the case of GCRs. The analysis, however, holds under certain conditions only. Next, we illustrate how the analysis might fail. Subsequently, we state the Generalized Concurrency Theorem.

Example 6. When *Ref2Gen_GCR* is applied at a match that maps Class 2,6 to a node with incoming or outgoing references or generalizations beyond the two references required by the match, the dangling edge condition prevents the applicability of the underlying first rule *removeMiddleMan* at the induced match. The deletion of Class 2 would not be possible because of these additional adjacent edges. Hence, the analysis of the application of *Ref2Gen_GCR* into sequential applications of *removeMiddleMan* and *extractSubclass* fails in that situation.

Theorem 1 (Generalized Concurrency Theorem). *Let $(\mathcal{C}, \mathcal{M})$ be an $\mathcal{M}$-adhesive category with $\mathcal{M}$-effective unions.*

Synthesis. *For each E-related transformation sequence $G_0 \Rightarrow_{\rho_1,m_1} G_1 \Rightarrow_{\rho_2,m_2} G_2$ there exists a direct transformation $G_0 \Rightarrow_{\rho_1*_{E,k}\rho_2,m} G_2$.*

Analysis. *Given a direct transformation $G_0 \Rightarrow_{\rho_1*_{E,k}\rho_2,m} G_2$, there exists an E-related transformation sequence $G_0 \Rightarrow_{\rho_1,m_1} G_1 \Rightarrow_{\rho_2,m_2} G_2$ with $m_1 = m \circ e'_1$ if and only if $G_0 \Rightarrow_{\rho_1,m_1} G_1$ exists, i.e., if and only if ρ_1 is applicable at m_1.*

Remark 1. At least for the case of $\mathcal{M}$-matching and in the presence of $\mathcal{M}$-effective unions, the Concurrency Theorem indeed becomes a corollary to our Generalized Concurrency Theorem. This is due to the observation that a GCR is a concurrent rule if and only if k is an isomorphism (Proposition 1). Similarly, the Generalized Concurrency Theorem subsumes the Short-cut Theorem [14, Theorem 7].

In the case of graphs, the Generalized Concurrency Theorem ensures that the situation illustrated in Example 6 is the only situation in which the analysis of the application of a GCR fails: When restricting to injective matching, a violation of the dangling edge condition is known to be the only possible obstacle to an application of a graph transformation rule [5, Fact 3.11].

6   Related Work

In this paper, we present a construction for generalized concurrent rules (GCRs) based on the double-pushout approach to rewriting. We compare it with existing constructions of concurrent rules that use some categorical setting.

Concerning double-pushout rewriting, after presenting Concurrency Theorems in specific categories (such as [11] for the case of graphs), such a theorem in a rather general categorical setting was obtained by Ehrig et al. [8]. In that work, spans $R_1 \leftarrow D \rightarrow L_2$ are used to encode the information about rule dependency. After (variants of) adhesive categories had been established as an axiomatic basis for double-pushout rewriting [21], the construction of concurrent rules and an according Concurrency Theorem was lifted to that setting: directly in [21] with the dependency information still encoded as span and by Ehrig et al. [5] with the dependency information now encoded as a co-span $R_1 \rightarrow E \leftarrow L_2$. Finally, the last construction, addressing plain rules, has been extended to the case of rules with general application conditions [9]. It is this construction that we present in Definition 5. A generalization of this construction to enable the reuse of graph elements (or object parts in general) as we present it in this paper is new.

Sequential composition of rules has also been presented for other categorical approaches to rewriting, for example, for *single-pushout rewriting* [22], *sesqui-pushout rewriting* [1,23], or for *double-pushout rewriting in context* (at least for special cases) [24]. The theory is most mature in the sesqui-pushout case, where Behr [1] has established a construction of concurrent rules in the setting of $\mathcal{M}$-adhesive categories for rules with general application conditions. It seems that our construction of GCRs would be similarly applicable to sesqui-pushout rewriting; however, applied as is, it would have a restricted expressivity in that context: In the category **Graph**, for instance, a rule application according to the sesqui-pushout semantics implies to (implicitly) delete all edges that are incident to a deleted node. Our construction would have to be extended for being able to identify such implicitly deleted items such that they can be preserved.

In the *cospan DPO approach*, rules are cospans (instead of spans) and the order of computation is switched compared to classical DPO rewriting, i.e., creation precedes deletion of elements [10]. This approach has been used, for example, to simultaneously rewrite a model and its meta-model [25] or to formalize the rewriting of graphs that are typed over a chain of type graphs [26] (here actually as cospan sesqui-pushout rewriting). As creation precedes deletion, the cospan DPO approach intrinsically offers support for certain kinds of information preservation. For example, attribute values of nodes that are to be deleted can be passed to newly created nodes first. However, in the category of graphs, cospan DPO rewriting is subject to virtually the same dangling edge condition as classical DPO rewriting (see [10]). This means, employing the cospan DPO approach instead of classical DPO rewriting does not address the problem of the applicability of rules. Moreover, modeling the kind of information preservation we are interested in (regarding elements deleted by one rule as recreated by a second) would require a specific form of rule composition also in the cospan approach. We do not expect this to be essentially simpler than the construction we provide for the classical DPO approach in this paper.

Behr and Sobociński [3] proved that, in the case of $\mathcal{M}$-matching, the concurrent rule construction is associative (in a certain technical sense; the same holds true for the sesqui-pushout case [1]). Based on that result, they presented the construction of a rule algebra that captures interesting properties of a given grammar. This has served, for example, as a starting point for a static analysis of stochastic rewrite systems [2]. Considering our GCR construction, it is future work to determine whether it is associative as well.

Kehrer et al. [19] addressed the automated generation of edit rules for models based on a given meta-model. Besides basic rules that create or delete model elements, they also generated *move* and *change rules*. It turns out that these can be built as GCRs from their basic rules but not as mere concurrent rules. Additionally, we introduced short-cut rules for more effective model synchronization [13,15,16] and showed here that short-cut rules are special GCRs. Hence, these works suggest that GCRs can capture typical properties of model edit operators. A systematic study of which edit operators can be captured as GCRs (and which GCRs capture typical model edit operators) remains future work.

7   Conclusion

In this paper, we present *generalized concurrent rules* (GCRs) as a construction that generalizes the constructions of concurrent and short-cut rules. We develop our theory of GCRs in the setting of double-pushout rewriting in $\mathcal{M}$-adhesive categories using rules with application conditions applied at $\mathcal{M}$-matches only. In contrast to concurrent rules, GCRs allow reusing elements that are deleted in the first rule and created in the second. As a central result (Theorem 1), we generalize the classical Concurrency Theorem.

From a theoretical point of view, it would be interesting to develop similar kinds of rule composition in the context of other categorical approaches to rewriting like the single- or the sesqui-pushout approach. Considering practical application scenarios, we are most interested in classifying the derivable GCRs of a given pair of rules according to their use and computing those that are relevant for certain applications in an efficient way.

Acknowledgments. This work was partially funded by the German Research Foundation (DFG), project TA294/17-1.

References

1. Behr, N.: Sesqui-pushout rewriting: concurrency, associativity and rule algebra framework. In: Echahed, R., Plump, D. (eds.) Proceedings Tenth International Workshop on Graph Computation Models, GCM@STAF 2019, Eindhoven, The Netherlands, 17th July 2019. EPTCS, vol. 309, pp. 23–52 (2019). https://doi.org/10.4204/EPTCS.309.2
2. Behr, N., Danos, V., Garnier, I.: Combinatorial conversion and moment bisimulation for stochastic rewriting systems. Log. Methods Comput. Sci. **16**(3) (2020). https://lmcs.episciences.org/6628
3. Behr, N., Sobocinski, P.: Rule algebras for adhesive categories. Log. Methods Comput. Sci. **16**(3) (2020). https://lmcs.episciences.org/6615
4. Boehm, P., Fonio, H.-R., Habel, A.: Amalgamation of graph transformations with applications to synchronization. In: Ehrig, H., Floyd, C., Nivat, M., Thatcher, J. (eds.) CAAP 1985. LNCS, vol. 185, pp. 267–283. Springer, Heidelberg (1985). https://doi.org/10.1007/3-540-15198-2_17
5. Ehrig, H., Ehrig, K., Prange, U., Taentzer, G.: Fundamentals of Algebraic Graph Transformation. MTCSAES. Springer, Heidelberg (2006). https://doi.org/10.1007/3-540-31188-2
6. Ehrig, H., Ermel, C., Golas, U., Hermann, F.: Graph and Model Transformation - General Framework and Applications. MTCSAES. Springer, Cham (2015). https://doi.org/10.1007/978-3-662-47980-3
7. Ehrig, H., Golas, U., Habel, A., Lambers, L., Orejas, F.: $\mathcal{M}$-adhesive transformation systems with nested application conditions. Part 1: parallelism, concurrency and amalgamation. Math. Struct. Comput. Sci. **24**(4), 240406 (2014). https://doi.org/10.1017/S0960129512000357
8. Ehrig, H., Habel, A., Kreowski, H.J., Parisi-Presicce, F.: Parallelism and concurrency in high-level replacement systems. Math. Struct. Comput. Sci. **1**(3), 361–404 (1991). https://doi.org/10.1017/S0960129500001353
9. Ehrig, H., Habel, A., Lambers, L.: Parallelism and concurrency theorems for rules with nested application conditions. Electron. Commun. Eur. Assoc. Softw. Sci. Technol. **26** (2010). https://doi.org/10.14279/tuj.eceasst.26.363
10. Ehrig, H., Hermann, F., Prange, U.: Cospan DPO approach: an alternative for DPO graph transformations. Bull. EATCS **98**, 139–149 (2009)
11. Ehrig, H., Rosen, B.K.: Parallelism and concurrency of graph manipulations. Theoret. Comput. Sci. **11**(3), 247–275 (1980). https://doi.org/10.1016/0304-3975(80)90016-X

12. Fowler, M.: Refactoring - Improving the Design of Existing Code. Addison Wesley object technology series. Addison-Wesley (1999). http://martinfowler.com/books/refactoring.html

13. Fritsche, L., Kosiol, J., Möller, A., Schürr, A., Taentzer, G.: A precedence-driven approach for concurrent model synchronization scenarios using triple graph grammars. In: de Lara, J., Tratt, L. (eds.) Proceedings of the 13th ACM SIGPLAN International Conference on Software Language Engineering (SLE 2020), November 16–17, 2020, Virtual, USA. ACM (2020). https://doi.org/10.1145/3426425.3426931

14. Fritsche, L., Kosiol, J., Schürr, A., Taentzer, G.: Short-Cut Rules - sequential composition of rules avoiding unnecessary deletions. In: Mazzara, M., Ober, I., Salaün, G. (eds.) STAF 2018. LNCS, vol. 11176, pp. 415–430. Springer, Cham (2018). https://doi.org/10.1007/978-3-030-04771-9_30

15. Fritsche, L., Kosiol, J., Schürr, A., Taentzer, G.: Efficient model synchronization by automatically constructed repair processes. In: Hähnle, R., van der Aalst, W. (eds.) FASE 2019. LNCS, vol. 11424, pp. 116–133. Springer, Cham (2019). https://doi.org/10.1007/978-3-030-16722-6_7

16. Fritsche, L., Kosiol, J., Schürr, A., Taentzer, G.: Avoiding unnecessary information loss: correct and efficient model synchronization based on triple graph grammars. Int. J. Software Tools Technol. Transf. (2020). https://doi.org/10.1007/s10009-020-00588-7

17. Golas, U., Habel, A., Ehrig, H.: Multi-amalgamation of rules with application conditions in $\mathcal{M}$-adhesive categories. Math. Struct. Comput. Sci. **24**(4) (2014). https://doi.org/10.1017/S0960129512000345

18. Habel, A., Pennemann, K.H.: Correctness of high-level transformation systems relative to nested conditions. Math. Struct. Comput. Sci. **19**(2), 245–296 (2009). https://doi.org/10.1017/S0960129508007202

19. Kehrer, T., Taentzer, G., Rindt, M., Kelter, U.: Automatically deriving the specification of model editing operations from meta-models. In: Van Van Gorp, P., Engels, G. (eds.) ICMT 2016. LNCS, vol. 9765, pp. 173–188. Springer, Cham (2016). https://doi.org/10.1007/978-3-319-42064-6_12

20. Kosiol, J., Taentzer, G.: A Generalized Concurrent Rule Construction for Double-Pushout Rewriting (2021). https://arxiv.org/abs/2105.02309

21. Lack, S., Sobociński, P.: Adhesive and quasiadhesive categories. Theoret. Inform. Appl. **39**(3), 511–545 (2005). https://doi.org/10.1051/ita:2005028

22. Löwe, M.: Algebraic approach to single-pushout graph transformation. Theor. Comput. Sci. **109**(1&2), 181–224 (1993). https://doi.org/10.1016/0304-3975(93)90068-5

23. Löwe, M.: Polymorphic Sesqui-Pushout graph rewriting. In: Parisi-Presicce, F., Westfechtel, B. (eds.) ICGT 2015. LNCS, vol. 9151, pp. 3–18. Springer, Cham (2015). https://doi.org/10.1007/978-3-319-21145-9_1

24. Löwe, M.: Double-Pushout rewriting in context. In: Guerra, E., Orejas, F. (eds.) ICGT 2019. LNCS, vol. 11629, pp. 21–37. Springer, Cham (2019). https://doi.org/10.1007/978-3-030-23611-3_2

25. Mantz, F., Taentzer, G., Lamo, Y., Wolter, U.: Co-evolving meta-models and their instance models: a formal approach based on graph transformation. Sci. Comput. Program. **104**, 2–43 (2015). https://doi.org/10.1016/j.scico.2015.01.002

26. Wolter, U., Macías, F., Rutle, A.: Multilevel typed graph transformations. In: Gadducci, F., Kehrer, T. (eds.) ICGT 2020. LNCS, vol. 12150, pp. 163–182. Springer, Cham (2020). https://doi.org/10.1007/978-3-030-51372-6_10

Transformations of Reaction Systems Over Categories by Means of Epi-Mono Factorization and Functors

Aaron Lye[✉]

Department of Computer Science, University of Bremen,
P.O.Box 33 04 40, 28334 Bremen, Germany
lye@uni-bremen.de

Abstract. A categorical approach to reaction systems is a generalization and unification of the intensely studied set-based and graph-based reaction systems such that a wider spectrum of data structures becomes available on which reaction systems can be based. Many types of graphs, hypergraphs, and graph-like structures are covered. As a class of suitable categories, *eiu*-categories have been introduced, which are closely related to well-known adhesive categories. In this paper, transformations of reaction systems over *eiu*-categories by means of epi-mono factorization and functors are investigated.

1 Introduction

A particular research strand of natural computing is the investigation of information processing taking place in nature [1]. One research topic is the computational nature of biochemical reactions. In 2007, the seminal concept of set-based reaction systems was introduced by Ehrenfeucht and Rozenberg in [2] to provide a formal framework for the modeling of biochemical processes (taking place in the living cell). The underlying idea is that interactions as well as the functioning of reactions are based on the mechanisms of facilitation and inhibition. Since then the framework has been intensely studied (see, e.g., [3–10]) and reaction systems turned out to be a novel paradigm of interactive and massively parallel computation suitable for modeling information processing in various fields beyond biochemistry.

A set-based reaction system consists of a finite background set B and a set of reactions A each of which is a triple of subsets of B called reactant, inhibitor and product respectively. A reaction is enabled on a state (being a subset of B) if the reactant is inside the state and the inhibitor outside. All enabled reactions of A are applied to some state in parallel yielding the union of all their products as results. Starting from initial states, the iterated applications of enabled reactions of A define the dynamic semantics of a reaction system where, before each step, a context set can be added to the current state making the processes interactive in this way.

© Springer Nature Switzerland AG 2021
F. Gadducci and T. Kehrer (Eds.): ICGT 2021, LNCS 12741, pp. 40–59, 2021.
https://doi.org/10.1007/978-3-030-78946-6_3

In [11,12] Kreowski and Rozenberg introduced graph surfing in graph-based reaction systems as a novel kind of graph transformation. They consider simple edge-labeled directed graphs.

But there are many further structures on which reaction systems can be based in a meaningful way. When the same kind of constructs and constructions can be considered for a spectrum of underlying structures, it may be worthwhile to come up with a categorical framework. In this way, the notions of interest can be defined once and for all and then used whenever certain structures form a category fitting into the framework. Recently, Kreowski and the author proposed such a categorical approach to reaction systems (cf. [13]) as a generalization and unification of the two approaches. In this framework, a wider spectrum of data structures becomes available on which reaction systems can be based. Many types of graphs, hypergraphs, and graph-like structures are covered. As a class of suitable categories, *eiu*-categories have been introduced, which are slightly more general than the well-known adhesive categories (cf. [14]).

Whenever one has a class of entities, one may try to use them as objects of a category by choosing suitable morphisms. Therefore, one may ask how reaction systems over a category may be provided with a meaningful notion of morphisms. In [15] it is shown that monomorphisms between backgrounds provide morphisms between reaction systems. In this paper, we generalize this idea by proposing a transformation based on epi-mono factorization. Hence, arbitrary morphisms between backgrounds give morphisms between reaction systems provided that the *eiu*-category has an epi-mono factorization. Furthermore, we introduce the notion of *eiu*-preserving functors and analyze the behavior of mapped reaction systems and interactive processes by *eiu*-preserving embedding functors.

The paper is organized as follows. Section 2 provides the notion of an *eiu*-category. In Sect. 3, we recall the notion of reaction systems over *eiu*-categories. In Sect. 4 and 5, we present the transformations of reaction systems over *eiu*-categories based on epi-mono factorization and functors, respectively. Section 6 concludes the paper.

2   The Categorical Prerequisites

In this section, the categorical prerequisites are provided that allow us to define reaction systems over an *eiu*-category in the next section. For the well-known categorical notions including subobjects, finite objects, initial objects, pullbacks, and special colimits cf., e.g., [13,15–18].

A category $\mathbf{C}$ is an *eiu-category* if $\mathbf{C}$ has an initial object $INIT$, and for every finite object B, pullbacks of pairs of subobjects of B, as well as colimits of the sets of all pairwise pullbacks of sets of subobjects of every finite object B subject to the following conditions: (1) $INIT$ has only itself as subobject and the initial morphism into B is a monomorphism and (2) the universal morphism from $COLIMIT(PB(S))$ into B for every set S of subobjects of B is a monomorphism, where $PB(S)$ is the set of all pairwise pullbacks and $COLIMIT(PB(S))$ the colimit object of the colimit of $PB(S)$.

We use the following notions and notations for eiu-categories and every of its finite objects B.

1. The subobject represented by the initial morphism into B is called *empty subobject* of B and denoted by $empty_B \colon INIT \to B$.
2. As pullbacks are stable under monomorphisms, the pullback morphisms $p_i' \colon PB(p_1, p_2) \to P_i$ of two subobjects $p_i \colon P_i \to B$ for $i = 1, 2$ are monomorphisms. Further, because monomorphisms are closed under composition, $p_1' \circ p_1 = p_2' \circ p_2$ represents a subobject of B called *intersection* of p_1 and p_2 which is denoted by $p_1 \cap p_2 \colon P_1 \cap P_2 \to B$.
3. Given a set S of subobjects of B, the universal morphism from $COLIMIT(PB(S))$ into B represents a subobject of B called *union* of S which is denoted by $union(S) \colon UNION(S) \to B$. We may write $p_1 \cup p_2$ for the binary (effective) $union(\{p_1, p_2\})$.

The initials e, i, and u of the three concepts are used to name the category. Intersection is a standard notion. With the terminology *empty subobject* and *union* we emphasize that we consider special initial morphisms and colimits.

Well-known and often used categories as well as certain diagram categories are eiu-categories, provided that the underlying category is an eiu-category. The following categories are examples for eiu-categories.

1. The category **Sets** of sets.
2. The category Σ-**Sets** of Σ-labeled sets for some alphabet Σ.
3. The category **Pos** of partially ordered sets.
4. The category **Graphs** of directed (unlabeled) graphs.
5. The category Σ-**Graphs** of Σ-graphs for some alphabet Σ.
6. The category (Σ_V, Σ_E)-**Graphs** of directed vertex- and edge-labeled graphs.
7. The category **BipartiteGraphs** of bipartite directed graphs.
8. The category Σ-**Hypergraphs** of Σ-hypergraphs for some alphabet Σ.
9. The category **Graphs**$^{\bullet \to TG}$ of *TG-typed graphs* for some *type graph TG*.

With the exception of **Pos** these categories are adhesive. In fact, every adhesive category with empty subobjects and pushouts being binary union is an eiu-category.

3   Reaction Systems Over eiu-Categories

In this section, we recall the notion of reaction systems over an eiu-category. This can be done in a straightforward way by replacing every occurrence of "(sub)set/(sub)graph" in the definition of set/graph-based reaction systems by "(sub)object" with one exception: the enabledness with respect to the inhibitor. The graph-based inhibitor (consisting of sets of vertices and edges) has not a direct counterpart as categorical objects do not provide explicit internal information like vertices and edges of graphs. Therefore, we replace it by a subobject $i \colon I \to B$ of the background like reactant and product accompanied by a subobject $i_0 \colon I_0 \to I$. This allows to require that the intersection of i and a current state is included in i_0 so that the "complement" of i and i_0 is forbidden.

3.1 Reaction Systems Over C

Let **C** be an *eiu*-category. Then we can define reaction systems over **C** in a way analogous to set-based and graph-based reaction systems.

Definition 1. 1. Let B be a finite object in **C**. A *reaction* over B is a triple $a = (r \colon R \to B, (i \colon I \to B, i_0 \colon I_0 \to I), p \colon P \to B)$ where r and p are non-empty subobjects of B, i is a subobject of B and i_0 is a subobject of I. The subobject r is called *reactant*, the pair (i, i_0) is called *inhibitor*, and p is called *product*.

2. A *state* $t \colon T \to B$ is a subobject of B.

3. A reaction $a = (r, (i, i_0), p)$ is *enabled* on a state t, denoted by $en_a(t)$, if $r \subseteq t$ and $t \cap i \subseteq i \circ i_0$, i.e., there is a monomorphism $s \colon R \to T$ with $r = t \circ s$ and, for the intersection $(T \cap I, i', t')$ of t and i, there is a monomorphism $s' \colon T \cap I \to I_0$ with $t \cap i = i \circ i_0 \circ s'$. The situation is illustrated in the following diagram.

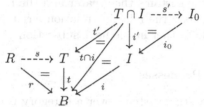

If a reaction is *disabled*, then this is denoted by $\overline{en}_a(t)$.

4. The *result* of a reaction a on a state t is $res_a(t) = p_a$ for $en_a(t)$ and $res_a(t) = empty_B$ otherwise.

5. The *result* of a set of reactions A on a state t is $res_A(t) = union(\{res_a(t) \mid a \in A\})$.

6. A *reaction system over* **C** is a pair $\mathcal{A} = (B, A)$ consisting of some finite object B in **C**, called *background*, and a finite set A of reactions over B. We may write $(B, A)_\mathbf{C}$ to indicate the underlying category.

7. The *result* of $\mathcal{A}$ on a state t is the result of A on t. It is denoted by $res_\mathcal{A}(t)$.

Remark 1. Some basic properties of enabledness and results which are known for set- and graph-based reaction systems carry over to reaction systems over a category.

1. A current state vanishes completely. But it or some subobject of it may be reproduced by the products of enabled reactions.

2. $res_\mathcal{A}(t)$ is uniquely defined for every state t so that $res_\mathcal{A}(t)$ is a function on the set of states of B.

3. All reactions contribute to $res_\mathcal{A}(t)$ in a maximally parallel and cumulative way. There is never any conflict.

4. As the addition of the empty subobject to a union of subobjects does not change the union, $res_\mathcal{A}(t) = res_{\{a \in A \mid en_a(t)\}}(t)$ holds for all states t.

5. As the intersection of a subobject and the empty subobject is empty, a reaction $a = (r, (empty_B, 1_{INIT}), p)$ is enabled on a state t if $r \subseteq t$. The empty inhibitor, denoted by $-$, has no effect. Therefore, the reaction is called *uninhibited*.
6. Let $a = (r, (i, i_0), p)$ be a reaction. If $r \cap i \not\subseteq i \circ i_0$, then a is never enabled.

Example 1. Consider the reaction system $\mathcal{A}^> = (B^>, A^>)$ over **Pos**, where $B^> = (V, E)$ for some finite set V and a parital order $E \subseteq V \times V$ (reflexive, antisymmetric and transitive) and $A^>$ consists of two types of uninhibited reactions:

1. $((\{x, y, z\}, \langle (x, y), (y, z) \rangle), -, (\{y, z\}, \langle (y, z) \rangle))$ for $x, y, z \in V, x \neq y, x \neq z, y \neq z$
2. $((\{y, z\}, \langle (y, z) \rangle), -, (\{z\}, \{(z, z)\}))$ for $y, z \in V, y \neq z$

where $\langle S \rangle$ denotes the partial order generated by the closure of the relation $S \subseteq E$. The first type sustains the greater and the respective reflexive relation; the second sustains only the reflexive relation wrt the second element. We will discuss this example further in the next subsection.

3.2   Interactive Processes

The definition of reaction systems over a category is chosen in such a way that the semantic notion of interactive processes can be carried over directly from the set-based and graph-based cases. Starting from initial states, the iterated applications of enabled reactions of A define the dynamic semantics of a reaction system where, before each step, a context can be added to the current state making the processes interactive.

Definition 2. 1. An *interactive process* $\pi = (\gamma, \delta)$ on $\mathcal{A} = (B, A)_{\mathbf{C}}$ consists of two sequences of subobjects of B $\gamma = c_0, \ldots, c_n$ and $\delta = d_0, \ldots, d_n$ for some $n \geq 1$ such that $d_i = res_A(c_{i-1} \cup d_{i-1})$ for $i = 1, \ldots, n$. The sequence γ is called *context sequence*, the sequence δ is called *result sequence* where d_0 is called *start*, and the sequence $\tau = t_0, \ldots, t_n$ with $t_i = c_i \cup d_i$ for $i = 0, \ldots, n$ is called *state sequence*.
2. π is called *context-independent* if $c_i \subseteq d_i$ for $i = 0, \ldots, n$.
3. τ is *repetition-free* if $t_i \neq t_j$ for all i, j with $0 \leq i < j \leq n$.

Example 2. Reconsider the reaction system $\mathcal{A}^>$ presented in Example 1. Let d_0 be an arbitrary subposet of B and let $c_i = (\emptyset, \emptyset)$ for each $i \in \mathbb{N}$. Then every state t_i in the state sequence $\tau^> = t_0, \ldots, t_n$ is just the result of applying all enabled reactions to the previous state, i.e., $t_{i+1} = res_{\mathcal{A}^>}(t_i)$. These successive applications of all the reactions produce maximal elements.

There are many situations where posets arise. For example B may be seen as a finite set of natural numbers equipped with the relation of divisibility or a finite vertex set of a directed acyclic graph ordered by reachability. When considering the latter, this interactive process produces finally reached vertices.

4 Transformation of Reaction Systems by Means of Epi-Mono Factorization

In this section, we show that, given a reaction system $\mathcal{A} = (B, A)$ over $\mathbf{C}$, where $\mathbf{C}$ has epi-mono factorization, a morphism $f \colon B \to B'$ induces a reaction system $f(\mathcal{A})$ by composing all the components of reactions with f, constructing the epi-mono factorization of each composed morphism and using the resulting monomorphisms of the factorization as components in the new reactions.

In Subsect. 4.1, we recall the notion of an epi-mono factorization. In Subsect. 4.2, we show that in *eiu*-categories with epi-mono factorization the factorization behaves well with respect to subobject inclusion, intersection and union. Afterwards, in Subsect. 4.3, we apply the transformation to reaction systems over *eiu*-categories with epi-mono factorization. In Subsect. 4.4, we present a result for the semantic notion of transformed interactive processes. Finally, in Subsect. 4.5, we define the category of reaction systems over *eiu*-categories with epi-mono factorization. This generalizes Theorem 2 in [15], where $f \colon B \to B'$ was restricted to be a monomorphism.

4.1 Epi-Mono Factorization

In general, a factorization of a morphism decomposes it into morphisms with special properties. In an epi-mono factorization, these morphisms are an epimorphism and a monomorphism.

Definition 3. 1. Let $f \colon A \to B, e \colon A \to C$, and $m \colon C \to B$ with $f = m \circ e$. If e is an epimorphism and m is a monomorphism, then e and m are called *epi-mono factorization* of f.
2. Let $\mathbf{C}$ be a category. If for every morphism in $Mor_{\mathbf{C}}$ such an epi-mono factorization exists and this decomposition is unique up to isomorphism, then $\mathbf{C}$ is said to have an epi-mono factorization.

As a naming convention, we denote the epimorphism by e_f, the monomorphism by m_f and the intermediate object by EM_f.

It is well-known that, the categories **Sets**, Σ-**Graphs**, and **TypedGraphs** have epi-mono factorizations. But factorization system exists on any (elementary) topos, i.e., any category which has finite limits, is Cartesian closed, and has a subobject classifier. This definition is quite close to the definition of an *eiu*-category. Indeed, factorization system exists on any pretopos.

4.2 Epi-Mono Factorization of Composed Morphisms

In particular, we are interested in the epi-mono factorization of composed morphisms $f \circ p$ where p is a monomorphism. The situation is illustrated in the following diagram.

$$
\begin{array}{ccc}
P & \xrightarrow{\ e_{f \circ p}\ } & EM_{f \circ p} \\
{\scriptstyle p}\downarrow & \searrow{\scriptstyle f \circ p} & \downarrow{\scriptstyle m_{f \circ p}} \\
B & \xrightarrow{\ f\ } & B'
\end{array}
$$

The epi-mono factorizations of such composed morphisms have useful properties with respect to subobject inclusion, intersection and union.

Lemma 1. *Let $f: B \to B'$ be a morphism.*

1. *Let $p_1: P_1 \to B, p_2: P_2 \to B$ be two subobjects of B. Then*
   (a) $p_1 \subseteq p_2$ implies $m_{f \circ p_1} \subseteq m_{f \circ p_2}$. More specifically, $p_1 = p_2 \circ s$ implies
   $$m_{f \circ p_1} \cong m_{f \circ p_2} \circ m_{e_{f \circ p_2} \circ s}.$$
   (b) $m_{f \circ (p_1 \cap p_2)} \subseteq m_{f \circ p_1} \cap m_{f \circ p_2}$.
2. *Let S be a set of subobjects of B, let $f(S) = \{f \circ p \mid p \in S\}$ and $m_f(S) = \{m_x \mid x \in f(S)\}$. Then $union(m_f(S))\}) \cong m_{f \circ union(S)}$.*

Proof. 1a. Let $P_i \xrightarrow{e_{f \circ p_i}} EM_{f \circ p_i} \xrightarrow{m_{f \circ p_i}} B'$ for $i = 1, 2$ be the epi-mono factorizations of the composed morphisms. Then the epi-mono factorization of $e_{f \circ p_2} \circ s$, the equations $f \circ p_1 = m_{f \circ p_1} \circ e_{f \circ p_1} = m_{f \circ p_2} \circ m_{e_{f \circ p_2} \circ s} \circ e_{e_{f \circ p_2} \circ s}$ and the uniqueness of epi-mono factorizations up to isomorphism gives us an isomorphism $z: EM_{e_{f \circ p_2} \circ s} \to EM_{f \circ p_1}$ such that $e_{f \circ p_1} \circ z^{-1} = e_{e_{f \circ p_2} \circ s}$ and $m_{f \circ p_1} \circ z = m_{f \circ p_2} \circ m_{e_{f \circ p_2} \circ s}$. Consequently, $m_{f \circ p_2} \subseteq m_{f \circ p_1}$. The situation is illustrated in the following diagram.

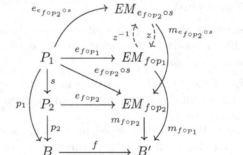

1b follows from 1a, the universal property of pullbacks (yielding the morphism $u: EM_{f \circ (p_1 \cap p_2)} \to EM_{f \circ p_1} \cap EM_{f \circ p_2}$) and the fact that $m_{e_{f \circ p_i} \circ p_i'}$ being a monomorphism implies that u is a monomorphism. The situation is illustrated in the following diagram.

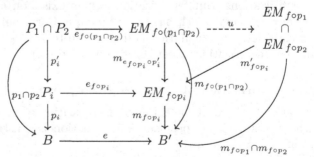

The proof of Point 2 is more complicated but uses similar arguments.

For $union(S)$, for each $p \in S$ and for each pullback $(PB(p_i, p_j), p_i', p_j') \in PB(S)$, we construct the epi-mono factorizations of the composed morphisms.

Point 1 gives respective subobject inclusions for the monomorphisms of the epi-mono factorizations. Then we construct $union(m_f(S))$ with injections m''_{fop} for each $m_{fop} \in m_f(S)$ and pairwise pullbacks $(PB(m_{fop_i}, m_{fop_j}), m'_{fop_i}, m'_{fop_j}) \in PB(m_f(S))$.

By definition $m_{fop_i} \cap m_{fop_j} = m_{fop_i} \circ m'_{fop_i}$ and for p_j analogously. Let $x_i \cong m_{e_{founion(S)} \circ p''_i}$ and x_j analogously. Then $m_{fop_i} = m_{founion(S)} \circ x_i$. Therefore, $m_{founion(S)} \circ x_i \circ m'_{fop_i} = m_{founion(S)} \circ x_j \circ m'_{fop_j}$, and because $m_{founion(S)}$ is a monomorphism, $x_i \circ m'_{fop_i} = x_j \circ m'_{fop_j}$. The universal property of union, therefore, gives us a unique morphism $y \colon UNION(m_f(S)) \to EM_{founion(S)}$, which is a monomorphism because $union(m_f(S))$ is. This proves $union(m_{foS}\}) \subseteq m_{founion(S)}$.

Altogether the situation is illustrated in the following diagram.

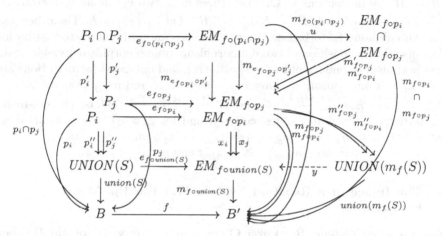

Let $\hat{p}_i = m''_{fop_i} \circ e_{fop_i}$ and $\hat{p}_j = m''_{fop_j} \circ e_{fop_j}$. Then $f \circ (p_i \cap p_j) = union(m_f(S)) \circ \hat{p}_i \circ p'_i = union(m_f(S)) \circ \hat{p}_j \circ p'_j$. Further, $union(m_f(S))$ being a monomorphism implies $\hat{p}_i \circ p'_i = \hat{p}_j \circ p'_j$. Consequently, the universal property of the colimit gives a morphism $z \colon UNION(S) \to UNION(m_f(S))$.

Let $UNION(S) \xrightarrow{e_z} EM_z \xrightarrow{m_z} UNION(m_f(S))$ be the epi-mono factorization of z. Then $union(m_f(S)) \circ m_z \circ e_z = m_{founion(S)} \circ e_{founion(S)}$. Further, because epi-mono factorizations are unique up to isomorphisms, there is an isomorphism $x \colon EM_{founion(S)} \to EM_z$. The composition $m_z \circ x$ is a monomorphism, meaning $m_{founion(S)} \subseteq union(m_f(S))$. Together with $union(m_f(S)) \subseteq m_{founion(S)}$ this implies $union(m_f(S)) \cong m_{founion(S)}$.

Remark 2. For the epi-mono factorization of the composition $f \circ empty_B$ ($= init_{B'}$ by definition) we assume the following two properties: $init_{B'} = empty_{B'}$ and $e_{init_{B'}}$ is an isomorphism. In many categories a morphism, which is both a monomorphism and an epimorphism, is an isomorphism. In these categories the first statement implies the second.

The factorization also behaves well with respect to the composition of epi-mono factorizations as the following lemma shows.

Lemma 2. *1. Epi-mono factorizations are closed under composition.*

2. *Let* $P \xrightarrow{p} B \xrightarrow{f} B' \xrightarrow{g} B''$ *be a composition of morphisms, where* p *is a monomorphism. Then* $m_{g \circ f \circ p} \cong m_{g \circ m_{f \circ p}}$. *The situation is illustrated in the following diagram.*

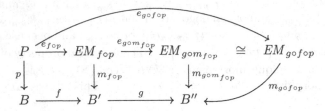

Proof. 1. Let $\mathbf{C}$ be a category with epi-mono factorization. Let $f \colon B \to B'$ and $g \colon B' \to B''$ be morphisms in $\mathbf{C}$. Then there exist two epi-mono factorizations $B \xrightarrow{e_f} EM_f \xrightarrow{m_f} B'$ and $B' \xrightarrow{e_g} EM_g \xrightarrow{m_g} B''$. Let $e_g \circ m_f = h$. Then, because $\mathbf{C}$ has an epi-mono factorization, there exist also an epi-mono factorization for h. Because the composition of two epimorphisms (monomorphisms) yields again an epimorphism (monomorphism, respectively), and epi-mono factorizations are unique up to isomorphism, we have $m_{g \circ f} \cong m_g \circ m_h$ and $m_{g \circ f} \cong e_h \circ e_f$.

2. Let $P \xrightarrow{e_{f \circ p}} EM_{f \circ p} \xrightarrow{m_{f \circ p}} B'$ and $B' \xrightarrow{e_g} EM_g \xrightarrow{m_g} B''$ be the epi-mono factorizations of $f \circ p$ and g, respectively, and let $e_g \circ m_{f \circ p} = h = m_h \circ e_h$. Then $g \circ m_{f \circ p} = m_g \circ h = m_g \circ (m_h \circ e_h) = (m_g \circ m_h) \circ e_h = m_{g \circ f \circ p} \circ e_h = m_{g \circ f \circ p} \circ e_{e_g \circ m_{f \circ p}}$. Hence, $m_{g \circ m_{f \circ p}} \cong m_{g \circ f \circ p}$ and $e_{g \circ m_{f \circ p}} \cong e_{e_g \circ m_{f \circ p}}$.

4.3 The Image of a Reaction System Wrt the Epi-Mono Factorization

Given a reaction system (B, A) over $\mathbf{C}$ and a morphism with domain B. Then the componentwise composition and epi-mono factorization of all components of reactions in A gives us a new reaction system.

Definition 4. Let $\mathcal{A} = (B, A)_{\mathbf{C}}$ be a reaction system and let $f \colon B \to B'$ be morphism satisfying $f \circ empty_B = empty_{B'}$.

1. Let $a = (r, (i, i_0), p) \in A$. Then $\overline{m}_f(a) = (m_{f \circ r}, (m_{f \circ i}, m_{e_{f \circ i} \circ i_0}), m_{f \circ p})$ is the *image of the reaction* wrt the epi-mono factorization.
2. $\overline{m}_f(a)$ is called *consistent* if $m_{f \circ r} \cap m_{f \circ i} \subseteq m_{f \circ i \circ i_0}$.
3. Let $\overline{m}_f(A) = \{\overline{m}_f(a) \mid a \in A\}$. Then $\overline{m}_f(\mathcal{A}) = (B', \overline{m}_f(A))$ is the *image of the reaction system* wrt the epi-mono factorization.
4. Let $\widetilde{m}_f(A) = \{\overline{m}_f(a) \mid m_{f \circ r} \cap m_{f \circ i} \subseteq m_{f \circ i \circ i_0}\} \subseteq \overline{m}_f(A)$. Then $\widetilde{m}_f(\mathcal{A}) = (B', \widetilde{m}_f(A))$ is the *consistent image of the reaction system* wrt the epi-mono factorization.

Example 3. Reconsider the reaction system $\mathcal{A}^>$ presented in Example 1. Let $f \colon (V, E) \to (V', E')$ be a poset morphism induced by an underlying non-injective morphism $f \colon V \to V'$ satisfying $(f(x), f(y)) \in E'$ for all $(a, b) \in E$. Then the image of $\mathcal{A}^>$ wrt the epi-mono factorization is as follows: $\overline{m}_f(\mathcal{A}^>) =$

$((f(V), f(E)), \overline{m}_f(A^>))$, where $f(V) = \{f(V) \mid v \in V\}$, $f(E) = \{(f(x), f(y)) \mid (x, y) \in E\}$ and for the images of reactions in $\overline{m}_f(A^>)$ several cases occur. Let $f(V_{xyz}) = \{f(x), f(y), f(z)\}$ and $f(V_{xy}) = \{f(x), f(y)\}$ for $a, b, c \in V$. Then for reactions of the first type we have four new types:

- $((f(V_{xyz}), \langle (f(x), f(y)), (f(y), f(z)) \rangle), -, (f(V_{yz}), \langle (f(y), f(z)) \rangle))$
  for $x, y, z \in V$ with $f(x) \neq f(y), f(x) \neq f(z), f(y) \neq f(z)$
- $((f(V_{yz}), \langle (f(y), f(z)) \rangle), -, (f(V_{yz}), \langle (f(y), f(z)) \rangle))$
  for $x, y, z \in V$ with $f(x) = f(y) \neq f(z)$.
- $((f(V_{xy}), \langle (f(x), f(y)) \rangle), -, (\{f(y)\}, \langle (f(y), f(y)) \rangle))$
  for $x, y, z \in V$ with $f(x) \neq f(y) = f(z)$.
- $((\{f(x)\}), \langle (f(x), f(x)) \rangle), -, (\{f(x)\}, \langle (f(x), f(x)) \rangle))$
  for $x, y, z \in V$ with $f(x) = f(y) = f(z)$.

The first is like before; the second sustains the image of the second element in the sequence $(x, y), (y, z)$ if $f(x) = f(y)$; the third acts like the reaction of the second type in the original reaction sytem; and the fouth sustains reflexivity.

For reactions of the second type we have two new types:

- $((f(V_{yz}), \langle (f(y), f(z)) \rangle), -, (f(V_{yz}), \langle (f(z), f(z)) \rangle))$
  for $y, z \in V$ with $f(y) \neq f(z)$
- $((\{f(z)\}, \langle (f(z), f(z)) \rangle), -, (\{f(z)\}, \langle (f(z), f(z)) \rangle))$
  for $y, z \in V$ with $f(y) = f(z)$.

The first is like before and the second only sustains reflexitivy.

The behaviour of the reaction system is hence very different to the original one, e.g. when considered as vertices and edges of a directed acyclic graph, then outgoing and incoming edges of vertices which are merged by f arc sustained.

Note that, because every reaction is uninhibited, the image of this reaction system is consistent.

In general, $\overline{m}_f(A)$ and $\tilde{m}_f(A)$ have the following properties.

Lemma 3. *Let* $t: T \to B$ *be a state of* A *and* $m_{fot}: EM_{fot} \to B'$ *the corresponding state in* $\overline{m}_f(A)$. *Then the following holds.*

1. *If* $\overline{m}_f(a) \in \overline{m}_f(A) \setminus \tilde{m}_f(A)$, *then* $\overline{m}_f(a)$ *is never enabled*,
2. *If* $\overline{m}_f(a) \in \tilde{m}_f(A)$, *then* $en_a(t)$ *implies* $en_{\overline{m}_f(a)}(m_{fot})$,
3. $m_{foresa(t)} \subseteq res_{\overline{m}_f(a)}(m_{fot})$ *for* $\overline{m}_f(a) \in \tilde{m}_f(A)$.

Proof. 1. Let $\overline{m}_f((r, (i, i_0), p)) \in \overline{m}_f(A) \setminus \tilde{m}_f(A)$. Then $m_{for} \cap m_{foi} \not\subseteq m_{foioi_0}$. Consequently, $m_{foioi_0} \subseteq m_{fot} \cap m_{foi}$ for any state m_{fot} satisfying $m_{for} \subseteq m_{fot}$. This means the reaction is disabled because of the inhibitor. For any state m_{fot} satisfying $m_{fot} \subseteq m_{for}$ the reaction is disabled because of the reactant.

2. Given a reaction $a = (r, (i, i_0), p)$ and a state t in A, $en_a(t)$ means $r \subseteq t$ and $t \cap i \subseteq i \circ i_0$. By Lemma 1 Point 1a, we have that $r \subseteq t$ implies $m_{for} \subseteq m_{fot}$, $t \cap i \subseteq i \circ i_0$ implies $m_{fo(t \cap i)} \subseteq m_{fo(io i_0)}$, and $m_{fo(io i_0)} \cong m_{foi} \circ m_{efoioi_0}$. Lemma 1 Point 1b gives us $m_{fot} \cap m_{foi} \subseteq m_{fo(t \cap i)}$. Hence, $m_{fot} \cap m_{foi} \subseteq m_{foi} \circ m_{efoioi_0}$. Moreover, by assuming $\overline{m}_f(a) \in \overline{m}_f(A)$, $m_{for} \cap m_{foi} \subseteq m_{foioi_0}$. Therefore, $en_{\overline{m}_f(a)}(m_{fot})$.

3. Let $\overline{m}_f(a) \in \tilde{m}_f(A)$. There are two cases to consider using the definition of results: $m_{fores_a(t)} = m_{fop}$ if $en_a(t)$ and $m_{fores_a(t)} = m_{foemptyB} = emptyB'$ otherwise. However, $res_{\overline{m}_f(a)}(m_{fot}) = m_{fop}$ if $en_{\overline{m}_f(a)}(f \circ t)$ and $res_{\overline{m}_f(a)}(m_{fot}) = emptyB'$ otherwise. Whenever, $en_a(t)$ implies $en_{\overline{m}_f(a)}(f \circ t)$, the subobjects are equal. If $en_{\overline{m}_f(a)}(f \circ t)$ but $\overline{en}_a(t)$, then $res_{\overline{m}_f(a)}(m_{fot}) = m_{fop}$ and $m_{fores_a(t)} = emptyB'$.

Corollary 1. $m_{fores_a(t)} = res_{\overline{m}_f(a)}(m_{fot})$ *if enabledness is reflected.*

If enabledness is reflected, then precise conclusions can be drawn for the transformed system. However, in general we get the following result.

Theorem 1. $m_{fores_A(t)} \subseteq res_{\tilde{m}_f(A)}(m_{fot}) = res_{\overline{m}_f(A)}(m_{fot}).$

Proof. Using the definition of results of reaction systems and sets of reactions as well as Point 2 of Lemma 1, one gets as stated: $m_{fores_A(t)} = m_{fores_A(t)} = m_{founion(\{res_a(t)|a \in A\})} \cong union(\{m_{fores_a(t)} \mid a \in A\}) \subseteq union(\{res_{\overline{m}_f(a)}(m_{fot}) \mid \overline{m}_f(a) \in \tilde{m}_f(A)\}) = res_{\tilde{m}_f(A)}(m_{fot}) = res_{\overline{m}_f(A)}(m_{fot}) = res_{\overline{m}_f(A)}(m_{fot}).$

Definition 5. *If* $m_{fores_A(t)} \cong res_{\tilde{m}_f(A)}(m_{fot}) = res_{\overline{m}_f(A)}(m_{fot})$, *then* f *is called* strong.

A strong non-injective morphism can be seen as minimizing a reaction system. In this way the research presented here generalizes concepts presented in [4] where the notion of minimal reaction systems over **Sets** (i.e., reaction systems with reactions using the minimal number of reactants, or the minimal number of inhibitors, or both) have been introduced. Furthermore, it relates to enabling equivalence of sets of reactions discussed in [5].

Example 4. Kreowski and Rozenberg demonstrated in [11] that graph-based reaction systems can simulate deterministic finite state automata (DFA) in such a way that recognition of strings is modeled by certain interactive processes that run on the corresponding state graphs of the automata.

Let $\mathcal{F} = (Q, \Sigma, \phi, s_0, F)$ be a DFA with the set of states Q, the set of input symbols Σ, the state transition function $\phi \colon Q \times \Sigma \to Q$, the initial state $s_0 \in Q$, and the set of final states $F \subseteq Q$. Let $\Gamma = \Sigma \cup \{run, fin\}$. Then the corresponding reaction system $\mathcal{A}(\mathcal{F}) = (B(\mathcal{F}), A(\mathcal{F}))$ over Γ-**Graph** is constructed as follows (simplified variant of the construction given in [11], edges are triples consisting of the source vertex, the target vertex and the label).

The background graph extends the state graph of $\mathcal{F}$ by a *run*-loop at each vertex and an extra vertex with a loop for each input symbol. Formally, let $\overline{Q} = Q \cup \{in\}$, $E_1 = \{(s, \phi(s, x), x) \mid s \in Q, x \in \Sigma\}$, $E_2 = \{(s_0, s_0, run)\}$, $E_3 = \{(s'', s'', fin) \mid s'' \in F\}$, $E_4 = \{(s, s, run) \mid s \in Q\}$, and $E_5 = \{(in, in, x) \mid x \in \Sigma\}$. Then $B(\mathcal{F}) = (\overline{Q}, \Gamma, E_1 \cup E_3 \cup E_4 \cup E_5)$ is the background graph, $gr(\mathcal{F}) = (Q, \Gamma, E_1 \cup E_2 \cup E_3)$ is the state graph of $\mathcal{F}$ and $gr(\mathcal{F})^- = (Q, \Gamma, E_1 \cup E_3)$.

The vertex *in* with its loops represents the input alphabet. The *run*-loops are used in the interactive processes. The set of reactions $A(\mathcal{F})$ consists of the following uninhibited reactions, where the symbol "$-$" is a shortcut for the inhibitor $(empty_{B(\mathcal{F})}, 1_\emptyset)$, where $\emptyset$ denotes the empty graph.

1. sustain the state graph $gr(\mathcal{F})^-$ with the reaction $(gr(\mathcal{F})^-, -, gr(\mathcal{F})^-)$.
2. moving along transition edges due to input symbol:

$(\; x \;\circlearrowleft\!(in)\;\; run \;\circlearrowleft\!(s) \xrightarrow{\;x\;} (s'),\; -,\; run \;\circlearrowleft\!(s')\;)$

for $s \in Q, x \in \Sigma$ with $s \neq s' = \phi(s,x)$,
3. moving along transition loops due to input symbol:

$(\; x \;\circlearrowleft\!(in)\;\; run \;\circlearrowleft\!(s)\!\circlearrowright x,\; -,\; run \;\circlearrowleft\!(s)\;)$

for $s \in Q, x \in \Sigma$ with $s = \phi(s,x)$.

Now let $f \colon B(\mathcal{F}) \to B'(\mathcal{F})$ be a non-injective Γ-graphmorphism which merges only states of the state graph and afterwards parallel edges with the same label. Then $f \circ empty_{B(\mathcal{F})} = empty_{B'(\mathcal{F})}$. Moreover, $B'(\mathcal{F})$ is a background graph of a DFA $\mathcal{F}'$. Furthermore, because every reaction is uninhibited, $\overline{m}_f(A) = \widetilde{m}_f(A)$ holds. Clearly, enabledness is not reflected. Consider some state graph with at least three states, where two have outgoing edges labeled x to the third state and one of the two states carries a *run*-loop. Assume that these two states and the two edges are merged. If the *in* state carries an x-loop, then the reaction corresponding to the merged state and the label x is enabled. However, in the original state graph only one of two reactions is enabled due to the lack of a second *run*-loop.

We will discuss this example further in the next subsection.

4.4 Transformation of Interactive Processes

The componentwise composition and epi-mono factorization of the subobjects specifying an interactive process give us again an interactive process.

Definition 6. Let $\mathbf{C}$ be an *eiu*-category with epi-mono factorization. Let $\pi = (\gamma, \delta)$ be an interactive process on some reaction system $\mathcal{A} = (B, A)_{\mathbf{C}}$ given by $\gamma = c_0, \ldots, c_n$ and d_0 and let $f \colon B \to B'$ be a morphism satisfying $f \circ empty_B = empty_{B'}$. Then $\overline{m}_f(\pi)$ given by $\overline{m}_f(\gamma) = (m_{f \circ c_0}, \ldots, m_{f \circ c_n})$ and $m_{f \circ d_0}$ is the *image of the interactive process* π *wrt the epi-mono factorization.*

Example 5. Meaningful interactive processes for the reaction system discussed in Example 4 are context sequences of the form $cs(x_1 \cdots x_n) = x_1 \;\circlearrowleft\!(in), \ldots,$ $x_n \;\circlearrowleft\!(in), \emptyset$ for $n \in \mathbb{N}^+$ and $x_i \in \Sigma$ for $i = 1, \ldots, n$. These context sequences are in one-to-one correspondence to words over Σ. Let $\pi(x_1 \cdots x_n)$ denote the interactive process that has $cs(x_1 \cdots x_n)$, for $n \in \mathbb{N}^+$ and $x_i \in \Sigma$ for $i = 1, \ldots, n$, as its context sequence and $D_0, \ldots, D_{n+1}$ with $D_0 = gr(F)$ as its result sequence. In all reaction steps, $gr(F)^-$ is a subgraph of D_i for $i = 1, \ldots, n+1$. In each reaction step i for $i = 1, \ldots, n$, the result graph D_{i+1} has a single *run*-loop at some vertex s_i and is accompanied by the context graph $x_i \;\circlearrowleft\!(in)$ so that exactly one of the *run*-reactions is enabled moving the *run*-loop at vertex $\phi(s_i, x_i)$. In this way, recognition processes in $\mathcal{F}$ correspond one-to-one to specific processes in $\mathcal{A}(\mathcal{F})$.

Reconsidering the two non-equivalent DFA $\mathcal{F} = (Q, \Sigma, \phi, s_0, F)$ and $\mathcal{F}' = (Q', \Sigma, \phi', s_0', F')$ over the same alphabet related by the morphism $f \colon B(\mathcal{F}) \to B(\mathcal{F}')$ as described in Example 4, we can observe the following. Because f acts as identity wrt the vertex in and its attached loops, the context sequences of the two interactive processes are the same.

If the corresponding interactive processes behave the same in the sense that the semantic of $\mathcal{F}$ is preserved and reflected by f, then f is a transformation in the usual sense of DFA minimization, i.e., the task of transforming a given DFA into an equivalent DFA that has a smaller/minimum number of states, where two DFAs are called equivalent if they recognize the same regular language.

As a direct consequence of Lemma 1 and Theorem 1 we get the following result for interactive processes.

Corollary 2. *1.* $m_{f \circ d_i} = m_{f \circ res_A(c_{i-1} \cup d_{i-1})} \cong res_{\overline{m}_f(A)}(m_{f \circ c_{i-1}} \cup m_{f \circ d_{i-1}})$ *for* $i = 1, \ldots, n$ *if f is strong; and* $m_{f \circ res_A(c_{i-1} \cup d_{i-1})} \subseteq res_{\overline{m}_f(A)}(m_{f \circ c_{i-1}} \cup m_{f \circ d_{i-1}})$ *for* $i = 1, \ldots, n$ *otherwise.*
2. If π is context independent, then $\overline{m}_f(\pi)$ is context independent.
3. Let $\tau = t_1, \ldots, t_n$ be the state sequence of π and $\overline{m}_f(\tau) = (m_{f \circ t_1}, \ldots, m_{f \circ t_n})$ the image of the state sequence. Then τ being repetition-free, does not imply $\overline{m}_f(\tau)$ being repetition-free (proof by simple a counter example where f maps everything to INIT and 1_{INIT}).

4.5   The Category of Reaction Systems Over an *eiu*-Category with Epi-Mono Factorization

Now we can define the category of reaction systems over an *eiu*-category with epi-mono factorization. This category is a generalization of the category $\mathbf{RS}(\mathbf{C})$ introduced in [15] provided that $\mathbf{C}$ has an epi-mono factorization.

A direct consequence of Theorem 1 is the following result.

Corollary 3. *Let $\mathbf{C}$ be an eiu-category with epi-mono factorization. Let $\mathcal{A} = (B, A)$ and $\mathcal{A}' = (B', A')$ be two reaction systems over $\mathbf{C}$ and let $f \colon B \to B'$ be a morphism satisfying $f \circ empty_B = empty_{B'}$. If $\widetilde{m}_f(A) \subseteq A'$, then $res_{\widetilde{m}_f(A)}(m_{f \circ t}) \subseteq res_{A'}(m_{f \circ t})$ for all states $t \colon T \to B$.*

Remark 3. This gives us a morphism between reaction systems over $\mathbf{C}$. The restriction to the smaller set of consistent reactions $\widetilde{m}_f(A)$ is meaningful because $res_{\overline{m}_f(A)}(m_{f \circ t}) = res_{\widetilde{m}_f(A)}(m_{f \circ t})$.

Definition 7. Let $\mathbf{C}$ be an *eiu*-category with epi-mono factorization. The category $\mathbf{RS}(\mathbf{C})_{\mathbf{em}}$ is defined as follows. Its objects are reactions systems over $\mathbf{C}$. Given two reaction systems $\mathcal{A} = (B, A)$ and $\mathcal{A}' = (B', A')$ over $\mathbf{C}$, a morphisms $f \colon \mathcal{A} \to \mathcal{A}'$ is given by a morphism $f \colon B \to B'$ satisfying $f \circ empty_B = empty_{B'}$ provided that $\widetilde{m}_f(A) \subseteq A'$ such that $f(A) = \widetilde{m}_f(A)$. Compositions and identities are given by the underlying morphisms.

The definition of composition and identities is meaningful as, for reaction systems $\mathcal{A} = (B, A)$, $\mathcal{A}' = (B', A')$ and $\mathcal{A}'' = (B'', A'')$ and for morphisms $f\colon \mathcal{A} \to \mathcal{A}'$ and $g\colon \mathcal{A}' \to \mathcal{A}''$, $(g \circ f)(A) = g(f(A)) \subseteq g(A') \subseteq A''$ (applying Lemma 2) and $1_B(A) = A$.

5 Transformation of Reaction Systems by Means of Functors

In this section, we discuss which functors preserve reaction systems. A functor relates two categories by mapping the objects and morphisms of one category to the objects and morphisms of the other category respectively in such a way that compositions and identities are preserved.

Definition 8. A *functor* $F\colon \mathbf{C} \to \mathbf{C}'$ is a pair of maps $F = (F_{Ob}, F_{Mor})$ with $F_{Ob}\colon Ob_{\mathbf{C}} \to Ob_{\mathbf{C}'}$ and $F_{Mor}(A, B)\colon Mor_{\mathbf{C}}(A, B) \to Mor_{\mathbf{C}'}(F_{Ob}(A), F_{Ob}(B))$ for each pair of objects $A, B \in Ob_{\mathbf{C}}$ such that $F(g \circ f) = F(g) \circ F(f)$ for each pair of morphisms $(f\colon A \to B)$ and $(g\colon B \to C) \in Mor_{\mathbf{C}}$, and $F(1_A) = 1_{F(A)}$ for each $A \in Ob_{\mathbf{C}}$.

For instance, the usual embedding of Σ-graphs into Σ-hypergraphs induces such a functor. The other way round, the usual transformation of a hypergraph into a graph can be extended to morphisms. The question is which properties of a functor $F\colon \mathbf{C} \to \mathbf{C}'$ are sufficient such that a reaction system $\mathcal{A}$ over $\mathbf{C}$ is translated into a reaction system $F(\mathcal{A})$ over $\mathbf{C}'$. Whenever this works, one can compare reaction systems over different categories.

Subsection 5.1 introduces the notion of *eiu*-preserving functors. In Subsect. 5.2 we start the investigation of functors and reaction systems for the case of *eiu*-preserving embedding functors. In Subsect. 5.3 the mapping of interactive processes are discussed. Finally, in Subsect. 5.4 we show that certain *eiu*-preserving embedding functors induce functors in the category of reaction systems such that properties of the reaction systems and the morphisms between them are preserved.

5.1 *eiu*-Preserving Functors

From the definition of *eiu*-categories we can deduce a notion of functors preserving the needed concepts.

Definition 9. A functor $F\colon \mathbf{C} \to \mathbf{C}'$ is *eiu-preserving* if it preserves monomorphisms, pullbacks along monomorphisms, colimits and finiteness of objects.

For two *eiu*-categories $\mathbf{C}$ and $\mathbf{C}'$ and an *eiu*-preserving functor $F\colon \mathbf{C} \to \mathbf{C}'$ the following holds.

Properties 1. *1. Let $p_1\colon P_1 \to B, p_2\colon P_2 \to B$ be two subobjects of B in $Mor_{\mathbf{C}}$. Then*
  (a) $p_1 \subseteq p_2$ implies $F(p_1) \subseteq F(p_2)$;

(b) $F(p_1 \cap p_2) = F(p_1) \cap F(p_2)$.

2. Let S be a set of subobjects of B in $Mor_{\mathbf{C}}$ and let $F(S) = \{F(p) \mid p \in S\}$. Then $union(F(S)) = F(union(S))$.

3. $F(empty_B) = empty_{F(B)}$

Points 1 and 2 hold by definition. Point 3 uses the fact that every colimit-preserving functor maps initial objects to initial objects, i.e., $F(INIT) = \overline{INIT}$, where $INIT$ and $\overline{INIT}$ are the initial objects in $\mathbf{C}, \mathbf{C'}$, respectively. Then, because $\mathbf{C'}$ is an eiu-category and initial morphisms are unique, this implies $F(empty_B) = empty_{F(B)}$.

Example 6. Because of the componentwise constructions of morphisms, pull-backs and colimits in both categories, the following functors are eiu-preserving.

– The *relabeling functor* $F\colon \Sigma_1\text{-}\mathbf{Graphs} \to \Sigma_2\text{-}\mathbf{Graphs}$ which is defined by $F_{Ob} = (1_V, 1_E, 1_{E \to V}, 1_{E \to V}, m\colon \Sigma_1 \to \Sigma_2)$ and $F_{Mor} = (1_{f_V}, 1_{f_E})$.

– The *forgetful functor* $F\colon \Sigma\text{-}\mathbf{Graphs} \to \mathbf{Set}$ which maps only the vertex set.

– The *embedding functor* $F\colon \Sigma\text{-}\mathbf{Graphs} \to \Sigma\text{-}\mathbf{Hypergraphs}$ which is defined by $F_{Ob}((V, E, s, t, l)) = (V, E, att, l)$, where $att(e) = s(e)t(e)$ for each $e \in E$, and $F_{Mor} = (1_{f_V}, 1_{f_E})$.

– The *embedding functor* $F\colon \mathbf{Hypergraphs} \to \mathbf{BipartiteGraphs}$ which is defined by $F_{Ob}((V, E, att)) = (V, E, Att(E), \emptyset, s_{Att(E)}, \emptyset, t_{Att(E)}, \emptyset)$, where $Att(E) = \{(e, i) \mid e \in E, i \in [k(e)]\}$, $k(e)$ is the type of e, $s_{Att(E)}((e, i)) = e$, $t_{Att(E)}((e, i)) = a_i(e)$, where $att(e) = a_1(e) \cdots a_{k(e)}(e)$, and $F_{Mor}((f_V, f_E) = (f_V, f_E, (f_E, f_V), \emptyset)$ because for $f = (f_V, f_E)\colon H_1 \to H_2$, $H_i = (V_i, E_i, att_i)$,

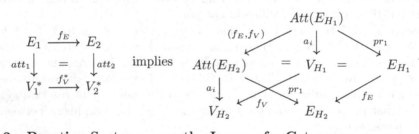

5.2 Reaction Systems over the Image of a Category

In order to define reaction systems over the image of a category, we need the following result for the image of an eiu-preserving functor.

Let $\mathbf{C}$ and $\mathbf{C'}$ be two eiu-categories and let $F\colon \mathbf{C} \to \mathbf{C'}$ be an eiu-preserving functor. If the image of F, denoted $F(\mathbf{C})$, is a category, then it is a subcategory of $\mathbf{C'}$. In particular, $F(\mathbf{C})$ is then an eiu-category, because F is eiu-preserving.

The image of a full functor always yields a category. For arbitrary functors this is not always the case. The composition of arrows in the image of the functor must also have a preimage. However, requiring a full functor is usually too strong.

Because a reaction system over $\mathbf{C}$ is a background object together with a collection of monomorphisms in $\mathbf{C}$ and eiu-preserving functors preserve monomorphisms, the mapping yields again a reaction system. In other words, F gives us a mapping for objects in the categories $\mathbf{RS}(\mathbf{C})$ and $\mathbf{RS}(\mathbf{C})_{\mathbf{em}}$ to objects in the categories $\mathbf{RS}(F(\mathbf{C}))$ and $\mathbf{RS}(F(\mathbf{C}))_{\mathbf{em}}$, respectively.

Definition 10. Let $\mathcal{A} = (B, A)_{\mathbf{C}} \in Ob_{\mathbf{RS(C)}}$ and let $F\colon \mathbf{C} \to \mathbf{C}'$ be an *eiu*-preserving functor satisfying that $F(\mathbf{C})$ be an *eiu*-category. For $a = (r, (i, i_0), p) \in A$, let $F(a) = (F(r), (F(i), F(i_0)), F(p))$. Then the *image* of $\mathcal{A}$ under F, denoted by $F(\mathcal{A})$, is $(F(B), F(A))_{\mathbf{C}'} \in Ob_{\mathbf{RS(C')}}$, where $F(A) = \{F(a) \mid a \in A\}$.

The properties of $F(\mathcal{A})$ depend on further properties of the functor. We start the investigation for the case of *eiu*-preserving embedding functors.

Lemma 4. *Let $F\colon \mathbf{C} \to \mathbf{C}'$ be eiu-preserving and an embedding. Then $F(\mathcal{A})$ has the following properties.*

1. $en_a(t)$ *implies* $en_{F(a)}(F(t))$ *for every state* $t\colon T \to B$.
2. $F(res_a(t)) \subseteq res_{F(a)}(F(t))$.
3. $F(res_{\mathcal{A}}(t)) \subseteq res_{F(\mathcal{A})}(F(t))$.

Proof. 1. $r \subseteq t$ implies $F(r) \subseteq F(t)$ and $i \cap t \subseteq i \circ i_0$ implies $F(i) \cap F(t) = F(i \cap t) \subseteq F(i \circ i_0) = F(i) \circ F(i_0)$ by Properties 1 Point 1. Because F is an embedding, it is injective on morphisms. Hence, $en_{F(a)}(F(t))$.

2. $F(res_a(t)) = F(p)$ if $en_a(t)$ and $F(empty_B) = empty_{F(B)}$ otherwise. But $res_{F(a)}(F(t))) = F(p)$ if $en_{F(a)}(F(t))$ (even if $\overline{en}_a(t)$) and $empty_{F(B)}$ otherwise.

3. $F(res_{\mathcal{A}}(t)) = F(res_{\mathcal{A}}(t)) = F(union(\{res_a(t) \mid a \in A\})) = union(F(\{res_a(t) \mid a \in A\})) \subseteq union(\{F(res_a(t)) \mid a \in A\}) \subseteq union(\{res_{F(a)}(F(t))) \mid F(a) \in F(A)\} = F'(res_{F(\mathcal{A})}(F'(t))) = F'(res_{F(\mathcal{A})}(F(t)))$.

Remark 4. If all monomorphisms in $F(\mathbf{C})$ have as preimage a monomorphism, then the implication in Point 1 becomes an equivalence, and Point 2 and 3 become equations.

5.3 The Image of an Interactive Process

The componentwise mapping of morphisms specifying an interactive process give us again an interactive process.

Definition 11. Let $\mathbf{C}$ be an *eiu*-category and $F\colon \mathbf{C} \to \mathbf{C}'$ an *eiu*-preserving functor. Let $\pi = (\gamma, \delta)$ be an interactive process on some reaction system $\mathcal{A} = (B, A)_{\mathbf{C}} \in Ob_{\mathbf{RS(C)}}$ given by $\gamma = c_0, \ldots, c_n$ and d_0. Then $F(\pi)$ given by $F(\gamma) = (F(c_0), \ldots, F(c_n))$ and $F(d_0)$ is the *image of the interactive process* π under F.

As a direct consequence of Properties 1, Theorem 4 and Remark 4 we get the following result for interactive processes.

Corollary 4. *1. Let $F\colon \mathbf{C} \to \mathbf{C}'$ be an embedding and eiu-preserving functor. Then $F(d_i) = F(res_{\mathcal{A}}(c_{i-1} \cup d_{i-1})) \subseteq res_{F(\mathcal{A})}(F(c_{i-1}) \cup F(d_{i-1}))$.*
2. If π is context-independent, then $F(\pi)$ is context-independent.

Remark 5. If $\mathbf{C}'$ is also an *eiu*-category, then interactive processes on $F(\mathcal{A})$ behave well outside $F(\mathbf{C})$.

Example 7. In [15] we modeled a vertex-coverability test by a family of reaction systems over the category Σ-**Hypergraphs**. A set of vertices X is a vertex cover of some hypergraph if each hyperedge has some attachment vertex in X.

Let $H = (V, E, att, l)$ be a Σ-hypergraph with $l(e) = *$ for some label $* \in \Sigma$ for all $e \in E$ (this means that all hyperedges are equally labeled and, hence, can be considered as unlabeled). Then $X \subseteq V$ is a *vertex cover* of H if each hyperedge has some attachment vertex in X. H is *k-vertex-coverable* for some $k \in \mathbb{N}$ if there is a hyperedge vertex cover of H with k elements.

The k-vertex-coverability test employs the reaction system $\mathcal{A}_{m,n} = (B_{m,n}, A_{m,n})$ for some $m, n \in \mathbb{N}$ with $m \leq n$ defined as follows. Let $\begin{bmatrix} n \\ m \end{bmatrix}$ be the set of all strings over $[n]$ of lengths up to m. Then the *complete hypergraph with twins* is defined by $CH_{m,n}^{(2)} = ([n], \begin{bmatrix} n \\ m \end{bmatrix} \times \{*, +\}, attach, lab)$ with $attach(u, *) = attach(u, +) = u$ and $lab(u, *) = *$ and $lab(u, +) = +$ for all $u \in \begin{bmatrix} n \\ m \end{bmatrix}$. The two parallel hyperedges $(u, *)$ and $(u, +)$ for $u \in \begin{bmatrix} n \\ m \end{bmatrix}$ are called

twins. The background hypergraph $B_{m,n}$ is $CH_{m,n}^{(2)}$ extended by a $*$-flag (type-1 hyperedge) at each vertex. The set of reactions $A_{m,n}$ contains the following elements, where, due to the one-to-one correspondence of categorial subobjects of a Σ-hypergraph and sub-Σ-hypergraphs, the subobjects are represented by the domain objects of the inclusion morphisms. The symbol "$-$" is a shortcut for the inhibitor $(empty_{B_{m,n}}, 1_{MPT})$.

1. $(\textcircled{\scriptsize j}, -, \textcircled{\scriptsize j})$ for all $j \in [n]$.

2. $(e^\bullet, -, e^\bullet)$ for all $e \in \begin{bmatrix} n \\ m \end{bmatrix} \times \{*, +\}$ where $e^\bullet$ is the sub-Σ-hypergraph of $B_{m,n}$ induced by e, i.e., $e^\bullet = (\{v_1, \dots, v_l\}, \{e\}, attach|_{\{e\}}, lab|_{\{e\}})$ with $attach(e) = v_1 \cdots v_l$, $v_j \in [n]$ for $j = 1, \dots, l$.

3. $(\textcircled{\scriptsize j}\!-\!\!\overset{1}{\boxed{*}}, -, \textcircled{\scriptsize j}\!-\!\!\overset{1}{\boxed{*}})$ for all $j \in [n]$.

4. $((u, *)^\bullet \cup v^\bullet, -, (u, +)^\bullet)$ for all $u \in \begin{bmatrix} n \\ m \end{bmatrix}$ and $v \in V$ occurring in u where $v^\bullet$ is the sub-Σ-hypergraph of $B_{m,n}$ with the vertex v and a $*$-flag at v.

The first three types of reactions applied to a state make sure that the state is sustained. The only changing reactions are of the fourth type. They add a $+$-labeled twin hyperedge whenever some attachment vertex of a $*$-labeled hyperedge has a $*$-flag. In the drawings, a circle represents a vertex and a box a flag. The label is inside the box, and a line from a box to a circle represents the attachment.

Let $H \subseteq CH_{m,n}^{(2)}$ be a sub-Σ-hypergraph with $*$-labeled hyperedges only. Let $i_1, \dots, i_k$ be a combination of k elements of $[n]$ for some $k \in \mathbb{N}$. Then one can consider the interactive process $\pi(H, i_1 \cdots i_k) = (\gamma(H, i_1 \cdots i_k), \delta(H, i_1 \cdots i_k))$ with $\gamma(H, i_1 \cdots i_k) = \textcircled{\scriptsize i_1}\!-\!\!\overset{1}{\boxed{*}}, \dots, \textcircled{\scriptsize i_k}\!-\!\!\overset{1}{\boxed{*}}, MPT$ and H as start. Then $\{i_1, \dots, i_k\}$ is a k-vertex-cover of H if and only if each hyperedge of H has a twin in the final result. Consequently, to test whether H is k-vertex-coverable, one may run the interactive process $\pi(H, i_1 \cdots i_k)$ for all combinations of k elements of $[n]$.

Using similar arguments as in Example 6, the embedding and eiu-preserving functor $F\colon \Sigma\text{-}\mathbf{Hypergraphs} \to \Sigma\text{-}\mathbf{BipartiteGraphs}$, where the labeling in the latter is defined for the second vertex set, gives us a corresponding family of reaction systems over the category $\Sigma\text{-}\mathbf{BipartiteGraphs}$. The vertex-coverability test performed by the interactive process in $\Sigma\text{-}\mathbf{Hypergraphs}$ translates directly to an interactive process in $\Sigma\text{-}\mathbf{BipartiteGraphs}$ such that $c_1, \ldots, c_n$ is a vertex cover for d_0 if and only if $F(c_1), \ldots, F(c_n)$ is a vertex cover for $F(d_0)$ in the following sense. A subset of vertices X of the set of vertices with labels is a vertex cover of some bipartite Σ-graph if and only if every vertex of the unlabeled set has at least one target in X for its incident edges.

5.4 The Induced Functor Between Categories of Reaction Systems

Moreover, because a morphism $(f_{RS}\colon (B, A)_\mathbf{C} \to (B', A')_\mathbf{C}) \in Mor_{\mathbf{RS(C)}}$ is defined in [15] by a monomorphisms $f\colon B \to B'$ provided that $f(A) = \{f(a) \mid a \in A\} \subseteq A'$, an eiu-preserving embedding functor with domain $\mathbf{C}$ where all monomorphisms in $F(\mathbf{C})$ have as preimage a monomorphism induces a well-defined mapping for $\mathbf{RS(C)}$. The property $f \circ res_A(t) \subseteq res_{A'}(f \circ t)$ for all states $t\colon T \to B$ is preserved.

Theorem 2. $F\colon \mathbf{C} \to \mathbf{C'}$ *being an embedding and eiu-preserving satisfying $F(\mathbf{C})$ being an eiu-category where all monomorphisms in $F(\mathbf{C})$ have as preimage a monomorphism induces an embedding functor $F_{RS}\colon \mathbf{RS(C)} \to \mathbf{RS(F(C))}$. If $(f_{RS}\colon \mathcal{A} \to \mathcal{A}') \in Mor_{\mathbf{RS(C)}}$, then $(F(f_{RS})\colon F(\mathcal{A}) \to F(\mathcal{A}')) \in Mor_{\mathbf{RS(F(C))}}$ satisfies $F(f) \circ res_{F(\mathcal{A})}(F(t)) \subseteq res_{F(\mathcal{A}')}(F(f) \circ F(t))$ for all states $t\colon T \to B$.*

Proof. Let $\mathcal{A} = (B, A)_\mathbf{C}, \mathcal{A}' = (B', A')_\mathbf{C} \in Ob_{\mathbf{RS(C)}}$. If $(f_{RS}\colon \mathcal{A} \to \mathcal{A}') \in Mor_{\mathbf{RS(C)}}$ is a monomorphism. Then by definition $f\colon B \to B'$ is a monomorphism and $f(A) \subseteq A'$. Because F preserves monomorphisms, it follows that $F(f)\colon F(B) \to F(B')$ as well as all morphisms in $F(f(A))$ and $F(A')$ are monomorphisms. $F(f(A)) \subseteq F(A')$ because F is an embedding. Clearly, $f \circ res_A(t) \subseteq res_{A'}(f \circ t)$ for all states $t\colon T \to B$ implies $F(f) \circ res_{F(\mathcal{A})}(F(t)) \subseteq res_{F(\mathcal{A}')}(F(f) \circ F(t))$ because $F(f \circ res_A(t)) = F(f) \circ F(res_A(t)) = F(f) \circ res_{F(A}(F(t))$ and $F(res_{A'}(f \circ t)) = res_{F(A')}(F(f \circ t)) = res_{F(A')}(F(f) \circ F(t))$ for all states $t\colon T \to B$.

6 Conclusion

In this paper, we have continued the research on reaction systems over eiu-categories. In this framework a wide spectrum of data structures become available on which reaction systems can be based. In particular, we have proposed two transformations of reaction systems over eiu-categories based on epimono factorization and functors. The first transformation generalizes the morphisms between reaction systems over eiu-categories (compared to [15] where only monomorphisms have been considered) provided that the underlying category has an epi-mono factorization. The second transformation is the first step to related reaction systems over different categories properly in the categorial framework. However, to shed more light on the significance of the framework, the investigation should be continued including the following topics.

1. In some categories epi-mono factorization exists only for a class of epimorphisms $\mathcal{E}$ and a class of monomorphisms $\mathcal{M}$. Some of such categories have been analyzed in the area of graph transformation. It may be worth investigating reaction systems over these categories.
2. The research in Sect. 5 may be continued by analyzing non-injective functors or functors preserving epi-mono factorizations in order to transform reaction system in the category $\mathbf{RS(C)_{em}}$.
3. How do evolutions of reaction systems over **Sets** presented in [5] and functors relate?
4. It would be interesting to clarify the relationship between *eiu*-categories and the well-studied adhesive categories that are successfully applied in the area of graph transformation in various variants (cf., e.g., [14,16,18–20]).
5. We used one standard notion of finite objects, where an object is finite if its set of subobjects is finite. However, it is worth considering a different notion such as compact objects, finitely presentable objects or finite objects in topoi.
6. In [15] we have shown that diagram categories provide a reservoir of *eiu*-categories. Another way to find appropriate categories is the restriction of *eiu*-categories to subcategories. For example, if one restricts the category Σ-**Graphs** to simple graphs, then this category is closed under empty subobjects, intersections and unions so that this category inherits all reaction systems over Σ-**Graphs** if the background graph is simple. How do general restriction principles look like that yield such subcategories?
7. Most of the example categories in this paper have graph-like structures as objects. But also monoids and partially ordered sets fit into the framework. Hence, one may like to know which kinds of algebraic structures form proper categories and how interesting reaction systems over such algebraic structures look like.

References

1. Kari, L., Rozenberg, G.: The many facets of natural computing. Commun. ACM **51**(10), 72–83 (2008)
2. Ehrenfeucht, A., Rozenberg, G.: Reaction systems. Fund. Inform. **75**(1–4), 263–280 (2007)
3. Brijder, R., Ehrenfeucht, A., Main, M.G., Rozenberg, G.: A tour of reaction systems. Int. J. Found. Comput. Sci. **22**(7), 1499–1517 (2011)
4. Ehrenfeucht, A., Kleijn, J., Koutny, M., Rozenberg, G.: Minimal reaction systems. Trans. Comp. Sys. Biology **14**, 102–122 (2012)
5. Ehrenfeucht, A., Kleijn, J., Koutny, M., Rozenberg, G.: Evolving reaction systems. Theor. Comput. Sci. **682**, 79–99 (2017)

6. Ehrenfeucht, A., Main, M.G., Rozenberg, G.: Functions defined by reaction systems. Int. J. Found. Comput. Sci. **22**(1), 167–178 (2011)

7. Ehrenfeucht, A., Petre, I., Rozenberg, G.: Reaction systems: A model of computation inspired by the functioning of the living cell. In: Konstantinidis, S., Moreira, N., Reis, R., Shallit, J. (eds.) The Role of Theory in Computing, pp. 11–32. World Scientific Publishing Co., Singapore (2017)

8. Ehrenfeucht, A., Rozenberg, G.: Introducing time in reaction systems. Theoret. Comput. Sci. **410**(4–5), 310–322 (2009)

9. Formenti, E., Manzoni, L., Porreca, A.E.: Fixed points and attractors of reaction systems. In: Beckmann, A., Csuhaj-Varjú, E., Meer, K. (eds.) CiE 2014. LNCS, vol. 8493, pp. 194–203. Springer, Cham (2014). https://doi.org/10.1007/978-3-319-08019-2_20

10. Salomaa, A.: Functions and sequences generated by reaction systems. Theoret. Comput. Sci. **466**(4–5), 87–96 (2012)

11. Kreowski, H.-J., Rozenberg, G.: Graph surfing by reaction systems. In: Lambers, L., Weber, J. (eds.) ICGT 2018. LNCS, vol. 10887, pp. 45–62. Springer, Cham (2018). https://doi.org/10.1007/978-3-319-92991-0_4

12. Kreowski, H., Rozenberg, G.: Graph transformation through graph surfing in reaction systems. J. Logical Algebr. Methods Program.**109**(100481) (2019)

13. Kreowski, H., Lye, A.: A categorial approach to reaction systems: First steps. Theoret. Comput. Sci. (2020)

14. Lack, S., Sobociński, P.: Adhesive and quasiadhesive categories. RAIRO Theoret. Inform. Appl **39**(3), 511–545 (2005)

15. Kreowski, H.-J., Lye, A.: Graph surfing in reaction systems from a categorial perspective. In: Hoffmann, B., Minas, M. (eds.) Proceedings of the 11th International Workshop on Graph Computation Models, (GCM 2020), Electronic Proceedings in Theoretical Computer Science (EPTCS), vol. 330, pp. 71–87. Open Publishing Association (2020)

16. Ehrig, H., Ehrig, K., Prange, U., Taentzer, G.: Fundamentals of Algebraic Graph Transformation. MTCSAES. Springer, Heidelberg (2006). https://doi.org/10.1007/3-540-31188-2

17. Adámek, J., Herrlich, H., Strecker, G.E.: Abstract and Concrete Categories - The Joy of Cats. Dover Publications (2009)

18. Ehrig, H., Ermel, C., Golas, U., Hermann, F.: Graph and Model Transformation - General Framework and Applications. Monographs in Theoretical Computer Science. An EATCS Series. Springer, Heidelberg (2015). https://doi.org/10.1007/978-3-662-47980-3

19. Corradini, A., Hermann, F., Sobociński, P.: Subobject transformation systems. Appl. Categ. Struct. **16**(3), 389–419 (2008)

20. Braatz, B., Ehrig, H., Gabriel, K., Golas, U.: Finitary $\mathcal{M}$-adhesive categories. In: Ehrig, H., Rensink, A., Rozenberg, G., Schürr, A. (eds.) ICGT 2010. LNCS, vol. 6372, pp. 234–249. Springer, Heidelberg (2010). https://doi.org/10.1007/978-3-642-15928-2_16

Graph Rewriting and Relabeling
with PBPO+

Roy Overbeek[(✉)], Jörg Endrullis, and Aloïs Rosset

Vrije Universiteit Amsterdam, Amsterdam, The Netherlands
{r.overbeek,j.endrullis,a.rosset}@vu.nl

Abstract. We extend the powerful Pullback-Pushout (PBPO) approach for graph rewriting with strong matching. Our approach, called PBPO+, exerts more control over the embedding of the pattern in the host graph, which is important for a large class of graph rewrite systems. In addition, we show that PBPO+ is well-suited for rewriting labeled graphs and certain classes of attributed graphs. For this purpose, we employ a lattice structure on the label set and use order-preserving graph morphisms. We argue that our approach is simpler and more general than related relabeling approaches in the literature.

1 Introduction

Injectively matching a graph pattern P into a host graph G induces a classification of G into three parts: (i) a *match graph* M, the image of P; (ii) a *context graph* C, the largest subgraph disjoint from M; and (iii) a *patch* J, the set of edges that are in neither M nor C. For example, if P and G are respectively

and

then M, C and J are indicated in green, black and red (and dotted), respectively. We call this kind of classification a *patch decomposition*.

Guided by the notion of patch decomposition, we recently introduced the expressive Patch Graph Rewriting (PGR) formalism [1]. Like most graph rewriting formalisms, PGR rules specify a replacement of a left-hand side (lhs) pattern L by a right-hand side (rhs) R. Unlike most rewriting formalisms, however, PGR rules allow one to (a) constrain the permitted shapes of patches around a match for L, and (b) specify how the permitted patches should be transformed, where transformations include rearrangement, deletion and duplication of patch edges.

Whereas PGR is defined set-theoretically, in this paper we propose a more sophisticated categorical approach, inspired by the same ideas. Such an approach is valuable for at least three reasons: (i) the classes of structures the method can be applied to is vastly generalized, (ii) typical meta-properties of interest (such as parallelism and concurrency) are more easily studied on the categorical level, and (iii) it makes it easier to compare to existing categorical frameworks.

F. Gadducci and T. Kehrer (Eds.): ICGT 2021, LNCS 12741, pp. 60–80, 2021.
https://doi.org/10.1007/978-3-030-78946-6_4

The two main contributions of this paper are as follows. First, we extend the Pullback Pushout (PBPO) approach by Corradini et al. [2] by strengthening the matching mechanism (Sect. 3). We call the resulting approach *PBPO with strong matching*, or *PBPO$^+$* for short. We argue that PBPO$^+$ is preferable over PBPO in situations where matching is nondeterministic, such as when specifying generative grammars or modeling execution. Moreover, we show that in certain categories (including toposes), any PBPO rule can be modeled by a set of PBPO$^+$ rules (and even a single rule when matching is monic), while the converse does not hold (Sect. 4).

Second, we show that PBPO$^+$ easily lends itself for rewriting labeled graphs and certain attributed graphs. To this end, we define a generalization of the usual category of labeled graphs, $\mathbf{Graph}^{(\mathcal{L}, \leq)}$, in which the set of labels forms a complete lattice $(\mathcal{L}, \leq)$ (Sect. 5). Not only does the combination of PBPO$^+$ and $\mathbf{Graph}^{(\mathcal{L}, \leq)}$ enable constraining and transforming the patch graph in flexible ways, it also provides natural support for modeling notions of relabeling, variables and sorts in rewrite rules. As we will clarify in the Discussion (Sect. 6), such mechanisms have typically been studied in the context of Double Pushout (DPO) rewriting [3], where the requirement to construct a pushout complement leads to technical complications and restrictions.

2 Preliminaries

We assume familiarity with various basic categorical notions, notations and results, including morphisms $X \to Y$, pullbacks and pushouts, monomorphisms (monos) $X \rightarrowtail Y$, identities $1_X : X \rightarrowtail X$ and the pullback lemma [4,5].

Definition 1 (Graph Notions). *A (labeled) graph G consists of a set of vertices V, a set of edges E, source and target functions $s, t : E \to V$, and label functions $\ell^V : V \to \mathcal{L}$ and $\ell^E : E \to \mathcal{L}$ for some label set $\mathcal{L}$.*

A graph is unlabeled *if $\mathcal{L}$ is a singleton.*

A premorphism between graphs G and G'' is a pair of maps $\phi = (\phi_V : V_G \to V_{G'}, \phi_E : E_G \to E_{G'})$ satisfying $(s_{G'}, t_{G'}) \circ \phi_E = \phi_V \circ (s_G, t_G)$.

A homomorphism is a label-preserving premorphism ϕ, i.e., a premorphism satisfying $\ell^V_{G'} \circ \phi_V = \ell^V_G$ and $\ell^E_{G'} \circ \phi_E = \ell^E_G$.

Definition 2 (Category Graph [6]). *The category $\mathbf{Graph}$ has graphs as objects, parameterized over some global (and usually implicit) label set $\mathcal{L}$, and homomorphisms as arrows.*

Although we will point out similarities with PGR, an understanding of PGR is not required for understanding this paper. However, the following PGR terminology will prove useful (see also the opening paragraph of Sect. 1).

Definition 3 (Patch Decomposition). *Given a premorphism $x : X \to G$, we call the image $M = im(x)$ of x the* match graph *in G, $G - M$ the* context graph *C induced by x (i.e., C is the largest subgraph disjoint from M), and the set of edges $E_G - E_M - E_C$ the set of* patch edges *(or simply,* patch*) induced by x. We refer to this decomposition induced by x as a* patch decomposition.

3   PBPO⁺

We introduce PBPO⁺, which strengthens the matching mechanism of PBPO [2]. In the next section, we compare the two approaches and elaborate on the expressiveness of PBPO⁺.

Definition 4 (PBPO⁺ Rewrite Rule). *A PBPO⁺ rewrite rule ρ is a collection of objects and morphisms, arranged as follows around a pullback square:*

$$\rho = \quad \begin{array}{ccc} L & \xleftarrow{l} K & \xrightarrow{r} R \\ t_L \downarrow & \text{PB} \uparrow t_K & \\ L' & \xleftarrow{l'} K' & \end{array}$$

L is the lhs pattern of the rule, L' its type graph and t_L the typing of L. Similarly for the interface K. R is the rhs pattern or replacement for L.

Remark 5 (A Mental Model for **Graph***).* In **Graph**, K' can be viewed as a collection of components, where every component is a (possibly generalized) subgraph of L', as indicated by l'. By "generalized" we mean that the components may unfold loops and duplicate elements. K is the restriction of K' to those elements that are also in the image of t_L.

We often depict the pushout $K' \xrightarrow{r'} R' \xleftarrow{t_R} R$ for span $K' \xleftarrow{t_K} K \xrightarrow{r} R$, because it shows the schematic effect of applying the rewrite rule. We reduce the opacity of R' to emphasize that it is not part of the rule definition.

Example 6 (Rewrite Rule in **Graph***).* A simple example of a rule for unlabeled graphs is the following:

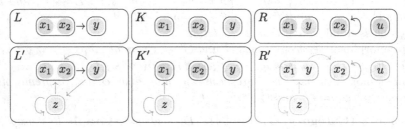

In this and subsequent examples, a vertex is a non-empty set $\{x_1, \ldots, x_n\}$ represented by a box $\boxed{x_1 \cdots x_n}$, and each morphism $f = (\phi_V, \phi_E) : G \to G'$ is the unique morphism satisfying $S \subseteq f(S)$ for all $S \in V_G$. For instance, for $\{x_1\}, \{x_2\} \in V_K$, $l(\{x_1\}) = l(\{x_2\}) = \{x_1, x_2\} \in V_L$. We will use examples that ensure uniqueness of each f (in particular, we ensure that ϕ_E is uniquely determined). Colors are purely supplementary.

Definition 7 (Strong Match). *A match morphism m and an adherence morphism α form a strong match for a typing t_L, denoted* strong(t_L, m, α), *if the square on the right is a pullback square.*

$$\begin{array}{ccc} L & \xrightarrow{m} & G_L \\ 1_L \downarrow & \text{PB} & \alpha \downarrow \\ L & \xrightarrow{t_L} & L' \end{array}$$

Remark 8 (Preimage Interpretation). In **Set**-like categories (such as **Graph**), the match diagram states that the preimage of $t_L(L)$ under $\alpha : G_L \to L'$ is L itself. So each element of $t_L(L)$ is the α-image of exactly one element of G_L.

In practice it is natural to first fix a match m, and to subsequently verify whether it can be extended into a suitable adherence morphism α.

Definition 9 (PBPO$^+$ Rewrite Step). *A PBPO$^+$ rewrite rule ρ (left) and adherence morphism $\alpha : G_L \to L'$ induce a rewrite step $G_L \Rightarrow_\rho^\alpha G_R$ on arbitrary G_L and G_R if the properties indicated by the commuting diagram (right)*

$$
\rho = \begin{array}{c} L \xleftarrow{l} K \xrightarrow{r} R \\ t_L \downarrow \quad \text{PB} \quad \downarrow t_K \\ L' \xleftarrow{l'} K' \end{array}
\qquad
\begin{array}{c}
K \xrightarrow{r} R \\
!u \downarrow \qquad \text{PO} \downarrow w \\
L \xrightarrowtail{m} G_L \xleftarrow{g_L} G_K \xrightarrow{g_R} G_R \\
1_L \downarrow \text{PB} \quad \alpha \downarrow \quad \text{PB} \downarrow u' \nearrow t_K \\
L \xrightarrowtail{t_L} L' \xleftarrow{l'} K'
\end{array}
$$

hold, where $u : K \to G_K$ is the unique (and necessarily monic) morphism satisfying $t_K = u' \circ u$. We write $G_L \Rightarrow_\rho G_R$ if $G_L \Rightarrow_\rho^\alpha G_R$ for some α.

It can be seen that the rewrite step diagram consists of a match square, a pullback square for extracting (and possibly duplicating) parts of G_L, and finally a pushout square for gluing these parts along pattern R.

The following lemma establishes the existence of a monic u by constructing a witness, and Lemma 11 establishes uniqueness.

Lemma 10 (Top-Left Pullback). *In the rewrite step diagram of Definition 9, there exists a morphism $u : K \to G_K$ such that $L \xleftarrow{l} K \xrightarrow{u} G_K$ is a pullback for $L \xrightarrow{m} G_L \xleftarrow{g_L} G_K$, $t_K = u' \circ u$, and u is monic.*   ⊛[1]

Lemma 11 (Uniqueness of u). *In the rewrite step diagram of Definition 9 (and in any category), there is a unique $v : K \to G_K$ such that $t_K = u' \circ v$.*   ⊛

Lemma 12 (Bottom-Right Pushout). *Let $K' \xrightarrow{r'} R' \xleftarrow{t_R} R$ be a pushout for cospan $R \xleftarrow{r} K \xrightarrow{t_K} K'$ of rule ρ in Definition 9. Then in the rewrite step diagram, there exists a morphism $w' : G_R \to R'$ such that $t_R = w' \circ w$, and $K' \xrightarrow{r'} R' \xleftarrow{w'} G_R$ is a pushout for $K' \xleftarrow{u'} G_K \xrightarrow{g_R} G_R$.*   ⊛

Lemmas 10 and 12 show that a PBPO$^+$ step defines a commuting diagram similar to the PBPO definition (Definition 16):

$$
\begin{array}{c}
L \xleftarrow{l} K \xrightarrow{r} R \\
m \downarrow \quad \text{PB} \quad !u \downarrow \quad \text{PO} \quad \downarrow w \\
L \xrightarrowtail{m} G_L \xleftarrow{g_L} G_K \xrightarrow{g_R} G_R \\
1_L \downarrow \text{PB} \downarrow \alpha \, / t_L \quad \text{PB} \quad \downarrow u' / t_K \text{ PO} \quad \downarrow w' / t_R \\
L \xrightarrowtail{t_L} L' \xleftarrow{l'} K' \xleftarrow{r'} R'
\end{array}
$$

We will omit the match diagram in depictions of steps.

---

[1] We use ⊛ instead of □ when the proof is available in the Appendix.

Example 13 (Rewrite Step). Applying the rule given in Example 6 to G_L (as depicted below) has the following effect:

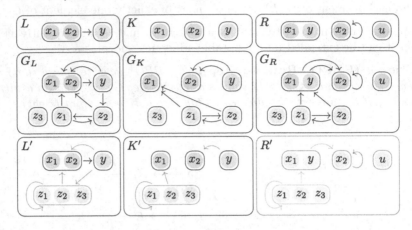

This example illustrates (i) how permitted patches can be constrained (e.g., L' forbids patch edges targeting y), (ii) how patch edge endpoints that lie in the image of t_L can be redefined, and (iii) how patch edges can be deleted.

In the examples of this section, we have restricted our attention to unlabeled graphs. In Sect. 5, we show that the category $\mathbf{Graph}^{(\mathcal{L}, \leq)}$ is more suitable than $\mathbf{Graph}$ for rewriting labeled graphs using PBPO⁺.

4   Expressiveness of PBPO⁺

The set of PBPO⁺ rules is a strict subset of the set of PBPO rules, and for any PBPO⁺ rule ρ, we have $\Rightarrow_\rho^{\mathrm{PBPO^+}} \subseteq \Rightarrow_\rho^{\mathrm{PBPO}}$ for the generated rewrite relations. Nevertheless, we will show that under certain assumptions, any PBPO rule can be modeled by a set of PBPO⁺ rules, but not vice versa. Thus, in many categories of interest (such as toposes), PBPO⁺ can define strictly more expressive grammars than PBPO. This result may be likened to Habel et al.'s result that restricting DPO to monic matching increases expressive power [7].

In Sect. 4.1, we recall and compare the PBPO definitions for rule, match and step, clarifying why PBPO⁺ is shorthand for *PBPO with strong matching*. We then argue why strong matching is usually desirable in Sect. 4.2. Finally, we prove a number of novel results on PBPO in Sect. 4.3, relating to monic matching, monic rules and strong matching. Our claim about PBPO⁺'s expressiveness follows as a consequence.

4.1   PBPO: Rule, Match and Step

Definition 14 (PBPO Rule [2]). *A PBPO rule ρ is a commutative diagram as shown on the right. The bottom span can be regarded as a typing for the top span. The rule is in* canonical form *if the left square is a pullback and the right square is a pushout.*

$$
\begin{array}{ccccc}
L & \xleftarrow{\;l\;} & K & \xrightarrow{\;r\;} & R \\
{\scriptstyle t_L}\downarrow & = & \downarrow{\scriptstyle t_K} & = & \downarrow{\scriptstyle t_R} \\
L' & \xleftarrow{\;l'\;} & K' & \xrightarrow{\;r'\;} & R'
\end{array}
$$

Every PBPO rule is equivalent to a rule in canonical form [2], and in PBPO$^+$, rules are limited to those in canonical form. The only important difference between a canonical PBPO rule and a PBPO$^+$ rule, then, is that a PBPO$^+$ rule requires monicity of t_L (and hence also of t_K).

Definition 15 (PBPO Match [2]). *A PBPO match for a typing $t_L : L \to L'$ is a pair of morphisms $(m : L \to G, \alpha : G \to L')$ such that $t_L = \alpha \circ m$.*

The pullback construction used to establish a match in PBPO$^+$ implies $t_L = \alpha \circ m$. Thus PBPO matches are more general than the strong match used in PBPO$^+$ (Definition 7). More specifically for **Graph**, PBPO allows mapping elements of the host graph G_L not in the image of $m : L \to G_L$ onto the image of t_L, whereas PBPO$^+$ forbids this. In the next subsection, we will argue why it is often desirable to forbid such mappings.

Definition 16 (PBPO Rewrite Step [2]).
A PBPO rule ρ (as in Definition 14) induces a PBPO step $G_L \Rightarrow_\rho^{m,\alpha} G_R$ shown on the right, where (i) $u : K \to G_K$ is uniquely determined by the universal property of pullbacks and makes the top-left square commuting, (ii) $w' : G_R \to R'$ is uniquely determined by the universal property of pushouts and makes the bottom-right square commuting, and $t_L = \alpha \circ m$.

$$
\begin{pmatrix}
L & \xleftarrow{\;\;l\;\;} & K & \xrightarrow{\;\;r\;\;} & R \\
{\scriptstyle m}\downarrow & = & \downarrow{\scriptstyle u}\ \text{PO} & & \downarrow{\scriptstyle w} \\
G_L & \xleftarrow{g_L} & G_K & \xrightarrow{g_R} & G_R \\
{\scriptstyle t_L}\downarrow{\scriptstyle \alpha}\ \text{PB} & & {\scriptstyle t_K}\downarrow{\scriptstyle u'} & = & {\scriptstyle t_R}\downarrow{\scriptstyle w'} \\
L' & \xrightarrow{\;\;l'\;\;} & K' & \xrightarrow{\;\;r'\;\;} & R'
\end{pmatrix}
$$

We write $G_L \Rightarrow_\rho G_R$ if $G_L \Rightarrow_\rho^{m,\alpha} G_R$ for some m and α.

The match square of PBPO$^+$ allows simplifying the characterization of u, as shown in the proof to Lemma 11. This simplification is not possible for PBPO (see Remark 17). The bottom-right square is omitted in the definition of a PBPO$^+$ rewrite step, but can be reconstructed through a pushout (modulo isomorphism). So this difference is not essential.

Remark 17. In a PBPO rewrite step, not every morphism $u : K \to G_K$ satisfying $u' \circ u = t_K$ corresponds to the arrow uniquely determined by the top-left pullback. Thus Lemma 11 does not hold for PBPO. This can be seen in the example of a (canonical) PBPO rewrite rule and step depicted in Fig. 1. Because our previous notational convention breaks for this example, we indicate two morphisms by dotted arrows. The others can be inferred.

Morphism $u : K \to G_K$ (as determined by the top-left pullback) is indicated, but it can be seen that three other morphisms $v : K \to G_K$ satisfy $u' \circ v = t_K$, because every $x \in V_{K'}$ has two elements in its preimage in G_K.

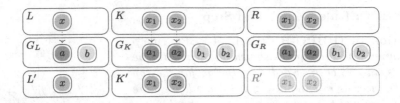

Fig. 1. Failure of Lemma 11 for PBPO.

Remark 18 (PBPO+ and AGREE). AGREE [8] by Corradini et al. is a rewriting approach closely related to PBPO. AGREE's match square can be regarded as a specialization of PBPO$^+$'s match square, since AGREE fixes the type morphism $t_L : L \rightarrowtail L'$ of a rule as the partial map classifier for L. Thus, PBPO$^+$ can also be regarded as combining PBPO's rewriting mechanism with a generalization of AGREE's strong matching mechanism.

4.2 The Case for Strong Matching

The two following examples serve to illustrate why we find it necessary to strengthen the matching criterion when matching is nondeterministic.

Example 19. In PBPO$^+$, an application of the rule

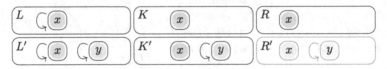

in an unlabeled graph G_L removes a loop from an isolated vertex that has a single loop, and preserves everything else. In PBPO, a match is allowed to map all of G_L into the component determined by vertex $\{x\}$, so that the rule deletes all of G_L's edges at once. (Before studying the next example, the reader is invited to consider what the effect of the PBPO rule is if R and R' are replaced by L and L', respectively.)

Example 20. Consider the following PBPO rule, and its application to host graph G_L (the morphisms are defined in the obvious way) shown in Fig. 2. Intuitively, host graph G_L is spiralled over the pattern of L'. The pullback then duplicates all elements mapped onto $x \in V_{L'}$ and any incident edges directed at a node mapped into $y \in V_{L'}$. The pushout, by contrast, affects only the image of $u : K \to G_K$.

The two examples show how locality of transformations cannot be enforced using PBPO. They also illustrate how it can be difficult to characterize the class of host graphs G_L and adherences α that establish a match, even for trivial left-hand sides. Finally, Example 20 in particular highlights an asymmetry that

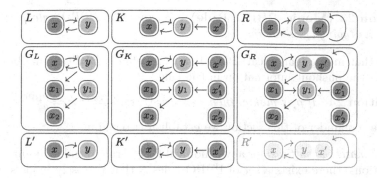

Fig. 2. The effects of PBPO rules can be difficult to oversee.

we find unintuitive: if one duplicates and then merges/extends pattern elements of L', the duplication affects all elements in the α-preimage of $t_L(L)$ (which could even consist of multiple components isomorphic to $t_L(L)$), whereas the pushout affects only $u(K) \subseteq G_K$. In PBPO$^+$, by contrast, transformations of the pattern affect the pattern only, and the overall applicability of a rule is easy to understand if the context graph is relatively simple (e.g., as in Example 13).

Remark 21 (Γ-preservation). A locality notion has been defined for PBPO called Γ-*preservation* [2]. Γ is some subobject of L', and a rewrite step $G_L \Rightarrow_\rho^{m,\alpha} G_R$ is said to be Γ-*preserving* if the $\alpha : G_L \to L'$ preimage of $\Gamma \subseteq L'$ is preserved from G_L to G_R (roughly meaning that this preimage is neither modified nor duplicated). Similarly, a rule is Γ-preserving if the rewrite steps it gives rise to are Γ-preserving. If one chooses Γ to be the context graph (the right component) of L' in Example 19, then the rule, interpreted as a PBPO rule, is Γ-preserving. However, the effect of the rule is not local in our understanding of the word, since PPBO does not prevent mapping parts of the context graph of G_L onto the image of t_L which usually *is* modified.

4.3 Modeling PBPO with PBPO$^+$

We will now prove that in many categories of interest (including locally small toposes), any PBPO rule can be modeled by a set of PBPO$^+$ rules; and even by a single rule if PBPO matches are restricted to monic matches. We do this by proving a number of novel results about PBPO.

Since any PBPO rule is equivalent to a canonical PBPO rule [2], we restrict attention to canonical rules in this section.

Definition 22. *We define the restrictions*

- $\Rightarrow_\rho^{\rightarrow} = \{(G, G') \in \Rightarrow_\rho \mid \exists m\ \alpha.\ G \Rightarrow_\rho^{m,\alpha} G' \wedge m \text{ is monic}\}$, *and*
- $\Rightarrow_\rho^{\text{SM}} = \{(G, G') \in \Rightarrow_\rho \mid \exists m\ \alpha.\ G \Rightarrow_\rho^{m,\alpha} G' \wedge \text{strong}(t_L, m, \alpha)\}$

for canonical PBPO rules ρ.

Definition 23 (Monic PBPO Rule). *A canonical PBPO rule ρ is called monic if its typing t_L is monic.*

Note that monic (canonical) PBPO rules ρ define a PBPO$^+$ rule by simply forgetting the pushout information in the rule.

Proposition 24. *If ρ is monic, then $\Rightarrow_\rho = \Rightarrow_\rho^{\rightharpoonup}$ and $\Rightarrow_\rho^{SM} = \Rightarrow_\rho^{PBPO^+}$.* ☐

In the remainder of this section we establish two claims:

1. *Monic matching suffices*: for any canonical PBPO rule ρ and assuming certain conditions, there exists a set of PBPO rules S that precisely models ρ when restricting S to monic matching, i.e., $\Rightarrow_\rho = \bigcup\{\Rightarrow_\sigma^{\rightharpoonup} \mid \sigma \in S\}$ (Corollary 28);
2. *Strong matching can be modeled through rule adaptation*: for any canonical PBPO rule σ and assuming certain conditions, there exists a monic rule τ such that $\Rightarrow_\sigma^{\rightharpoonup} = \Rightarrow_\tau^{SM}$ (Lemma 32).

Because PBPO$^+$ rewriting boils down to using monic PBPO rules with a strong matching policy, from these facts and conditions it follows that any PBPO rule can be modeled by a set of PBPO$^+$ rules (Theorem 33).

The following definition defines a rule σ for every factorization of a type morphism t_L of a rule τ.

Definition 25 (Compacted Rule). *For any canonical PBPO rule ρ (on the left) and factorization $t_L = t_{L_c} \circ e$ where e is epic (note that t_{L_c} is uniquely determined since e is right-cancellative), the compacted rule ρ_e is defined as the lower half of the commuting diagram on the right:*

$$
\begin{array}{ccc}
L & \xleftarrow{\;l\;} K \xrightarrow{\;r\;} & R \\
t_L \downarrow & \text{PB} \;\; \downarrow t_K \;\; \text{PO}\downarrow & \downarrow t_R \\
L' & \xleftarrow{\;l'\;} K' \xrightarrow{\;r'\;} & R'
\end{array}
\qquad
t_L \left(
\begin{array}{ccc}
L & \xleftarrow{\;l\;} K \xrightarrow{\;r\;} & R \\
e\downarrow & \text{PB}\downarrow \quad \text{PO}\downarrow & \\
L_c & \xleftarrow{\quad} K_c \xrightarrow{\quad} & R_c \\
t_{L_c}\downarrow & \text{PB}\Big\downarrow\; t_K \; \text{PO}\downarrow & \\
L' & \xleftarrow{\;l'\;} K' \xrightarrow{\;r'\;} & R'
\end{array}
\right) t_R
$$

Proposition 26. *The properties implicitly asserted in Definition 25 hold.* ☐

Lemma 27. *Let ρ be a canonical PBPO rule, G_L an object, and $m' \circ e : L \to G_L$ a match morphism for a mono m' and epi e. We have:*

$$
G_L \Rightarrow_\rho^{(m' \circ e),\alpha} G_R \quad \Longleftrightarrow \quad G_L \Rightarrow_{\rho_e}^{m',\alpha} G_R .
$$

⊛

Recall that a category is locally small if the collection of morphisms between any two objects A and B (and so also all factorizations) forms a set.

Corollary 28. *In locally small categories in which any morphism can be factorized into an epi followed by a mono, for every canonical PBPO rule ρ, there exists a set of PBPO rules S such that $\Rightarrow_\rho = \bigcup\{\Rightarrow_\sigma^{\rightarrow} \mid \sigma \in S\}$.* □

Definition 29 (Amendable Category). *A category is* amendable *if for any $t_L : L \to L'$, there exists a factorization $L \overset{t'_L}{\rightarrowtail} L'' \overset{\beta}{\to} L'$ of t_L such that for any factorization $L \overset{m}{\rightarrowtail} G_L \overset{\alpha}{\to} L'$ of t_L, there exists an α' making the diagram*

$$
\begin{array}{ccc}
& t_L & \\
L \underset{m}{\rightarrowtail} G_L & \underset{\alpha}{\rightrightarrows} & L' \\
1_L \Big\downarrow \quad t'_L \quad \Big\downarrow \alpha' & \nearrow & \\
L \underset{t'_L}{\rightarrowtail} L'' & \beta &
\end{array}
$$

commute.

The category is strongly amendable *if there exists a factorization of t_L witnessing amendability that moreover makes the left square a pullback square.*

Strong amendability is intimately related to the concept of *materialization* [9]. Namely, if the factorization $L \overset{t'_L}{\rightarrowtail} L'' \overset{\beta}{\to} L'$ of t_L establishes the pullback square and is final (the α' morphisms not only exist, but they exist uniquely), then $\beta \circ t'_L$ is the materialization of t_L. In general we do not need finality, and for one statement (Lemma 32) we require weak amendability only.

We have the following sufficient condition for strong amendability.

Proposition 30. *If all slice categories $\mathbf{C}/X$ of a category $\mathbf{C}$ have partial map classifiers, then $\mathbf{C}$ is strongly amendable.*

Proof. Immediate from the fact that in this case all arrows have materializations [9, Proposition 8]. □

Corollary 31. *Any topos is strongly amendable.*

Proof. Toposes have partial map classifiers, and any slice category of a topos is a topos. □

Lemma 32. *In an amendable category $\mathcal{C}$, for any PBPO rule ρ, there exists a monic PBPO rule σ such that $\Rightarrow_\rho^{\rightarrow} = \Rightarrow_\sigma$. If $\mathcal{C}$ is moreover strongly amendable, then additionally $\Rightarrow_\rho^{\rightarrow} = \Rightarrow_\sigma^{SM}$.*

Proof. Given rule ρ on the left

$$
\begin{array}{ccc}
L \overset{l}{\longleftarrow} K \overset{r}{\longrightarrow} R \\
t_L \Big\downarrow \quad PB \quad \Big\downarrow t_K \quad \Big\downarrow t_R \\
 \quad\quad\quad PO \\
L' \overset{}{\longleftarrow} K' \overset{}{\longrightarrow} R' \\
 l' \quad\quad r'
\end{array}
\qquad
t_L \left(
\begin{array}{ccc}
L \overset{l}{\longleftarrow} K \overset{r}{\longrightarrow} R \\
t'_L \Big\downarrow \quad PB \Big\downarrow t'_K \quad PO \Big\downarrow t'_R \\
L'' \overset{l''}{\longleftarrow} K'' \overset{r''}{\longrightarrow} R'' \\
\beta \Big\downarrow \quad PB \Big\downarrow \quad PO \Big\downarrow \\
L' \overset{l'}{\longleftarrow} K' \overset{r'}{\longrightarrow} R'
\end{array}
\right.
$$

we can construct rule σ as the upper half of the diagram on the right, where $L \xrightarrow{t'_L} L'' \xrightarrow{\beta} L'$ is the factorization of t_L witnessing strong amendability. Then the first claim $\Rightarrow_\rho^{\rightarrow} = \Rightarrow_\sigma$ follows by considering the commuting diagram

$$
\begin{array}{c}
\end{array}
$$

and the second claim $\Rightarrow_\rho^{\rightarrow} = \Rightarrow_\sigma^{\text{SM}}$ for strongly amendable categories follows by observing that the square marked by † is a pullback square.   □

Theorem 33. *In locally small, strongly amendable categories in which every morphism f can be factored into an epi e followed by a mono m, any PBPO rule ρ can be modeled by a set of PBPO$^+$ rules.*   ⊛

Corollary 34. *In any locally small topos, any PBPO rule ρ can be modeled by a set of PBPO$^+$ rules.*

Proof. By Corollary 31 and the fact that toposes are epi-mono factorizable.   □

5 Category Graph$^{(\mathcal{L}, \leq)}$

Unless one employs a meta-notation or restricts to unlabeled graphs, as we did in Sect. 3, it is sometimes impractical to use PBPO$^+$ in the category **Graph**. The following example illustrates the problem.

Example 35. Suppose the set of labels is $\mathcal{L} = \{0, 1\}$. To be able to injectively match pattern $L = \xrightarrow{1} 0$ in any context, one must inject it into the type graph L' shown on the right in which every dotted loop represents two edges (one for each label), and every dotted non-loop represents four edges (one for each label, in either direction). In general, to allow any context, one needs to include $|\mathcal{L}|$ additional vertices in L', and $|\mathcal{L}|$ complete graphs over $V_{L'}$.

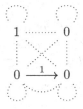

Beyond this example, and less easily alleviated with meta-notation, in **Graph** it is impractical or impossible to express rules that involve (i) arbitrary labels (or classes of labels) in the application condition; (ii) relabeling; or (iii) allowing and capturing arbitrary subgraphs (or classes of subgraphs) around a match graph. As we will discuss in Sect. 6, these features have been non-trivial to express in general for algebraic graph rewriting approaches.

We define a category which allows flexibly addressing all of these issues.

Definition 36 (Complete Lattice). *A* complete lattice $(\mathcal{L}, \leq)$ *is a poset such that all subsets S of $\mathcal{L}$ have a supremum (join) $\bigvee S$ and an infimum (meet) $\bigwedge S$.*

Definition 37 (Graph$^{(\mathcal{L}, \leq)}$). For a complete lattice $(\mathcal{L}, \leq)$, we define the category **Graph**$^{(\mathcal{L}, \leq)}$, where objects are graphs labeled from $\mathcal{L}$, and arrows are graph premorphisms $\phi : G \to G'$ that satisfy $\ell_G(x) \leq \ell_{G'}(\phi(x))$ for all $x \in V_G \cup E_G$.

In terms of graph structure, the pullbacks and pushouts in **Graph**$^{(\mathcal{L}, \leq)}$ are the usual pullbacks and pushouts in **Graph**. The only difference is that the labels that are identified by respectively the cospan and span are replaced by their meet and join, respectively.

The sufficient condition of Proposition 30 does not hold in **Graph**$^{(\mathcal{L}, \leq)}$. Nonetheless, we have the following result.

Lemma 38. Graph$^{(\mathcal{L}, \leq)}$ is strongly amendable. ⊛

One very simple but extremely useful complete lattice is the following.

Definition 39 (Flat Lattice). *Let $\mathcal{L}^{\perp, \top} = \mathcal{L} \uplus \{\perp, \top\}$. We define the flat lattice induced by $\mathcal{L}$ as the smallest poset $(\mathcal{L}^{\perp, \top}, \leq)$, which has $\perp$ as a global minimum and $\top$ as a global maximum (so in particular, the elements of $\mathcal{L}$ are incomparable). In this context, we refer to $\mathcal{L}$ as the* base label set.

One feature flat lattices provide is a kind of "wildcard element" $\top$.

Example 40 (Wildcards). Using flat lattices, L' of Example 35 can be fully expressed for any base label set $\mathcal{L} \ni 0, 1$ as shown on the right (node identities are omitted). The visual syntax and naming shorthands of PGR [1] (or variants thereof) could be leveraged to simplify the notation further.

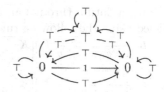

As the following example illustrates, the expressive power of a flat lattice stretches beyond wildcards: it also enables relabeling of graphs. (Henceforth, we will depict a node x with label u as x^u.)

Example 41 (Relabeling). As vertex labels we employ the flat lattice induced by the set $\{a, b, c, \ldots\}$, and assume edges are unlabeled for notational simplicity. The following diagram displays a rule (L, L', K, K', R) for overwriting an arbitrary vertex's label with c, in any context. We include an application to an example graph in the middle row:

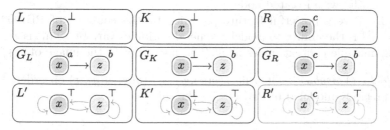

The example demonstrates how (i) labels in L serve as lower bounds for matching, (ii) labels in L' serve as upper bounds for matching, (iii) labels in K' can be used to decrease matched labels (so in particular, $\bot$ "instructs" to "erase" the label and overwrite it with $\bot$, and $\top$ "instructs" to preserve labels), and (iv) labels in R can be used to increase labels.

Complete lattices also support modeling sorts.

Example 42 (Sorts). Let $p_1, p_2, \ldots \in \mathbb{P}$ be a set of processes and $d_1, d_2, \ldots \in \mathbb{D}$ a set of data elements. Assume a complete lattice over labels $\mathbb{P} \cup \mathbb{D} \cup \{\mathbb{P}, \mathbb{D}, \triangleright, @\}$, arranged as in the diagram on the right.

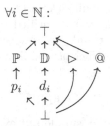

Moreover, assume that the vertices $x, y, \ldots$ in the graphs of interest are labeled with a p_i or d_i, and that edges are labeled with a $\triangleright$ or $@$. In such a graph,

- an edge $x^{d_i} \xrightarrow{@} y^{p_j}$ encodes that process p_j holds a local copy of datum d_i (x will have no other connections); and
- a chain of edges $x^{p_i} \xrightarrow{\triangleright} y^{d_k} \xrightarrow{\triangleright} z^{d_l} \xrightarrow{\triangleright} \cdots \xrightarrow{\triangleright} u^{p_j}$ encodes a directed FIFO channel from process p_i to process $p_j \neq p_i$, containing a sequence of elements $d_k, d_l, \ldots$. An empty channel is modeled as $x^{p_i} \xrightarrow{\triangleright} u^{p_j}$.

Receiving a datum through an incoming channel (and storing it locally) can be modeled using the following rule:

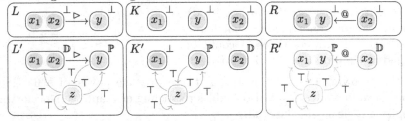

The rule illustrates how sorts can improve readability and provide type safety. For instance, the label $\mathbb{D}$ in L' prevents empty channels from being matched. More precisely, always the last element d of a non-empty channel is matched. K' duplicates the node holding d: for duplicate x_1, the label is forgotten but the connection to the context retained, allowing it to be fused with y; and for x_2, the connection is forgotten but the label retained, allowing it to be connected to y as an otherwise isolated node.

Finally, a very powerful feature provided by the coupling of PBPO$^+$ and **Graph**$^{(\mathcal{L}, \leq)}$ is the ability to model a general notion of variable. This is achieved by using multiple context nodes in L' (i.e., nodes not in the image of t_L).

Example 43 (Variables). The rule $f(g(x), y) \to h(g(x), g(y), x)$ on ordered trees can be precisely modeled in PBPO$^+$ by the rule

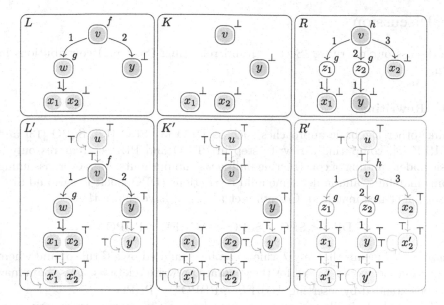

if one restricts the set of rewritten graphs to straightforward representations of trees: nodes are labeled by symbols, and edges are labeled by $n \in \mathbb{N}$, the position of its target (argument of the symbol).

Remark 44 (On Adhesivity). A category is *adhesive* [10] if (i) it has all pullbacks, (ii) it has all pushouts along monos, and (iii) pushouts along monos are stable and are pullbacks [11, Theorem 3.2]. Adhesivity implies the uniqueness of pushout complements (up to isomorphism) [10, Lemma 4.5], which in turn ensures that DPO rewriting in adhesive categories is deterministic. Moreover, certain meta-properties of interest (such as the Local Church-Rosser Theorem and the Concurrency Theorem) hold for DPO rewriting in adhesive categories (and many graphical structures are adhesive). For these reasons, DPO and adhesivity are closely related in the literature.

We make two observations in connection to adhesivity. First, PBPO$^+$ rewriting makes strictly weaker assumptions than DPO rewriting: it is enough to assume conditions (i) and (ii) above. Second, $\mathbf{Graph}^{(\mathcal{L}, \leq)}$ is non-adhesive for any choice of complete lattice in which the maximum $\top$ and minimum $\bot$ are distinct: the square

$$
\begin{array}{ccc}
\bot & \rightarrowtail & \top \\
\downarrow & & \downarrow \\
X & \rightarrowtail & \top
\end{array}
$$

is a pushout along a mono for $X \in \{\bot, \top\}$, and hence admits two pushout complements. Thus, not only can PBPO$^+$ be applied to non-adhesive categories, $\mathbf{Graph}^{(\mathcal{L}, \leq)}$ is a graphical non-adhesive category with practical relevance. To what extent meta-properties of interest carry over to PBPO$^+$/$\mathbf{Graph}^{(\mathcal{L}, \leq)}$ rewriting from DPO rewriting on adhesive categories is left for future work.

6 Discussion

We discuss our rewriting (Sect. 6.1) and relabeling (Sect. 6.2) contributions in turn.

6.1 Rewriting

Unlike other algebraic approaches such as DPO [3], SPO [12], SqPO [13] and AGREE [8], computing a rewrite step in PBPO and PBPO$^+$ requires only a basic understanding of constructing pullbacks and pushouts. Moreover, assuming monic matching, and under some mild restrictions (DPO is left-linear, and SPO uses conflict-free matches), Corradini et al. [2,13] have shown that

$$\text{DPO} < \text{SPO} < \text{SqPO} < \text{AGREE} < \text{PBPO}$$

where $\mathcal{F} < \mathcal{G}$ means that any $\mathcal{F}$ rule ρ can be simulated by a $\mathcal{G}$ rule σ, and where *simulation* means $\Rightarrow_\rho^{\mathcal{F}} \subseteq \Rightarrow_\sigma^{\mathcal{G}}$ for the generated rewrite relations. This chain may now be extended by inserting AGREE < PBPO$^+$ < PBPO.

If instead of simulation one uses *modeling* as the expressiveness criterion, i.e., $\subseteq$ is strengthened to $=$, then the situation is different. Writing $\prec$ for this expressiveness relation, we conjecture that in **Graph** and with monic matching (which implies conflict-freeness of SPO matches)

$$\text{SPO} \prec \text{SqPO} \prec \text{AGREE} \quad \overset{\prec}{} \quad \begin{matrix} \text{PBPO}^+ \\ \curlyvee \\ \text{PBPO} \end{matrix} \quad \overset{\prec}{} \quad \text{DPO}$$

holds,[2] and the other comparisons do not hold. PBPO $\prec$ PBPO$^+$ follows from Lemma 32 and the fact that **Graph** is strongly amendable, and PBPO does not stand in any other modeling relation due to uncontrolled global effects (in particular, a straightforward adaptation of Example 19 shows DPO $\not\prec$ PBPO).

Other graph rewriting approaches that bear certain similarities to PBPO$^+$ (see also the discussion in [2]) include the double-pullout graph rewriting approach by Kahl [14]; variants of the aforementioned formalisms, such as the cospan SqPO approach by Mantz [15, Section 4.5]; and the recent drag rewriting framework by Dershowitz and Jouannaud [16]. Double-pullout graph rewriting also uses pullbacks and pushouts to delete and duplicate parts of the context (extending DPO), but the approach is defined in the context of collagories [17], and to us it is not yet clear in what way the two approaches relate. Cospan SqPO can be understood as being almost dual to SqPO: rules are cospans, and transformation steps consists of a pushout followed by a final pullback complement. An interesting question is whether PBPO$^+$ can also model cospan SqPO. Drag rewriting is a non-categorical approach to generalizing term rewriting, and like

---

[2] The modeling of DPO and SPO rules for category **Graph** in PBPO$^+$ is similar to the approach described in our paper on PGR [1].

$PBPO^+$, allows relatively fine control over the interface between pattern and context, thereby avoiding issues related to dangling pointers and the construction of pushout complements. Because drag rewriting is non-categorical and drags have inherently more structure than graphs, it is difficult to relate $PBPO^+$ and drag rewriting precisely. These could all be topics for future investigation.

Finally, let us just note that the combination of $PBPO^+$ and $\mathbf{Graph}^{(\mathcal{L}, \leq)}$ does not provide a strict generalization of Patch Graph Rewriting (PGR) [1], our conceptual precursor to $PBPO^+$ (Sect. 1). This is because patch edge endpoints that lie in the context graph can be redefined in PGR (e.g., the direction of edges between context and pattern can be inverted), but not in $PBPO^+$. Beyond that, $PBPO^+$ is more general and expressive. Therefore, at this point we believe that the most distinguishing and redeeming feature of PGR is its visual syntax, which makes rewrite systems much easier to define and communicate. In order to combine the best of both worlds, our aim is to define a similar syntax for (a suitable restriction of) $PBPO^+$ in the future.

6.2 Relabeling

The coupling of $PBPO^+$ and $\mathbf{Graph}^{(\mathcal{L}, \leq)}$ allows relabeling and modeling sorts and variables with relative ease, and does not require a modification of the rewriting framework. Most existing approaches study these topics in the context of DPO, where the requirement to ensure the unique existence of a pushout complement requires restricting the method and proving non-trivial properties:

– Parisi-Presicce et al. [18] limit DPO rules $L \leftarrow K \rightarrow R$ to ones where $K \rightarrow R$ is monic (meaning merging is not possible), and where some set-theoretic consistency condition is satisfied. Moreover, the characterization of the existence of rewrite step has been shown to be incorrect [19], supporting our claim that pushout complements are not easy to reason about.
– Habel and Plump [19] study relabeling using the category of partially labeled graphs. They allow non-monic morphisms $K \rightarrow R$, but they nonetheless add two restrictions to the definition of a DPO rewrite rule. Among others, these conditions do not allow hard overwriting arbitrary labels as in Example 41. Moreover, the pushouts of the DPO rewrite step must be restricted to pushouts that are also pullbacks. Finally, unlike the approach suggested by Parisi-Presice et al., Habel and Plump's approach does not support modeling notions of sorts and variables.

  Our conjecture that $PBPO^+$ can model DPO in $\mathbf{Graph}$ extends to this relabeling approach in the following sense: given a DPO rule over graphs partially labeled from $\mathcal{L}$ that moreover satisfies the criteria of [19], we conjecture that there exists a $PBPO^+$ rule in $\mathbf{Graph}^{(\mathcal{L}^{\perp, \top}, \leq)}$ that models the same rewrite relation when restricting to graphs totally labeled over the base label set $\mathcal{L}$.

Later publications largely appear to build on the approach [19] by Habel and Plump. For example, Schneider [20] gives a non-trivial categorical formulation;

Hoffman [21] proposes a two-layered (set-theoretic) approach to support variables; and Habel and Plump [22] generalize their approach to $\mathcal{M}, \mathcal{N}$-adhesive systems (again restricting $K \to R$ to monic arrows).

The transformation of attributed structures has been explored in a very general setting by Corradini et al. [2], which involves a comma category construction and suitable restrictions of the PBPO notions of rewrite rule and rewrite step. We leave relating their and our approach to future work.

Acknowledgments. We thank Andrea Corradini and anonymous reviewers for useful discussions, suggestions and corrections. We would also like to thank Michael Shulman, who identified the sufficient conditions for amendability for us [23]. The authors received funding from the Netherlands Organization for Scientific Research (NWO) under the Innovational Research Incentives Scheme Vidi (project. No. VI.Vidi.192.004).

Appendix

A   PBPO⁺

We will need both directions of the well known pullback lemma.

Lemma 45 (Pullback Lemma). *Consider the diagram on the right. Suppose the right square is a pullback square and the left square commutes. Then the outer square is a pullback square iff the left square is a pullback square.*   □

$$
\begin{array}{ccccc}
A & \longrightarrow & B & \longrightarrow & C \\
\downarrow & & \downarrow \text{ PB} & & \downarrow \\
D & \longrightarrow & E & \longrightarrow & F
\end{array}
$$

Lemma 10 (Top-Left Pullback). *In the rewrite step diagram of Definition 9, there exists a morphism $u : K \to G_K$ such that $L \xleftarrow{l} K \xrightarrow{u} G_K$ is a pullback for $L \xrightarrow{m} G_L \xleftarrow{g_L} G_K$, $t_K = u' \circ u$, and u is monic.*

Proof. In the following diagram, u satisfying $t_K = u' \circ u$ and $m \circ l = g_L \circ u$ is inferred by using that G_K is a pullback and commutation of the outer square.

$$
t_L \left(
\begin{array}{ccc}
L & \xleftarrow{\;l\;} & K \\
\downarrow{\scriptstyle m} & & \vdots\, u \\
G_L & \xleftarrow{\;g_L\;} & G_K \\
\downarrow{\scriptstyle \alpha} & \text{PB} & \downarrow{\scriptstyle u'} \\
L' & \xleftarrow{\;l'\;} & K'
\end{array}
\right) t_K
$$

By direction $\Longrightarrow$ of the pullback lemma (Lemma 45), the created square is a pullback square, and so by stability of monos under pullbacks, u is monic.   □

Lemma 11 (Uniqueness of u). *In the rewrite step diagram of Definition 9 (and in any category), there is a unique $v : K \to G_K$ such that $t_K = u' \circ v$.*

Proof. In the following diagram, the top-right pullback is obtained using Lemma 10, and the top-left pullback is a rotation of the match diagram:

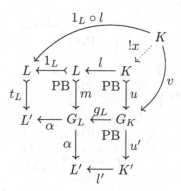

By direction $\Longleftarrow$ of the pullback lemma, $L \xleftarrow{1_L \circ l} K \xrightarrow{u} G_K$ is a pullback for the topmost outer square.

Now suppose that for a morphism $v : K \to G_K, t_K = u' \circ v$. Then $\alpha \circ g_L \circ u = l' \circ u' \circ u = l' \circ t_K = l' \circ u' \circ v = \alpha \circ g_L \circ v$. Hence both v and u make the topmost outer square commute. Hence there exists a unique x such that (simplifying) $l \circ x = l$ and $u \circ x = v$. From known equalities and monicity of t_K we then derive

$$u \circ x = v$$
$$u' \circ u \circ x = u' \circ v$$
$$t_K \circ x = t_K$$
$$t_K \circ x = t_K \circ 1_K$$
$$x = 1_K.$$

Hence $u = v$. $\square$

Lemma 12 (Bottom-Right Pushout). *Let* $K' \xrightarrow{r'} R' \xleftarrow{t_R} R$ *be a pushout for cospan* $R \xleftarrow{r} K \xrightarrow{t_K} K'$ *of rule* ρ *in Definition 9. Then in the rewrite step diagram, there exists a morphism* $w' : G_R \to R'$ *such that* $t_R = w' \circ w$, *and* $K' \xrightarrow{r'} R' \xleftarrow{w'} G_R$ *is a pushout for* $K' \xleftarrow{u'} G_K \xrightarrow{g_R} G_R$.

Proof. The argument is similar to the proof of Lemma 10, but now uses the dual statement of the pullback lemma. $\square$

B   Expressiveness of PBPO+

Lemma 27 *Let* ρ *be a canonical PBPO rule,* G_L *an object, and* $m' \circ e : L \to G_L$ *a match morphism for a mono* m' *and epi* e. *We have:*

$$G_L \Rightarrow_\rho^{(m' \circ e), \alpha} G_R \quad \Longleftrightarrow \quad G_L \Rightarrow_{\rho_e}^{m', \alpha} G_R .$$

Proof. By using the following commuting diagram:

$$\begin{array}{ccc}
L & \xleftarrow{\ l\ } K \xrightarrow{\ r\ } R \\
\end{array}$$

$$t_L \left(\begin{array}{ccc}
L & \xleftarrow{\ l\ } K \xrightarrow{\ r\ } R \\
\downarrow e \quad \mathrm{PB} \downarrow \qquad \mathrm{PO} \downarrow \\
L_c \longleftarrow K_c \longrightarrow R_c \\
\Upsilon m' \; \mathrm{PB} \Upsilon \qquad \mathrm{PO} \downarrow \\
t_{L_c} G_L \longleftarrow G_K \longrightarrow G_R \\
\alpha \downarrow \quad \mathrm{PB} \downarrow \qquad \mathrm{PO} \downarrow \\
L' \xleftarrow{\ l'\ } K' \xrightarrow{\ r'\ } R'
\end{array} \right)$$

□

Theorem 33. *In locally small, strongly amendable categories in which every morphism f can be factored into an epi e followed by a mono m, any PBPO rule ρ can be modeled by a set of PBPO⁺ rules.*

Proof. From Corollary 28 we obtain a set of PBPO rules S that collectively model ρ using monic matching, and by Lemma 32 each $\sigma \in S$ can be modeled by a monic rule τ_σ with a strong matching rewrite policy. By Proposition 24, the set $\{\tau_\sigma \mid \sigma \in S\}$ corresponds to a set of PBPO⁺ rules.   □

C   Category Graph$^{(\mathcal{L}, \leq)}$

Lemma 38. Graph$^{(\mathcal{L}, \leq)}$ is strongly amendable.

Proof. Let **UGraph** refer to the category of unlabeled graphs.

Given a $t_L : L \to L'$ in **Graph**$^{(\mathcal{L}, \leq)}$, momentarily forget about the labels and consider the unlabeled version (overloading names) in **UGraph**. Because **UGraph** is a topos, it is strongly amendable. Thus we can obtain a factorization $L \overset{t'_L}{\rightarrowtail} L'' \overset{\beta}{\to} L'$ on the level of **UGraph** that witnesses strong amendability, i.e., for any factorization of $L \overset{m}{\rightarrowtail} G \overset{\alpha}{\to} L'$ of t_L in **UGraph**, there exists an α' such that

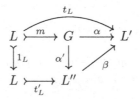

commutes and the left square is a pullback square.

The idea now is to lift the bottom unlabeled factorization into **Graph**$^{(\mathcal{L}, \leq)}$. As far as graph structure is concerned, we know that it is a suitable factorization candidate. Then all that needs to be verified are the order requirements $\leq$ on the labels.

For the lifting of L'', choose the graph in which every element has the same label as its image under β. Then clearly the lifting of β of **UGraph** into **Graph**$^{(\mathcal{L}, \leq)}$ is well-defined, and so is the lifting of t'_L (using that t_L is well-defined in **Graph**$^{(\mathcal{L}, \leq)}$).

Now given any factorization $L \xrightarrow{m} G \xrightarrow{\alpha} L'$ in **Graph**$^{(\mathcal{L}, \leq)}$, lift the α' that is obtained by considering the factorization on the level of **UGraph**. Then the lifting of α' is well-defined by $\alpha = \beta \circ \alpha'$ and well-definedness of α and β in **Graph**$^{(\mathcal{L}, \leq)}$. All that remains to be checked is that the left square is a pullback as far as the labels are concerned, i.e., whether for every $x \in V_L \cup E_L$, $\ell_L(x) = \ell_L(1_L(x)) \wedge \ell_G(m(x))$. This follows using $\ell_L(x) \leq \ell_G(m(x))$ and the complete lattice law $\forall a\ b.a \leq b \implies a \wedge b = a$. $\qquad\square$

References

1. Overbeek, R., Endrullis, J.: Patch graph rewriting. In: Gadducci, F., Kehrer, T. (eds.) ICGT 2020. LNCS, vol. 12150, pp. 128–145. Springer, Cham (2020). https://doi.org/10.1007/978-3-030-51372-6_8

2. Corradini, A., Duval, D., Echahed, R., Prost, F., Ribeiro, L.: The PBPO graph transformation approach. J. Log. Algebraic Methods Program. **103**, 213–231 (2019)

3. Ehrig, H., Pfender, M., Schneider, H.J.: Graph-grammars: an algebraic approach. In: Proceedings of Symposium on Switching and Automata Theory (SWAT), pp. 167–180. IEEE Computer Society (1973)

4. Mac Lane, S.: Categories for the Working Mathematician, vol. 5. Springer, New York (1971). https://doi.org/10.1007/978-1-4612-9839-7

5. Awodey, S.: Category Theory. Oxford University Press, Oxford (2006)

6. Ehrig, H., Ehrig, K., Prange, U., Taentzer, G.: Fundamentals of Algebraic Graph Transformation. Springer, Heidelberg (2006). https://doi.org/10.1007/3-540-31188-2

7. Habel, A., Müller, J., Plump, D.: Double-pushout graph transformation revisited. Math. Struct. Comput. Sci. **11**(5), 637–688 (2001)

8. Corradini, A., Duval, D., Echahed, R., Prost, F., Ribeiro, L.: AGREE – algebraic graph rewriting with controlled embedding. In: Parisi-Presicce, F., Westfechtel, B. (eds.) ICGT 2015. LNCS, vol. 9151, pp. 35–51. Springer, Cham (2015). https://doi.org/10.1007/978-3-319-21145-9_3

9. Corradini, A., Heindel, T., König, B., Nolte, D., Rensink, A.: Rewriting abstract structures: materialization explained categorically. In: Bojańczyk, M., Simpson, A. (eds.) FoSSaCS 2019. LNCS, vol. 11425, pp. 169–188. Springer, Cham (2019). https://doi.org/10.1007/978-3-030-17127-8_10

10. Lack, S., Sobociński, P.: Adhesive categories. In: Walukiewicz, I. (ed.) FoSSaCS 2004. LNCS, vol. 2987, pp. 273–288. Springer, Heidelberg (2004). https://doi.org/10.1007/978-3-540-24727-2_20

11. Garner, R., Lack, S.: On the axioms for adhesive and quasiadhesive categories. Theory Appl. Categories **27**(3), 27–46 (2012)

12. Löwe, M.: Algebraic approach to single-pushout graph transformation. Theor. Comput. Sci. **109**(1&2), 181–224 (1993)

13. Corradini, A., Heindel, T., Hermann, F., König, B.: Sesqui-pushout rewriting. In: Corradini, A., Ehrig, H., Montanari, U., Ribeiro, L., Rozenberg, G. (eds.) ICGT 2006. LNCS, vol. 4178, pp. 30–45. Springer, Heidelberg (2006). https://doi.org/10.1007/11841883_4

14. Kahl, W.: Amalgamating pushout and pullback graph transformation in collagories. In: Ehrig, H., Rensink, A., Rozenberg, G., Schürr, A. (eds.) ICGT 2010. LNCS, vol. 6372, pp. 362–378. Springer, Heidelberg (2010). https://doi.org/10.1007/978-3-642-15928-2_24

15. Mantz, F.: Coupled transformations of graph structures applied to model migration. Ph.D. thesis, University of Marburg (2014)

16. Dershowitz, N., Jouannaud, J.-P.: Drags: a compositional algebraic framework for graph rewriting. Theor. Comput. Sci. **777**, 204–231 (2019)

17. Kahl, W.: Collagories: relation-algebraic reasoning for gluing constructions. J. Log. Algebraic Methods Program. **80**(6), 297–338 (2011)

18. Parisi-Presicce, F., Ehrig, H., Montanari, U.: Graph rewriting with unification and composition. In: Ehrig, H., Nagl, M., Rozenberg, G., Rosenfeld, A. (eds.) Graph Grammars 1986. LNCS, vol. 291, pp. 496–514. Springer, Heidelberg (1987). https://doi.org/10.1007/3-540-18771-5_72

19. Habel, A., Plump, D.: Relabelling in graph transformation. In: Corradini, A., Ehrig, H., Kreowski, H.-J., Rozenberg, G. (eds.) ICGT 2002. LNCS, vol. 2505, pp. 135–147. Springer, Heidelberg (2002). https://doi.org/10.1007/3-540-45832-8_12

20. Schneider, H.J.: Changing labels in the double-pushout approach can be treated categorically. In: Kreowski, H.-J., Montanari, U., Orejas, F., Rozenberg, G., Taentzer, G. (eds.) Formal Methods in Software and Systems Modeling. LNCS, vol. 3393, pp. 134–149. Springer, Heidelberg (2005). https://doi.org/10.1007/978-3-540-31847-7_8

21. Hoffmann, B.: Graph transformation with variables. In: Kreowski, H.-J., Montanari, U., Orejas, F., Rozenberg, G., Taentzer, G. (eds.) Formal Methods in Software and Systems Modeling. LNCS, vol. 3393, pp. 101–115. Springer, Heidelberg (2005). https://doi.org/10.1007/978-3-540-31847-7_6

22. Habel, A., Plump, D.: $\mathcal{M}, \mathcal{N}$-adhesive transformation systems. In: Ehrig, H., Engels, G., Kreowski, H.-J., Rozenberg, G. (eds.) ICGT 2012. LNCS, vol. 7562, pp. 218–233. Springer, Heidelberg (2012). https://doi.org/10.1007/978-3-642-33654-6_15

23. Shulman, M.: Subobject- and factorization-preserving typings. MathOverflow. https://mathoverflow.net/q/381933. Accessed 22 Jan 2021

Incorrectness Logic for Graph Programs

Christopher M. Poskitt[✉]

Singapore Management University, Singapore, Singapore
cposkitt@smu.edu.sg

Abstract. Program logics typically reason about an over-approximation of program behaviour to prove the absence of bugs. Recently, program logics have been proposed that instead prove the *presence* of bugs by means of *under-approximate reasoning*, which has the promise of better scalability. In this paper, we present an under-approximate program logic for a nondeterministic graph programming language, and show how it can be used to reason deductively about program incorrectness, whether defined by the presence of forbidden graph structure or by finitely failing executions. We prove this 'incorrectness logic' to be sound and complete, and speculate on some possible future applications of it.

Keywords: Program logics · Under-approximate reasoning · Bugs

1 Introduction

Many problems in computer science and software engineering can be modelled in terms of rule-based graph transformations [13], motivating research into verifying the correctness of grammars and programs based on this unit of computation. Various approaches towards this goal have been proposed, with techniques including model checking [9], unfoldings [4,16], k-induction [29], weakest preconditions [10,11], abstract interpretation [17], and program logics [5,24,25].

Verification approaches based on program logics and proofs typically reason about over-approximations of program behaviours to prove the absence of bugs. For instance, proving a partial correctness specification $\{pre\}P\{post\}$ guarantees that for states satisfying *pre*, every terminating execution of P ends in a state satisfying *post*. Recently, authors have begun to investigate *under-approximate* program logics that instead prove the *presence* of bugs, motivated by the promise of better scalability that may result from reasoning only about the subset of paths that matter. De Vries and Koutavas [30] proposed the first program logic of this kind, using it to reason about state reachability for randomised nondeterministic algorithms. O'Hearn [21] extended the idea to an *incorrectness logic* that tracked both successful and erroneous executions. Under-approximate program logics have also been explored for local reasoning [28] and proving insecurity [18].

An under-approximate specification $[pres]P[res]$ specifies a reachability property in the reverse direction: that every state satisfying *res* ('result') is reachable by executing P on *some* state (not necessarily all) satisfying *pres* ('presumption'). In other words, *res* under-approximates the reachable states, allowing

© Springer Nature Switzerland AG 2021
F. Gadducci and T. Kehrer (Eds.): ICGT 2021, LNCS 12741, pp. 81–101, 2021.
https://doi.org/10.1007/978-3-030-78946-6_5

for sound reasoning about undesirable behaviours without any false positives, i.e. a formal logical basis for bug catching. This is one of many dualities under-approximate program logics have with Hoare logics [14]. Other important dual-ities include the inverted rule of consequence in which postconditions can be strenghtened (e.g. by dropping disjuncts/paths), as well as the completeness proof which relies on *weakest postconditions* rather than weakest preconditions.

In this paper, we present an under-approximate program logic for reasoning about the presence of bugs in nondeterministic attribute-manipulating graph programs. Following O'Hearn [21], we design it as an *incorrectness logic*, and show how it can be used to reason deductively about the presence of forbidden graph structures or finitely failing executions (e.g. due to the failure of find-ing a match for a rule). As our main technical result, we prove the soundness and relative completeness of our incorrectness logic with respect to a relational denotational semantics. The work in this paper is principally a theoretical expo-sition, but is motivated by some possible future applications, such as the use of incorrectness logic as a basis for sound reasoning in symbolic execution tools for graph and model transformations (e.g. [1,3,20]).

The paper is organised as follows. In Sect. 2 we provide preliminary defi-nitions of graphs and graph morphisms. In Sect. 3 we define graph programs using a relational denotational semantics, as well as an assertion language ('E-conditions') for specifying properties of program states. In Sect. 4, we present an incorrectness logic for graph programs and demonstrate it on some examples. In Sect. 5, we formally define the assertion transformations used in our incorrect-ness logic, and present our main soundness and completeness results. Finally, we review some related work in Sect. 6 before concluding in Sect. 7.

2 Preliminaries

We use a definition of graphs in which edges are directed, nodes (resp. edges) are partially (resp. totally) labelled, and parallel edges are allowed to exist. All graphs in this paper will be totally labelled except for the interface graphs in rule applications (for technical reasons to support relabelling [12]).

A *graph* over a label alphabet $\mathcal{C}$ is a system $G = \langle V_G, E_G, s_G, t_G, l_G, m_G \rangle$ comprising a finite set V_G of *nodes*, a finite set E_G of *edges*, *source* and *target* functions $s_G, t_G \colon E_G \to V_G$, a partial *node labelling function* $l_G \colon V_G \to \mathcal{C}$, and a total *edge labelling function* $m_G \colon E_G \to \{\Box\}$. If $V_G = \emptyset$, then G is the *empty graph*, which we denote by $\emptyset$. Given a node $v \in V_G$, we write $l_G(v) = \bot$ to express that $l_G(v)$ is undefined. A graph G is *totally labelled* if l_G is a total function. Note that for simplicity of presentation, in this paper, we label all edges with a 'blank' label denoted by $\Box$ and rendered as $\longrightarrow$ in diagrams. Note also that we use an undirected edge ⑧—⑧ to represent a pair of edges ⑧⇄⑧.

We write $\mathcal{G}(\mathcal{C}_\bot)$ (resp. $\mathcal{G}(\mathcal{C})$) to denote the *class* of all (resp. all totally labelled) graphs over label alphabet $\mathcal{C}$. Let $\mathcal{L}$ denote the label alphabet $\mathbb{Z}^+$, i.e. all non-empty sequences of integers. In diagrams we will delimit the integers of the sequence using colons, e.g. 5:6:7:8.

A *graph morphism* $g \colon G \to H$ between graphs G, H in $\mathcal{G}(\mathcal{C}_\perp)$ consists of two functions $g_V \colon V_G \to V_H$ and $g_E \colon E_G \to E_H$ that preserve sources, targets and labels; that is, $s_H \circ g_E = g_V \circ s_G$, $t_H \circ g_E = g_V \circ t_G$, $m_H \circ g_E = m_G$, and $l_H(g_V(v)) = l_G(v)$ for all nodes v for which $l_G(v) \neq \perp$. We call G, H respectively the *domain* and *codomain* of g.

A morphism g is *injective* (*surjective*) if g_V and g_E are injective (surjective). Injective morphisms are usually denoted by hooked arrows, $\hookrightarrow$. A morphism g is an *isomorphism* if it is injective, surjective, and satisfies $l_H(g_V(v)) = \perp$ for all nodes v with $l_G(v) = \perp$. In this case G and H are *isomorphic*, which is denoted by $G \cong H$. Finally, a morphism g is an *inclusion* if $g(x) = x$ for all nodes and edges x.

3   Graph Programs and Assertions

We begin by introducing the graph programs that will be the target of our incorrectness logic, as well as an assertion language ('E-conditions') that will be used for specifying properties of the program states (which consist of graphs). To allow for a self-contained presentation, our programs are a simplified 'core' of full-fledged graph programming languages (e.g. GP 2 [23]) which have several more features for practicality (e.g. additional types, negative application conditions).

First, we define the underlying unit of computation in graph programs: the application of a graph transformation rule with relabelling.

Definition 1 (Rule). A *(concrete) rule* $r \colon \langle L \hookleftarrow K \hookrightarrow R \rangle$ comprises totally labelled graphs $L, R \in \mathcal{G}(\mathcal{L})$, a partially labelled graph $K \in \mathcal{G}(\mathcal{L}_\perp)$, and inclusions $K \hookrightarrow L$, $K \hookrightarrow R$. We call L, R the *left-* and *right-hand graphs of* r, and K its *interface*. □

Intuitively, an application of a rule r to a graph $G \in \mathcal{G}(\mathcal{L})$ removes items in $L - K$, preserves those in K, adds the items in $R - K$, and relabels the unlabelled nodes in K. An injective morphism $g \colon L \hookrightarrow G$ is a *match* for r if it satisfies the dangling condition, i.e. no node in $g(L) - g(K)$ is incident to an edge in $G - g(L)$. In this case, G directly derives $H \in \mathcal{G}(\mathcal{L})$ with *comatch* $h \colon R \hookrightarrow H$, denoted $G \Rightarrow_{r,g,h} H$ (or just $G \Rightarrow_r H$), by: (1): removing all nodes and edges in $g(L) - g(K)$; (2) disjointly adding all nodes and edges from $R - K$, keeping their labels (for $e \in E_R - E_K$, $s_H(e)$ is $s_R(e)$ if $s_R(e) \in V_R - V_K$, otherwise $g_V(s_R(e))$; targets analogous); (3) for every node in K, $l_H(g_V(v))$ becomes $l_R(v)$. Semantically, direct derivations are constructed as two 'natural pushouts' (see [12] for the technical details).

In practical graph programming languages, we need a more powerful unit of computation—the rule schema—which describes (potentially) infinitely many concrete rules by labelling the graphs over expressions. We define a simple abstract syntax 'Exp' (Fig. 1) which derives a label alphabet of (lists of) integer expressions, including variables ('Var') of type integer.

A graph in $\mathcal{G}(\mathcal{L})$ can be obtained from a graph in $\mathcal{G}(\text{Exp})$ by means of an *interpretation*, which is a partial function $I \colon \text{Var} \to \mathbb{Z}$. We denote the domain of

Exp   ::= Integer | Integer ':' Exp
Integer   ::= Digit {Digit} | Var | '−' Integer | Integer ArithOp Integer
ArithOp ::= '+' | '−' | '*' | '/'

Fig. 1. Abstract syntax of rule schema labels

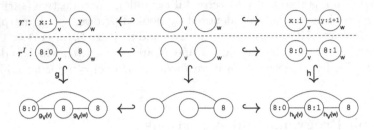

Fig. 2. Example rule schema application

I by dom(I), and the set of variables used in a graph $G \in \mathcal{G}(\text{Exp})$ by vars(G). If vars(G) $\subseteq$ dom(I), then $G^I \in \mathcal{G}(\mathcal{L})$ is the graph obtained by evaluating the expressions in the standard way, with variables x substituted for $I(\text{x})$. Interpretations may also be applied to morphisms, e.g. $p: P \hookrightarrow C$ becomes $p^I : P^I \hookrightarrow C^I$.

Definition 2 (Rule schema). A *rule schema* $r: \langle L \Rightarrow R \rangle$ with $L, R \in \mathcal{G}(\text{Exp})$ represents concrete rules $r^I : \langle L^I \hookleftarrow K \hookrightarrow R^I \rangle$ where dom(I) = vars(L) and K consists of the preserved nodes only (with all nodes unlabelled). Note that we assume for any rule schema, vars(R) $\subseteq$ vars(L). □

The application of a rule schema $r = \langle L \Rightarrow R \rangle$ to a graph $G \in \mathcal{G}(\mathcal{L})$ consists of the following steps: (1) choose an interpretation I with dom(I) = vars(L); (2) choose a *match*, i.e. a morphism $g: L^I \hookrightarrow G$ that satisfies the dangling condition with respect to $r^I : \langle L^I \hookleftarrow K \hookrightarrow R^I \rangle$; (3) apply r^I with match g. If a graph H with comatch $h: R^I \hookrightarrow H$ is derived from G via these steps, we write $G \Rightarrow_{r,g,h}$ (or just $G \Rightarrow_r H$). Moreover, if a graph H can be derived from a graph G via some r in a set of rule schemata $\mathcal{R}$, we write $G \Rightarrow_{\mathcal{R}} H$ (i.e. nondeterministic choice of rule schema). If no rule schema in the set has a match for G, we write $G \not\Rightarrow_{\mathcal{R}}$ (i.e. finite failure).

Example 1 (Rule schema application). Figure 2 displays a rule schema $r: \langle L \hookleftarrow K \hookrightarrow R \rangle$ with its interface (top row), a possible instantiation r^I where $I(\text{x}) = I(\text{y}) = 8$ and $I(\text{i}) = 0$ (middle row). Finally, the bottom row depicts a direct derivation from G (bottom left) to H (bottom right) via r^I. □

Definition 3 (Graph programs). *(Graph) programs* are defined inductively. Given a set of rule schemata $\mathcal{R}$, $\mathcal{R}$ and $\mathcal{R}!$ are programs. If P, Q are programs and $\mathcal{R}$ a set of rule schemata, then $P; Q$ and if $\mathcal{R}$ then P else Q are programs. □

$$
\begin{aligned}
[\![\mathcal{R}]\!]ok &= \{(G,H) \mid G \Rightarrow_\mathcal{R} H\} \\
[\![\mathcal{R}]\!]er &= \{(G,G) \mid G \not\Rightarrow_\mathcal{R}\} \\
[\![P;Q]\!]\epsilon &= \{(G,H) \mid \exists G'.(G,G') \in [\![P]\!]ok \text{ and } (G',H) \in [\![Q]\!]\epsilon\} \\
&\quad \cup \ (\text{if } \epsilon = er \text{ then } \{(G,H) \mid (G,H) \in [\![P]\!]er\}) \\
[\![\mathcal{R}!]\!]ok &= [\![\mathcal{R}]\!]er \cup [\![\mathcal{R};\mathcal{R}!]\!]ok \\
[\![\mathcal{R}!]\!]er &= \emptyset \\
[\![\text{if } \mathcal{R} \text{ then } P \text{ else } Q]\!]\epsilon &= \{(G,H) \mid \exists G'.(G,G') \in [\![\mathcal{R}]\!]ok \text{ and } (G,H) \in [\![P]\!]\epsilon\} \\
&\quad \cup \{(G,H) \mid (G,G) \in [\![\mathcal{R}]\!]er \text{ and } (G,H) \in [\![Q]\!]\epsilon\}
\end{aligned}
$$

Fig. 3. A relational denotational semantics for graph programs

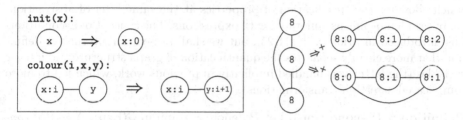

Fig. 4. Rules for the program init; colour! and two possible executions

Intuitively, $\mathcal{R}$ denotes a single nondeterministic application of a rule schemata set. This results in failure if none of the rules are applicable to the current graph. The program $\mathcal{R}!$ denotes as-long-as-possible iteration of $\mathcal{R}$, in which the iteration terminates the moment that $\mathcal{R}$ is no longer applicable to the current graph (the program never fails). Finally, the program $P;Q$ denotes sequential composition, and if $\mathcal{R}$ then P else Q denotes conditional branching, determined by testing the applicability of $\mathcal{R}$ (note that $\mathcal{R}$ will not transform the current graph).

Each graph program is given a simple relational denotational semantics (in the style of [21]). We associate each program P with two semantic functions, $[\![P]\!]ok$ and $[\![P]\!]er$, which respectively describe state (i.e. graph) transitions for successful and finitely failing computations. Unlike operational semantics for graph programs (e.g. [23]), we do not explicitly track a 'fail' state, but rather return pairs (G,H) where H is the last graph derived from G before the failure.

Definition 4 (Semantics). The *semantics* of a graph program P is given by a binary relation $[\![P]\!]\epsilon \subseteq \mathcal{G}(\mathcal{L}) \times \mathcal{G}(\mathcal{L})$, defined according to Fig. 3.   □

Note that divergence is treated in an implicit way: a program that always diverges is associated with empty relations. For example, $[\![\langle \emptyset \Rightarrow \emptyset \rangle!]\!]ok = \emptyset$.

Example 2 (Buggy colouring). Figure 4 contains an example graph program $P =$ init; colour! that purportedly computes a graph colouring, i.e. an association of integers ('colours') with nodes such that no two adjacent nodes are associated with the same colour. The program nondeterministically assigns a colour of '0' to a node, encoding it as the second element of the label's sequence, before iteratively matching adjacent pairs of coloured/uncoloured nodes and assigning a colour to the latter obtained by incrementing the colour of the former. Note that the edges are undirected for simplicity.

Two possible executions are shown in Fig. 4, the first of which leads to a correct colouring, and the second of which leads to an illegal one. Moreover, the program can finitely fail on input graphs for which init has no match. (We shall use incorrectness logic to logically prove the presence of such outcomes.)

Before we can define an incorrectness logic for graph programs, we require an assertion language for expressing properties of the states, i.e. graphs in $\mathcal{G}(\mathcal{L})$. For this purpose we shall use nested conditions with expressions ('E-conditions'), which allow for the specification of properties at the same level of abstraction, i.e. by graph morphisms annotated with expressions. The concept of E-conditions was introduced in prior work [24,25], but we shall present an alternative definition that more cleanly separates the quantification of graph structure and integer variables (the latter was handled implicitly in previous work, which led to more complicated assertion transformations).

Definition 5 (E-condition). Let P denote a graph in $\mathcal{G}(\text{Exp})$. A *nested condition with expressions* (short. *E-condition) over* P is of the form true, γ, $\exists x.c$, or $\exists a.c'$, where γ is an interpretation constraint (i.e. a Boolean expression over 'Exp'), x is a variable in Var, c is an E-condition over P, $a \colon P \hookrightarrow C$ is an injective graph morphism over $\mathcal{G}(\text{Exp})$, and c' is an E-condition over C. Moreover, $\neg c_1$, $c_1 \wedge c_2$, and $c_1 \vee c_2$ are E-conditions over P if c_1, c_2 are E-conditions over P. □

The *free variables* of an E-condition c, denoted $\text{FV}(c)$, are those variables present in node labels and interpretation constraints that are not bound by any variable quantifier (defined in the standard way). If c is defined over the empty graph $\emptyset$ and $\text{FV}(c) = \emptyset$, we call c an *E-constraint*. Furthermore, a mapping of free variables to expressions $\sigma = (x_1 \mapsto e_1, \cdots)$ is called a *substitution*, and c^σ denotes the E-condition c but with all free variables x substituted for $\sigma(x)$.

Definition 6 (Satisfaction of E-conditions). Let c denote an E-condition over P, I an interpretation with $\text{dom}(I) = \text{FV}(c)$, and $p \colon P^I \hookrightarrow G$ an injective morphism over $\mathcal{G}(\mathcal{L})$. The *satisfaction* relation $p \models^I c$ is defined inductively.

If c has the form true, then $p \models^I c$ always. If c is an interpretation constraint γ, then $p \models^I c$ if γ^I = true (defined in the standard way). If c has the form $\exists x.c'$ where c' is an E-condition over P, then $p \models^I c$ if $p \models^{I[x \mapsto v]} c'$ for some $v \in \mathbb{Z}$. If c has the form $\exists a \colon P \hookrightarrow C.c'$ where c' is an E-condition over C, then $p \models^I c$ if there exists an injective morphism $q \colon C^I \hookrightarrow G$ such that $q \circ a^I = p$ and $q \models^I c'$.

$$P^I \overset{a^I}{\hookrightarrow} C^I$$
$$p \searrow \underset{=}{\ } \swarrow q \models^I c'$$
$$G$$

Finally, the satisfaction of Boolean formulae over E-conditions is defined in the standard way. □

The satisfaction of E-constraints by graphs is defined as a special case of the general definition. That is, a graph $G \in \mathcal{G}(\mathcal{L})$ *satisfies* an E-constraint c, denoted $G \models c$, if $i_G : \emptyset \hookrightarrow G \models^{I_\emptyset} c$, where $I_\emptyset$ is the empty interpretation, i.e. with $\mathrm{dom}(I_\emptyset) = \emptyset$.

For brevity, we write `false` for $\neg\mathbf{true}$, $c \Longrightarrow d$ for $\neg c \vee d$, $\forall \mathbf{x}.c$ for $\neg \exists \mathbf{x}.\neg c$, $\forall a.c$ for $\neg \exists a.\neg c$, and $\exists \mathbf{x}_1, \cdots \mathbf{x}_n.c$ for $\exists \mathbf{x}_1. \cdots \exists \mathbf{x}_n.c$ (analogous for $\forall$). Furthermore, if the domain of a morphism can unambiguously be inferred from the context, we write only the codomain. For example, the E-constraint $\exists \emptyset \hookrightarrow C. \exists C \hookrightarrow C'. \mathbf{true}$ can be written as $\exists C. \exists C'$.

Example 3 (E-constraint). The following E-constraint expresses that for every pair of integer-labelled nodes, if the labels differ, then the nodes are adjacent:

$$\forall \mathbf{x}, \mathbf{y}. \; \forall \underset{v}{\textcircled{x}}, \underset{w}{\textcircled{y}} . \; \mathbf{x} \neq \mathbf{y} \Longrightarrow \exists \underset{v}{\textcircled{x}} {\rightarrow} \underset{w}{\textcircled{y}} \vee \exists \underset{v}{\textcircled{x}} {\leftarrow} \underset{w}{\textcircled{y}}$$

Note that v, w are node identifiers to indicate which nodes are the same along the chain of nested morphisms, as can be seen when denoting them in full:

$$\forall \mathbf{x}, \mathbf{y}. \; \forall \emptyset \hookrightarrow \underset{v}{\textcircled{x}}, \underset{w}{\textcircled{y}} . \; \mathbf{x} \neq \mathbf{y} \Longrightarrow \exists \underset{v}{\textcircled{x}}, \underset{w}{\textcircled{y}} \hookrightarrow \underset{v}{\textcircled{x}} {\rightarrow} \underset{w}{\textcircled{y}} \vee \exists \underset{v}{\textcircled{x}}, \underset{w}{\textcircled{y}} \hookrightarrow \underset{v}{\textcircled{x}} {\leftarrow} \underset{w}{\textcircled{y}}$$

These node identifiers may be omitted when the mappings are unambiguous.

4 Proving the Presence of Bugs

Before we define the proof rules of our incorrectness logic, it is important to define what an *incorrectness specification* is and what it means for it to be *valid*. In over-approximate program logics (e.g. [24,25]) a specification is given in the form of a triple, $\{c\}P\{d\}$, which under partial correctness expresses that if a graph satisfies precondition c, and program P successfully terminates on it, then the resulting graph will always satisfy d. The postcondition d over-approximates the graphs reachable upon termination of P from graphs satisfying c.

Incorrectness logic [21], however, is based on under-approximate reasoning, for which a specification $[c]P[d]$ has a rather different meaning (and thus a different notation). Here, we call the pre-assertion c a *presumption* and the post-assertion d a *result*. The triple specifies that if a graph satisfies d, then it can be derived from *some* graph satisfying c by executing P on it. In other words, d under-approximates the states reached as a result of executing P on graphs satisfying c. It does not specify that every graph satisfying c derives a graph satisfying d, and it does not preclude graphs satisfying $\neg c$ from deriving such graphs either.

The principal benefit of proving such triples is then proving the *presence of bugs*, and can be thought of as providing a possible formal foundation for static bug catchers, e.g. symbolic execution tools. In graph programs, this amounts to formal proofs of the presence of *illegal graph structure*, but it can also facilitate proofs of the presence of *finite failure*. To accommodate this, we adopt O'Hearn's approach [21] of tracking exit conditions ϵ in the result, $[c]P[\epsilon : d]$, using *ok* to represent normal executions and *er* to track finite failures.

$$\textsc{RuleSetSucc} \quad \vdash [c \wedge \text{App}(\mathcal{R})] \; \mathcal{R} \; [ok : \text{WPost}(\mathcal{R}, c)][er : \texttt{false}]$$

$$\textsc{RuleSetFail} \quad \vdash [c \wedge \neg\text{App}(\mathcal{R})] \; \mathcal{R} \; [ok : \texttt{false}][er : c \wedge \neg\text{App}(\mathcal{R})]$$

$$\textsc{SeqSucc} \; \frac{\vdash [c] \; P \; [ok : e] \qquad \vdash [e] \; Q \; [\epsilon : d]}{\vdash [c] \; P; Q \; [\epsilon : d]} \qquad \textsc{SeqFail} \; \frac{\vdash [c] \; P \; [er : d]}{\vdash [c] \; P; Q \; [er : d]}$$

$$\textsc{IfElse} \; \frac{\vdash [c \wedge \text{App}(\mathcal{R})] \; P \; [\epsilon : d] \qquad \vdash [c \wedge \neg\text{App}(\mathcal{R})] \; Q \; [\epsilon : d]}{\vdash [c] \; \texttt{if } \mathcal{R} \texttt{ then } P \texttt{ else } Q \; [\epsilon : d]}$$

$$\textsc{Cons} \; \frac{c \Longleftarrow c' \quad \vdash [c'] \; P \; [\epsilon : d'] \quad d' \Longleftarrow d}{\vdash [c] \; P \; [\epsilon : d]}$$

$$\textsc{IterZero} \quad \vdash [c \wedge \neg\text{App}(\mathcal{R})] \; \mathcal{R}! \; [ok : c \wedge \neg\text{App}(\mathcal{R})][er : \texttt{false}]$$

$$\textsc{Iter} \; \frac{\vdash [c \wedge \text{App}(\mathcal{R})] \; \mathcal{R}; \mathcal{R}! \; [ok : d \wedge \neg\text{App}(\mathcal{R})]}{\vdash [c \wedge \text{App}(\mathcal{R})] \; \mathcal{R}! \; [ok : d \wedge \neg\text{App}(\mathcal{R})]}$$

$$\textsc{IterVar} \; \frac{\vdash [c_{i-1}] \; \mathcal{R} \; [ok : c_i] \text{ for all } 0 < i \leq n, \text{ and } c_n \Longrightarrow \neg\text{App}(\mathcal{R})}{\vdash [c_0] \; \mathcal{R}! \; [ok : c_n]}$$

Fig. 5. Incorrectness axioms and proof rules for graph programs

Definition 7 (Under-approximate validity). Let c, d denote E-constraints, P a graph program, and ϵ an exit condition. A specification $[c] \; P \; [\epsilon : d]$ is *valid*, denoted $\models [c] \; P \; [\epsilon : d]$, if for every graph $H \in \mathcal{G}(\mathcal{L})$ such that $H \models d$, there exists a graph $G \in \mathcal{G}(\mathcal{L})$ such that $G \models c$ and $(G, H) \in [\![P]\!]\epsilon$. □

Figure 5 presents the axioms and proof rules of our incorrectness logic for graph programs, which are adapted from O'Hearn's incorrectness logic for imperative programs [21]. We say that a triple is *provable*, denoted $\vdash [c]P[\epsilon : d]$, if it can be instantiated from any axiom, or deduced as the consequent of any proof rule with provable antecedents. We use the notation $\vdash [c]P[ok : d_1][er : d_2]$ as shorthand for two separate triples, $\vdash [c]P[ok : d_1]$ and $\vdash [c]P[er : d_2]$.

Note that a number of axioms and proof rules rely on some transformations that we have not yet defined: $\text{App}(\mathcal{R})$, which expresses the existence of a match for $\mathcal{R}$, and $\text{WPost}(\mathcal{R}, c)$, which expresses the *weakest postcondition* that must be satisfied to guarantee the existence of a pre-state satisfying c. These transformations will be formally defined in Sect. 5.

The axioms RULESETSUCC and RULESETFAIL allow for reasoning about the most fundamental unit of graph programs: rule schema application. The former covers the successful case: if a graph satisfies the weakest postcondition for rule schemata set $\mathcal{R}$ and E-constraint c, then it can be derived from some graph satisfying the presumption $c \wedge \text{App}(\mathcal{R})$. The latter of the axioms covers the possibility that $\mathcal{R}$ cannot be applied: in this case, we have an exit condition of er to track its finite failure.

Sequential composition is handled by SEQSUCC as well as SEQFAIL (to cover the possibility of the first program resulting in failure). The conditional construct is covered by IFELSE: note that failure can only result from failure in the two branches, and not from the guard $\mathcal{R}$, which is simply tested to choose the branch.

It is important to highlight the rule of consequence, CONS, as the implications in the side conditions are reversed from those of the corresponding Hoare logic rule [2,14]. In incorrectness logic, we instead weaken the precondition and strengthen the postcondition. Intuitively, this allows us to soundly drop disjuncts in the result and thus reason about *fewer paths* in the post-state, which may support better scalability in tools [21].

For the iteration of rule schemata sets, we have a number of cases. The axiom ITERZERO covers the case when a rule schemata set is no longer applicable (note that this does not result in failure). The proof rule ITER unrolls a step of the iteration. Traditional loop invariants are less important in these proof rules than they are for Hoare logic, as we are reasoning about a subset of paths rather than *all* of them. To see this, consider the triple $\models [inv]\mathcal{R}![ok : inv \wedge \neg\text{App}(\mathcal{R})]$ with invariant inv. Under-approximate validity requires every graph H satisfying inv and $\neg\text{App}(\mathcal{R})$ to be derivable by applying $\mathcal{R}!$ to some graph G satisfying inv. One can always find such a graph by taking $G = H$.

Finally, ITERVAR combines ITERZERO and ITER into one rule. It expresses that a triple $\vdash [c_0]\mathcal{R}![ok : c_n]$ can be proven if: (1) c_n implies the termination of the iteration (i.e. the non-applicability of $\mathcal{R}$); and (2) if triples can be proven for the n iterations of $\mathcal{R}$. ITERVAR is a stricter version of the backwards variant rule for while-loops in [21,30]: had we adopted the rule in full, we would be able to prove triples such as $\vdash [c(0)]\mathcal{R}![ok : \exists n.n \geq 0.c(n) \wedge \neg\text{App}(\mathcal{R})]$. Here, $c(i)$ denotes a parameterised predicate, i.e. in our case, a function mapping expressions to E-constraints. Unfortunately, these are not possible to express using E-constraints, and including them would strictly increase their expressive power beyond first-order graph properties and the current capabilities of 'WPost'.

Example 4 (Colouring: finite failure). In our first example, we prove the incorrectness specification $\vdash [\neg\exists x.\exists \textcircled{x}]$ `init`; `colour!` $[er : \neg\exists x.\exists \textcircled{x}]$ for the program of Fig. 4. This triple specifies that if a graph does not contain any integer-labelled nodes, then it can be derived from another graph satisfying the same condition that the program finitely fails on. Since `init` would fail on any such graph, this specification is valid: the graph in the post-state is exactly the graph in the pre-state. Figure 6 proves this triple using incorrectness logic.

RuleSetFail $\dfrac{\square}{\vdash [\textbf{true} \wedge \neg\text{App}(\texttt{init})]\ \texttt{init}\ [er : \textbf{true} \wedge \neg\text{App}(\texttt{init})]}$
Cons $\dfrac{}{\vdash [\neg\text{App}(\texttt{init})]\ \texttt{init}\ [er : \neg\text{App}(\texttt{init})]}$
SeqFail $\dfrac{}{\vdash [\neg\text{App}(\texttt{init})]\ \texttt{init}; \texttt{colour!}\ [er : \neg\text{App}(\texttt{init})]}$

Fig. 6. Proving the presence of failure (E-constraints in Fig. 8)

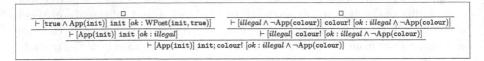

$\dfrac{\square}{\vdash [\textbf{true} \wedge \text{App}(\texttt{init})]\ \texttt{init}\ [ok : \text{WPost}(\texttt{init}, \textbf{true})]}$ $\dfrac{\square}{\vdash [illegal \wedge \neg\text{App}(\texttt{colour})]\ \texttt{colour!}\ [ok : illegal \wedge \neg\text{App}(\texttt{colour})]}$
$\dfrac{}{\vdash [\text{App}(\texttt{init})]\ \texttt{init}\ [ok : illegal]}$ $\dfrac{}{\vdash [illegal]\ \texttt{colour!}\ [ok : illegal \wedge \neg\text{App}(\texttt{colour})]}$
$\dfrac{}{\vdash [\text{App}(\texttt{init})]\ \texttt{init}; \texttt{colour!}\ [ok : illegal \wedge \neg\text{App}(\texttt{colour})]}$

Fig. 7. Proving the presence of an illegal graph (E-constraints in Fig. 8)

Example 5 (Colouring: illegal graph). While proving the presence of failure for the program of Fig. 4 is simple, there are some interesting subtleties involved in proving the presence of illegal graph structure. Let us consider:

$$\vdash [\exists \textbf{x}.\exists \textbf{x}]\ \texttt{init}; \texttt{colour!}\ [ok : (\exists \textbf{a},\textbf{b},\textbf{j}.\exists \text{(a:j)}\!\!-\!\!\text{(b:j)}) \wedge (\exists \textbf{x}.\exists \text{(x:0)}) \wedge (\neg\exists \textbf{x}.\exists \text{(x:i)}\!\!-\!\!\text{(y)})]$$

which specifies that if a graph has an illegal colouring, at least one node coloured '0', and `colouring` is no longer applicable, then it can be derived by applying the program to some graph containing an integer-labelled node (i.e. that `init` does not fail on). This triple is provable (Fig. 7) and valid, but not because of any problem with `colour`. Consider, for example, the graph $\text{(8:0)}\ \text{(8:8)}\!\!-\!\!\text{(8:8)}$. This is trivially reachable from graphs that already contain the illegal structure, e.g. $\text{(8)}\ \text{(8:8)}\!\!-\!\!\text{(8:8)}$, thus we are able to complete the proof using the IterZero rule.

Finally, we strengthen the condition on the result to try and prove the presence of an illegal colouring that is created by the program itself (see Fig. 8 for the E-constraints):

$$\vdash [c]\ \texttt{init}; \texttt{colour!}\ [ok : d \wedge \neg\text{App}(\texttt{colour})]$$

The E-constraint c expresses that there exists at least one node and that no node is coloured (instead of using conjunction, we express this more compactly using nesting). The E-constraint d expresses that there are three coloured nodes (with colours $0, 1, 1$). Together, the triple specifies that every graph satisfying $d \wedge \neg\text{App}(\texttt{colour})$ can be derived from at least one graph satisfying c. This triple is valid and provable (Fig. 9) as the illegal colouring is a logical possibility of some executions of `colour!`. Note that we cannot use an assertion such as $\exists \textbf{a},\textbf{b}.\exists \text{(a:1)}\!\!-\!\!\text{(b:1)}$ in place of d, as this is satisfied by the graph $\text{(8:1)}\!\!-\!\!\text{(8:1)}$ which is impossible to derive from any graph satisfying c.

As E-constraints are equivalent to first-order logic on graphs [24], we are precluded from proving a more general non-local condition, e.g. "there exists a cycle with an illegal colouring". However, there are more powerful logics equipped with similar transformations that may be possible to use instead [19,27].

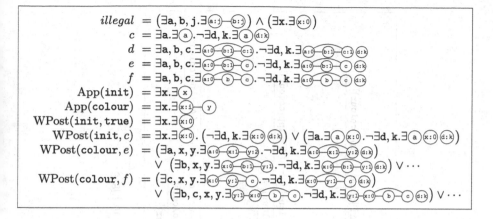

Fig. 8. E-constraints used in the proofs of Figs. 6, 7, and 9

5 Transformations, Soundness, and Completeness

This section presents formal definitions and characterisations of the transformations that are used in some of our incorrectness axioms and proof rules. Following this, we present our main technical result: the soundness and completeness of our incorrectness logic with respect to the denotational semantics.

First, we consider 'App', which transforms a set of rule schemata into an E-constraint that expresses the minimum requirements on a graph for at least one of the rules to be applicable. Intuitively, the E-constraint expresses the presence of a match for a left-hand side, i.e. a morphism that satisfies the dangling condition. This transformation is adapted from similar transformations in [10,24].

Proposition 1 (Applicability). For every graph $G \in \mathcal{G}(\mathcal{L})$ and set of rule schemata $\mathcal{R}$,

$$G \models \text{App}(\mathcal{R}) \text{ if and only if } \exists H.\ G \Rightarrow_{\mathcal{R}} H.$$

Construction. Define $\text{App}(\emptyset) = \texttt{false}$ and then $\text{App}(\{r_1, \cdots r_n\}) = \text{app}(r_1) \vee \cdots \text{app}(r_n)$. Given a rule schema $r = \langle L \hookleftarrow K \hookrightarrow R \rangle$ over variables $x_1, \cdots x_m$, define $\text{app}(r) = \exists x_1, \cdots x_m.\ \exists \emptyset \hookrightarrow L.\ \text{Dang}(r)$.

Finally, define $\text{Dang}(r) = \bigwedge_{a \in A} \neg \exists x_a. \exists a$ where the index set A ranges over all injective morphisms (equated up to isomorphic codomains) $a\colon L \hookrightarrow L^{\oplus}$ such that the pair $\langle K \hookrightarrow L, a \rangle$ has no natural pushout complement and each $L^{\oplus}$ is a graph that can be obtained from L by adding either: (1) a single loop with label $\square$; (2) a single edge with label $\square$ between distinct nodes; or (3) a single node labelled with fresh variable x_a and a non-looping edge incident to it with label $\square$. If the index set A is empty, then $\text{Dang}(r) = \texttt{true}$. $\qquad\square$

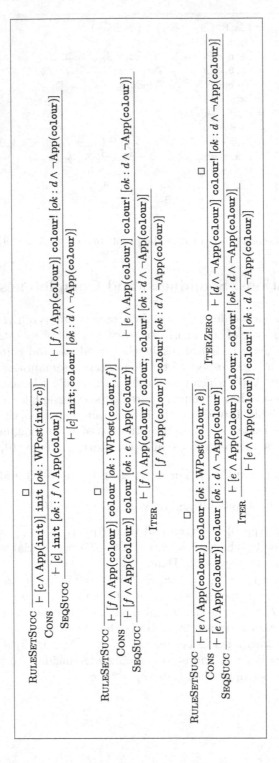

Fig. 9. Proving the presence of an illegal colouring (E-constraints in Fig. 8)

Next, we consider 'WPost', which transforms a set of rule schemata and a presumption into a weakest postcondition, i.e. the weakest property a graph must satisfy to guarantee the *existence* of a pre-state that satisfies the presumption. WPost is defined via two intermediate transformations: 'Shift' and 'Right'.

We begin by defining 'Shift', which can be used to transform an E-constraint c into an E-condition over the left-hand side of a rule L by considering all the ways that a 'match' can overlap with c. Our definition is adapted from the shifting constructions of [10,24] to handle the explicit quantification of label variables. Intuitively, this step is handled via a disjunction over all possible substitutions of a variable in c for integer expressions or variables in L, i.e. to account for interpretations in which they refer to the same values.

To facilitate this, we require that the labels in c are lists of variables that are distinct from those in L. This is a mild assumption, as an arbitrary expression can simply be replaced with a variable that is then equated with the original expression in an interpretation constraint.

Lemma 1 (E-constraint to left E-condition). Let r denote a rule schema and c an E-constraint labelled over lists of variables distinct from those in r. For every graph $G \in \mathcal{G}(\mathcal{L})$ and morphism $g\colon L^I \hookrightarrow G$ with $\mathrm{dom}(I) = \mathrm{vars}(L)$,

$$g\colon L^I \hookrightarrow G \models^I \mathrm{Shift}(r,c) \text{ if and only if } G \models c.$$

Construction. Let c denote an E-constraint and r a rule with left-hand side L. We define $\mathrm{Shift}(r,c) = \mathrm{Shift}'(\emptyset \hookrightarrow L, c)$. We define Shift' inductively for morphisms $p\colon P \hookrightarrow P'$ and E-conditions over P. Let $\mathrm{Shift}'(p,\mathtt{true}) = \mathtt{true}$ and $\mathrm{Shift}'(p,\gamma) = \gamma$. Then:

$$\mathrm{Shift}'(p, \exists \mathbf{x}.\ c) = \big(\exists \mathbf{x}.\ \mathrm{Shift}'(p,c)\big) \bigvee_{l \in \Sigma_{P'}} \mathrm{Shift}'(p, c^{(\mathbf{x}\mapsto l)})$$

$$\mathrm{Shift}'(p, \exists a\colon P \hookrightarrow C.\ c) = \bigvee_{e \in \varepsilon} \exists b\colon P' \hookrightarrow E.\ \mathrm{Shift}'(s\colon C \hookrightarrow E, c)$$

In the third case, $\Sigma_{P'}$ is the set of all variables and integer expressions present in the labels of $V_{P'}$. In the fourth case, construct pushout (1) of p and a as depicted in the diagram. The disjunction ranges over the set ε, which we define to contain every surjective morphism $e\colon C' \hookrightarrow E$ such that $b = e \circ a'$ and $s = e \circ q$ are injective morphisms. (We consider codomains of each e up to isomorphism, so the disjunction is finite.)

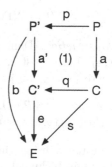

Shift and Shift' are defined for Boolean formulae over E-conditions in the standard way. $\qquad\square$

Example 6 (Shift). Consider the rule schema `init` (Fig. 4) and E-constraint c (Fig. 8). After simplification, the transformation $\mathrm{Shift}(\mathtt{init}, c)$ results in:

$$\big(\exists \textcircled{x} \hookrightarrow \textcircled{x}.\neg\exists d, k.\exists \textcircled{x} \hookrightarrow \textcircled{x}\ \textcircled{\scriptsize d:k}\big) \vee \big(\exists a.\exists \textcircled{x} \hookrightarrow \textcircled{a}\ \textcircled{x}.\neg\exists d, k.\exists \textcircled{a}\ \textcircled{x} \hookrightarrow \textcircled{a}\ \textcircled{x}\ \textcircled{\scriptsize d:k}\big)$$

The second intermediate transformation for 'WPost' is 'Right', which transforms an E-condition over the left-hand side of a rule to an E-condition over the right-hand side. This construction is based on transformation 'L' from [10,24] but in the reverse direction.

Lemma 2 (Left to right E-condition). Let $r = \langle L \hookleftarrow K \hookrightarrow R \rangle$ denote a rule schema and c an E-condition over L. Then for every direct derivation $G \Rightarrow_{r,g,h} H$ with $g \colon L^I \hookrightarrow G$ and $h \colon R^I \hookrightarrow H$,

$$g \colon L^I \hookrightarrow G \models^I c \text{ if and only if } h \colon R^I \hookrightarrow H \models^I \text{Right}(r, c).$$

Construction. We define $\text{Right}(r, \mathtt{true}) = \mathtt{true}$, Right $(r, \gamma) = \gamma$, and $\text{Right}(r, \exists \mathtt{x}.\ c) = \exists \mathtt{x}.\ \text{Right}(r, c)$. Let $\text{Right}(r, \exists a.\ c) = \exists b.\ \text{Right}(r^*, c)$ if $\langle K \hookrightarrow L, a \rangle$ has a natural pushout complement (1), where $r^* = \langle X \hookleftarrow Z \hookrightarrow Y \rangle$ denotes the rule 'derived' by also constructing natural pushout (2). If $\langle K \hookrightarrow L, a \rangle$ has no natural pushout complement, then $\text{Right}(r, \exists a.\ c) = \mathtt{false}$.

Right is defined for Boolean formulae over E-conditions as per usual. □

Example 7 (Right). Continuing from Example 6, applying the transformation $\text{Right}(\mathtt{init}, \text{Shift}(\mathtt{init}, c))$ results in the E-condition:

$$\left(\exists\, \fbox{x:0} \hookrightarrow \fbox{x:0}.\neg \exists \mathtt{d}, \mathtt{k}. \exists \fbox{x:0} \hookrightarrow \fbox{x:0}\ \fbox{d:k} \right) \vee \left(\exists a. \exists \fbox{x:0} \hookrightarrow \text{ⓐ}\ \fbox{x:0}.\neg \exists \mathtt{d}, \mathtt{k}. \exists\, \text{ⓐ}\ \fbox{x:0} \hookrightarrow \text{ⓐ}\ \fbox{x:0}\ \fbox{d:k} \right)$$

Next, we can give 'WPost' a simple definition based on the two intermediate transformations. Intuitively, it constructs a disjunction of E-constraints that demand the existence of some co-match that would result from applying the rule schema set to a graph satisfying the presumption.

Proposition 2 (Weakest postcondition). Let $\mathcal{R}$ denote a rule schemata set and c an E-constraint. Then for every graph $H \in \mathcal{G}(\mathcal{L})$,

$$H \models \text{WPost}(\mathcal{R}, c) \text{ if and only if } \exists G.\ G \models c \text{ and } G \Rightarrow_{\mathcal{R}} H.$$

Construction. Define $\text{WPost}(\emptyset, c) = \mathtt{false}$ and $\text{WPost}(\mathcal{R}, c) = \bigvee_{r \in \mathcal{R}} \text{wpost}$ (r, c). Let $\text{wpost}(r, c) = \exists \mathtt{x}_1, \cdots \mathtt{x}_n. \exists \emptyset \hookrightarrow R.\text{Dang}(r^{-1}) \wedge \text{Right}(r, \text{Shift}(r, c))$ where $\{\mathtt{x}_1, \cdots, \mathtt{x}_n\} = \text{vars}(R)$ and r^{-1} is the reversal of rule r. □

Example 8 (WPost). Continuing from Example 7, applying the transformation $\text{WPost}(\mathtt{init}, c)$ results in the E-constraint given in Fig. 8.

Finally, using the characterisations of 'App' and 'WPost', we can present the main technical results of our paper: the soundness and completeness of our incorrectness logic for graph programs. Soundness means that any triple provable in our logic is valid in the sense of Definition 7, i.e. that graphs satisfying the result are reachable from some graph satisfying the presumption. The proof of this theorem is by structural induction on triples.

Theorem 1 (Soundness). For all E-constraints c, d, graph programs P, and exit conditions ϵ,

$$\vdash [c] \ P \ [\epsilon : d] \text{ implies } \models [c] \ P \ [\epsilon : d].$$

$\square$

Completeness is the other side of the coin: it means that any valid triple can be proven using our logic. As is typical, we prove *relative completeness* [7] in which completeness is relative to the existence of an oracle for deciding the validity of assertions (as in CONS). The idea is to separate incompleteness due to the incorrectness logic from incompleteness in deducing valid assertions, and determine that no proof rules are missing. Our proof relies on some semantically (or extensionally) defined assertions, WPOST$[P, c]$, that characterise exactly the weakest postcondition of an arbitrary program P relative to an E-constraint c.

Theorem 2 (Relative completeness). For all E-constraints c, d, graph programs P, and exit conditions ϵ,

$$\models [c] \ P \ [\epsilon : d] \text{ implies } \vdash [c] \ P \ [\epsilon : d].$$

$\square$

It is important to remark that it is unknown whether E-constraints are *expressive* enough to specify precisely the assertion WPOST$[P, c]$ in general; in fact, there is evidence to suggest they may not be [31]. This is, however, a limitation of the logic and not the incorrectness proof rules, and expressiveness may not be a problem faced by stronger assertion languages for graphs, such as those supporting non-local properties [19,22,27].

6 Related Work

Over-approximate program logics for proving the absence of bugs have been studied extensively [2]. Our program logic differs by focusing on under-approximate reasoning, i.e. proofs about the presence of bugs (in our case, forbidden graph structure or finitely failing execution paths). The first under-approximate calculus of this kind was introduced by De Vries and Koutavas [30], who proposed the notion of under-approximate validity, and defined a 'Reverse Hoare Logic' for proving reachability specifications over the proper states of imperative randomised programs. O'Hearn's incorrectness logic [21] extended this program logic to support under-approximate reasoning about executions that result in errors, an idea we adopt to support reasoning about both successful computations (ok) and finitely failing executions (er). Both of these program logics use variants to reason about while-loop termination, but unlike standard Hoare logics, require that the variant decreases in the backwards direction. Our ITERVAR rule is similar, but requires the number of iterations to be known as E-conditions are not expressive enough to specify parameterised graph properties, for example, the existence of a cycle of length n.

Raad et al. [28] combined separation logic with incorrectness logic to facilitate proofs about the presence of bugs using local reasoning, i.e. specifications that focus only on the region of memory being accessed. They found that the original model of separation logic, which does not distinguish dangling pointers from pointers we have no knowledge about, to be incompatible with the under-approximate frame rule. This was resolved by refining the model with negative heap assertions that can specify that a location has been de-allocated.

Murray [18] proposed the first under-approximate relational logic, allowing for reasoning about the behaviours of pairs of programs. As many important security properties (e.g. noninterference, function sensitivity, refinement) can be specified as relational properties, Murray's program logic can be used to provably demonstrate the presence of insecurity.

Bruni et al. [6] incorporate incorrectness logic in a proof system for abstract interpretation that combines over- and under-approximation. Given an abstraction that is 'locally complete' (i.e. complete only for some specific inputs, rather than all possible inputs), they show that it is possible to prove both the presence as well as the absence of true alerts.

Incorrectness logics allow formal reasoning about reachability specifications—in our context, the presence of finite failure or forbidden graph structure. A complementary approach is to find counterexamples (i.e. instances of the forbidden structure) using model checkers such as GROOVE [9]. Analysing graph transformation systems can be challenging, however, as they often have infinite state spaces, but this can be mitigated by using bounded model checking [15].

7   Conclusion and Future Work

We proposed an incorrectness logic for under-approximate reasoning about graph programs, demonstrating that the deductive rules of Hoare logics can be 'reversed' to prove the presence of graph transformation bugs, such as the possibility of illegal graph substructures or finitely failing execution paths. In particular, we presented a calculus of incorrectness axioms and rules, proved them to be sound and relatively complete with respect to a denotational semantics of graph programs, and demonstrated their use to prove the presence of various bugs in a faulty node colouring program.

This paper was principally a theoretical exposition, but was motivated by some potentially interesting applications. One idea (suggested by O'Hearn [21]) is to recast static bug catchers in terms of finding under-approximation proofs. For instance, incorrectness logic might be able to provide soundness arguments for approaches that symbolically execute graph or model transformations (e.g. [1, 3,20]). Another idea is to use it to complement over-approximate proofs: if one is unable to prove a partial correctness specification or the absence of failure [26], switch to under-approximate proofs instead and reason about the circumstances that could cause some undesirable result to be reachable.

Beyond exploring these potential applications, future work should also extend our logic to a full-fledged graph programming language (e.g. GP 2 [23], or the recipes of GROOVE [8,9]). It is also important to investigate how to make incorrectness reasoning for graph programs easier. This could be in the form of guidelines on how to come up with incorrectness specifications (reasoning over a whole graph can be counter-intuitive, as Examples 4 and 5 demonstrate), or some derived proof rules for simplifying reasoning about common patterns.

Acknowledgements. I am grateful to the ICGT'21 referees for their detailed reviews and suggestions, which have helped to improve the quality of this paper.

Appendix

Proof (Proposition 1; Lemmata 1–2). By induction over the form of E-conditions, following the proof structure for transformations 'App', 'A', and 'L' for the similar assertion language in [24]. □

Proof (Proposition 2). $\Longrightarrow$. Assume that $H \models \mathrm{WPost}(\mathcal{R}, c)$. There exists some $r \in \mathcal{R}$ such that:

$$H \models \mathrm{wpost}(r, c) = \exists \mathbf{x}_1, \cdots \mathbf{x}_n.\exists\emptyset \hookrightarrow R.\mathrm{Dang}(r^{-1}) \wedge \mathrm{Right}(r, \mathrm{Shift}(r, c)).$$

There exists an $h \colon R^I \hookrightarrow G$ such that $h \models^I \mathrm{Dang}(r^{-1}) \wedge \mathrm{Right}(r, \mathrm{Shift}(r, c))$. Using Proposition 1, there exists a direct derivation from some graph G to H via $r = \langle L \Rightarrow R \rangle$, and by Lemma 2, there exists some $g \colon L^I \hookrightarrow G$ such that $g \models^I \mathrm{Shift}(r, c)$. By Lemma 1, $G \models c$.
$\Longleftarrow$. Assume that there exists a graph G such that $G \models c$ and $G \Rightarrow_{\mathcal{R}} H$. There exists some $r = \langle L \Rightarrow R \rangle \in \mathcal{R}$ such that $G \Rightarrow_r H$. By the definition of $\models$, Lemma 1, and Lemma 2, there exists some $h \colon R^I \hookrightarrow G \models^I \mathrm{Right}(r, \mathrm{Shift}(r, c))$. By the definition of direct derivations and Proposition 1, $h \models^I \mathrm{Dang}(r^{-1})$, and thus $h \models^I \mathrm{Dang}(r^{-1}) \wedge \mathrm{Right}(r, \mathrm{Shift}(r, c))$. By the definition of $\models$, $H \models \exists \mathbf{x}_1, \cdots \mathbf{x}_n.\exists R.\mathrm{Dang}(r^{-1}) \wedge \mathrm{Right}(r, \mathrm{Shift}(r, c))$, that is, $H \models \mathrm{wpost}(r, c)$. Being a disjunct of $\mathrm{WPost}(r, c)$, we derive the result $H \models \mathrm{WPost}(r, c)$. □

Proof (Theorem 1). Given $\vdash [c]P[\epsilon : d]$, we need to show that $\models [c]P[\epsilon : d]$. We consider each axiom and proof rule in turn and proceed by induction on proofs.

RULESETSUCC, RULESETFAIL. The validity of these axioms follows immediately from the definitions of $[\![\mathcal{R}]\!]ok$, $[\![\mathcal{R}]\!]er$, Proposition 1, and Proposition 2.

SEQSUCC. Suppose that $\vdash [c]P; Q[ok : d]$. By induction, we have $\models [c]P[ok : e]$ and $\models [e]Q[ok : d]$. By definition of $\models$, for all $H.H \models d$, there exists a $G'.G' \models e$ with $(G', G) \in [\![Q]\!]ok$, and for all $G'.G' \models e$, there exists a $G.G \models c$ with $(G, G') \in [\![P]\!]ok$. From the definition of $\models$ and $[\![P; Q]\!]ok$, it then follows that $\models [c]P; Q[ok : d]$. Analogous for case $\vdash [c]P; Q[er : d]$.

SEQFAIL. Suppose that $\vdash [c]P; Q[er : d]$. By induction, we have $\models [c]P[er : d]$. By definition of $\models$, for all $H.H \models d$, there exists a $G.G \models c$ with $(G, H) \in [\![P]\!]er$. By the definition of $[\![P; Q]\!]er$ and $\models$, it follows that $\models [c]P; Q[er : d]$.

IFELSE. Suppose that $\vdash [c]$if $\mathcal{R}$ then P else $Q[\epsilon : d]$. By induction, $\models [c \wedge \mathrm{App}(\mathcal{R})]P[\epsilon : d]$ and $\models [c \wedge \neg\mathrm{App}(\mathcal{R})]Q[\epsilon : d]$. From the definition of $\models$, $[\![$if $\mathcal{R}$ then P else $Q]\!]\epsilon$, and Proposition 1, we obtain the result that $\models [c]$if $\mathcal{R}$ then P else $Q[\epsilon : d]$.

CONS. Suppose that $\vdash [c]P[\epsilon : d]$. By induction, we have $\models [c']P[\epsilon : d']$, $\models d \implies d'$, and $\models c' \implies c$. It immediately follows that $\models [c]P[\epsilon : d]$.

ITERZERO. For every graph $G.G \models c \wedge \neg\mathrm{App}(\mathcal{R})$, by Proposition 1, $G \not\Rightarrow_{\mathcal{R}}$, $(G, G) \in [\![\mathcal{R}]\!]er$, and thus $(G, G) \in [\![\mathcal{R}!]\!]ok$. It immediately follows that $\models [c \wedge \neg\mathrm{App}(\mathcal{R})]\mathcal{R}![ok : c \wedge \neg\mathrm{App}(\mathcal{R})]$.

ITER. Suppose that $\vdash [c \wedge \mathrm{App}(\mathcal{R})]\mathcal{R}![ok : d \wedge \neg\mathrm{App}(\mathcal{R})]$. By induction, $\models [c \wedge \mathrm{App}(\mathcal{R})]\mathcal{R}; \mathcal{R}![ok : d \wedge \neg\mathrm{App}(\mathcal{R})]$. By definition of $\models$, for all $H.H \models d \wedge \neg\mathrm{App}(\mathcal{R})$, there exists some $G.G \models c \wedge \mathrm{App}(\mathcal{R})$ and $(G, H) \in [\![\mathcal{R}; \mathcal{R}!]\!]ok$. By the definition of $[\![\mathcal{R}!]\!]ok$ and $\models$, we obtain $\models [c \wedge \mathrm{App}(\mathcal{R})]\mathcal{R}![ok : d \wedge \neg\mathrm{App}(\mathcal{R})]$.

ITERVAR. Suppose that $\vdash [c_0]\mathcal{R}![ok : c_n]$. By induction, $\models [c_{i-1}]\mathcal{R}[ok : c_i]$ for every $0 < i \leq n$ and $\models c_n \implies \neg\mathrm{App}(\mathcal{R})$. By the definition of $\models$ and $[\![\mathcal{R}]\!]ok$, for every $G_i.G_i \models c_i$, there exists some $G_{i-1}.G_{i-1} \models c_{i-1}$ and $G_{i-1} \Rightarrow_{\mathcal{R}} G_i$. It follow that there is a sequence of derivations $G_0 \Rightarrow_{\mathcal{R}} \cdots \Rightarrow_{\mathcal{R}} G_n$ with $G_0 \models c_0$ and $G_n \models c_n$. By $\models c_n \implies \neg\mathrm{App}(\mathcal{R})$ and Proposition 1, we have $G_n \not\Rightarrow_{\mathcal{R}}$, i.e. $(G_n, G_n) \in [\![\mathcal{R}]\!]er$. Together with the definition of $[\![\mathcal{R}!]\!]ok$, it follows that $\models [c_0]\mathcal{R}![ok : c_n]$. $\square$

Proof (Theorem 2). We prove relative completeness extensionally by showing that for every program P, extensional assertion c, and exit condition $\epsilon \in \{ok, er\}$, $\vdash [c]P[\epsilon : \mathrm{WPOST}[P, c]]$, where $\mathrm{WPOST}[P, c]$ is an extensional assertion expressing the weakest postcondition relative to P and c, i.e. if $\models [c]P[\epsilon : d]$ for any d, then $d \implies \mathrm{WPOST}[P, c]$ is valid. Relative completeness is obtained by applying the rule of consequence to $\vdash [c]P[\epsilon : \mathrm{WPOST}[P, c]]$.

Rule Application ($\epsilon = ok$). Immediate from RULESETSUCC and CONS.

Rule Application ($\epsilon = er$). Immediate from RULESETFAIL, the definition of $[\![\mathcal{R}]\!]er$, and CONS.

Sequential Application ($\epsilon = ok$). In this case,

$$
\begin{aligned}
&H \models \mathrm{WPOST}[P; Q, c] \\
&\text{iff } \exists G.G \models c \text{ and } (G, H) \in [\![P; Q]\!]ok \\
&\text{iff } \exists G, G'.G \models c, (G, G') \in [\![P]\!]ok, \text{ and } (G', H) \in [\![Q]\!]ok \\
&\text{iff } \exists G'.G' \models \mathrm{WPOST}[P, c] \text{ and } (G', H) \in [\![Q]\!]ok \\
&\text{iff } H \models \mathrm{WPOST}[Q, \mathrm{WPOST}[P, c]]
\end{aligned}
$$

By induction we have $\vdash [\mathrm{WPOST}[P, c]]Q[ok : \mathrm{WPOST}[Q, \mathrm{WPOST}[P, c]]]$ and $\vdash [c]P[ok : \mathrm{WPOST}[P, c]]$. By SEQSUCC we derive the triple $\vdash [c]P; Q[ok : \mathrm{WPOST}[Q, \mathrm{WPOST}[P, c]]]$, and by CONS $\vdash [c]P; Q[ok : \mathrm{WPOST}[P; Q, c]]$.

Sequential Application ($\epsilon = er$). If the program $P; Q$ fails and the error occurs in Q, then the proof is analogous to the *ok* case. If the error occurs in P:

$$H \models \text{WPOST}[P; Q, c]$$
$$\text{iff } \exists G.G \models c \text{ and } (G, H) \in [\![P; Q]\!]er$$
$$\text{iff } \exists G.G \models c \text{ and } (G, H) \in [\![P]\!]er$$
$$\text{iff } H \models \text{WPOST}[P, c]$$

By induction we have $\vdash [c]P[er : \text{WPOST}[P, c]]$, and by SEQFAIL derive $\vdash [c]P; Q[er : \text{WPOST}[P, c]]$. With CONS we get $\vdash [c]P; Q[er : \text{WPOST}[P; Q, c]]$.

If-then-else. The proof for this case follows a similar structure to sequential composition but treating the two branches separately.

Iteration. Define c_i as $\text{WPOST}[\mathcal{R}, c_{i-1}]$ for every $0 < i \leq n$. We have:

$$G_n \models \text{WPOST}[\mathcal{R}!, c_0]$$
$$\text{iff } \exists G_0.G_0 \models c_0 \text{ and } (G_0, G_n) \in [\![\mathcal{R}!]\!]ok$$
$$\text{iff } \exists G_0, \cdots G_{n-1}.(G_{i-1}, G_i) \in [\![\mathcal{R}]\!]ok \text{ for all } 0 < i \leq n, \text{ and } (G_n, G_n) \in [\![\mathcal{R}]\!]er$$
$$\text{iff } \exists G_1, \cdots G_{n-1}.G_1 \models \text{WPOST}[\mathcal{R}, c_0], (G_{i-1}, G_i) \in [\![\mathcal{R}]\!]ok \text{ for all } 1 < i \leq n$$
$$\quad \text{and } (G_n, G_n) \in [\![\mathcal{R}]\!]er$$
$$\text{iff } G_n \models c_n \text{ and } c_n \implies \neg\text{App}(\mathcal{R})$$

By induction, $\vdash [c_{i-1}]\mathcal{R}[ok : \text{WPOST}[\mathcal{R}, c_{i-1}]]$ and thus $\vdash [c_{i-1}]\mathcal{R}[ok : c_i]$. By ITERVAR and CONS derive the result, $\vdash [c_0]\mathcal{R}![ok : \text{WPOST}[\mathcal{R}!, c_0]]$. $\qquad \square$

References

1. Al-Sibahi, A.S., Dimovski, A.S., Wasowski, A.: Symbolic execution of high-level transformations. In: SLE 2016, pp. 207–220. ACM (2016)
2. Apt, K.R., de Boer, F.S., Olderog, E.: Verification of Sequential and Concurrent Programs. Texts in Computer Science. Springer, London (2009). https://doi.org/10.1007/978-1-84882-745-5
3. Azizi, B., Zamani, B., Rahimi, S.K.: SEET: symbolic execution of ETL transformations. J. Syst. Softw. **168**, 110675 (2020)
4. Baldan, P., Corradini, A., König, B.: A framework for the verification of infinite-state graph transformation systems. Inf. Comput. **206**(7), 869–907 (2008)
5. Brenas, J.H., Echahed, R., Strecker, M.: Verifying graph transformation systems with description logics. In: Lambers, L., Weber, J. (eds.) ICGT 2018. LNCS, vol. 10887, pp. 155–170. Springer, Cham (2018). https://doi.org/10.1007/978-3-319-92991-0_10
6. Bruni, R., Giacobazzi, R., Gori, R., Ranzato, F.: A logic for locally complete abstract interpretations. In: LICS 2021. IEEE (2021, to appear)
7. Cook, S.A.: Soundness and completeness of an axiom system for program verification. SIAM J. Comput. **7**(1), 70–90 (1978)
8. Corrodi, C., Heußner, A., Poskitt, C.M.: A semantics comparison workbench for a concurrent, asynchronous, distributed programming language. Formal Aspects Comput. **30**(1), 163–192 (2018)

9. Ghamarian, A.H., de Mol, M., Rensink, A., Zambon, E., Zimakova, M.: Modelling and analysis using GROOVE. Int. J. Softw. Tools Technol. Transfer **14**(1), 15–40 (2012)
10. Habel, A., Pennemann, K.: Correctness of high-level transformation systems relative to nested conditions. Math. Struct. Comput. Sci. **19**(2), 245–296 (2009)
11. Habel, A., Pennemann, K.-H., Rensink, A.: Weakest preconditions for high-level programs. In: Corradini, A., Ehrig, H., Montanari, U., Ribeiro, L., Rozenberg, G. (eds.) ICGT 2006. LNCS, vol. 4178, pp. 445–460. Springer, Heidelberg (2006). https://doi.org/10.1007/11841883_31
12. Habel, A., Plump, D.: Relabelling in graph transformation. In: Corradini, A., Ehrig, H., Kreowski, H.-J., Rozenberg, G. (eds.) ICGT 2002. LNCS, vol. 2505, pp. 135–147. Springer, Heidelberg (2002). https://doi.org/10.1007/3-540-45832-8_12
13. Heckel, R., Taentzer, G.: Graph Transformation for Software Engineers - With Applications to Model-Based Development and Domain-Specific Language Engineering. Springer, Cham (2020). https://doi.org/10.1007/978-3-030-43916-3
14. Hoare, C.A.R.: An axiomatic basis for computer programming. Commun. ACM (CACM) **12**(10), 576–580 (1969)
15. Isenberg, T., Steenken, D., Wehrheim, H.: Bounded model checking of graph transformation systems via SMT solving. In: Beyer, D., Boreale, M. (eds.) FMOODS/FORTE -2013. LNCS, vol. 7892, pp. 178–192. Springer, Heidelberg (2013). https://doi.org/10.1007/978-3-642-38592-6_13
16. König, B., Esparza, J.: Verification of graph transformation systems with context-free specifications. In: Ehrig, H., Rensink, A., Rozenberg, G., Schürr, A. (eds.) ICGT 2010. LNCS, vol. 6372, pp. 107–122. Springer, Heidelberg (2010). https://doi.org/10.1007/978-3-642-15928-2_8
17. Makhlouf, A., Percebois, C., Tran, H.N.: Two-level reasoning about graph transformation programs. In: Guerra, E., Orejas, F. (eds.) ICGT 2019. LNCS, vol. 11629, pp. 111–127. Springer, Cham (2019). https://doi.org/10.1007/978-3-030-23611-3_7
18. Murray, T.: An under-approximate relational logic: heralding logics of insecurity, incorrect implementation & more. CoRR abs/2003.04791 (2020)
19. Navarro, M., Orejas, F., Pino, E., Lambers, L.: A navigational logic for reasoning about graph properties. J. Logical Algebraic Methods Program. **118**, 100616 (2021)
20. Oakes, B.J., Troya, J., Lúcio, L., Wimmer, M.: Full contract verification for ATL using symbolic execution. Softw. Syst. Model. **17**(3), 815–849 (2018)
21. O'Hearn, P.W.: Incorrectness logic. Proc. ACM Program. Lang. **4**(POPL), 10:1–10:32 (2020)
22. Orejas, F., Pino, E., Navarro, M., Lambers, L.: Institutions for navigational logics for graphical structures. Theoret. Comput. Sci. **741**, 19–24 (2018)
23. Plump, D.: The design of GP 2. In: WRS 2011. EPTCS, vol. 82, pp. 1–16 (2011)
24. Poskitt, C.M.: Verification of graph programs. Ph.D. thesis, University of York (2013)
25. Poskitt, C.M., Plump, D.: Hoare-style verification of graph programs. Fund. Inform. **118**(1–2), 135–175 (2012)
26. Poskitt, C.M., Plump, D.: Verifying total correctness of graph programs. ECE-ASST, vol. 61 (2013)
27. Poskitt, C.M., Plump, D.: Verifying monadic second-order properties of graph programs. In: Giese, H., König, B. (eds.) ICGT 2014. LNCS, vol. 8571, pp. 33–48. Springer, Cham (2014). https://doi.org/10.1007/978-3-319-09108-2_3

28. Raad, A., Berdine, J., Dang, H.-H., Dreyer, D., O'Hearn, P., Villard, J.: Local reasoning about the presence of bugs: incorrectness separation logic. In: Lahiri, S.K., Wang, C. (eds.) CAV 2020. LNCS, vol. 12225, pp. 225–252. Springer, Cham (2020). https://doi.org/10.1007/978-3-030-53291-8_14

29. Schneider, S., Dyck, J., Giese, H.: Formal verification of invariants for attributed graph transformation systems based on nested attributed graph conditions. In: Gadducci, F., Kehrer, T. (eds.) ICGT 2020. LNCS, vol. 12150, pp. 257–275. Springer, Cham (2020). https://doi.org/10.1007/978-3-030-51372-6_15

30. de Vries, E., Koutavas, V.: Reverse Hoare logic. In: Barthe, G., Pardo, A., Schneider, G. (eds.) SEFM 2011. LNCS, vol. 7041, pp. 155–171. Springer, Heidelberg (2011). https://doi.org/10.1007/978-3-642-24690-6_12

31. Wulandari, G.S., Plump, D.: Verifying graph programs with first-order logic. In: GCM 2020. EPTCS, vol. 330, pp. 181–200 (2020)

Powerful and NP-Complete: Hypergraph Lambek Grammars

Tikhon Pshenitsyn[✉][iD]

Department of Mathematical Logic and Theory of Algorithms,
Faculty of Mathematics and Mechanics, Lomonosov Moscow State University,
GSP-1, Leninskie Gory, Moscow 119991, Russian Federation
tpshenitsyn@lpcs.math.msu.su

Abstract. We consider two approaches to generating formal string languages: context-free grammars and Lambek grammars, which are based on the Lambek calculus. They are equivalent in the sense that they generate the same set of languages (disregarding the empty word). It is well known that context-free grammars can be generalized to hyperedge replacement grammars (HRGs) preserving their main principles and properties. In this paper, we study a generalization of the Lambek grammars to hypergraphs and investigate the recognizing power of the new formalism. We show how to define the hypergraph Lambek calculus (HL), and then introduce *hypergraph Lambek grammars* based on HL. It turns out that such grammars recognize all isolated-node bounded languages generated by HRGs. However, they are more powerful than HRGs: they recognize at least finite intersections of such languages. Thus the Pentus theorem along with the pumping lemma and the Parikh theorem have no place for hypergraph Lambek grammars. Besides, it can be shown that hypergraph Lambek grammars are NP-complete, so they constitute an attractive alternative to HRGs, which are also NP-complete.

1 Introduction

Formal grammars include two large classes opposed to each other: context-free grammars and categorial grammars. The former generate languages by means of productions: one starts with a single symbol and then applies productions to it, hence generating a string. On contrary, categorial grammars work in a "deductive way": they start with a whole string, assign types to strings and then check whether the resulting sequence of types is correct w.r.t. some uniform laws. There are many options of choosing such uniform laws, which lead to different formalisms. The laws are often provided by a logical calculus; in such cases, grammars are called type-logical.

The study was funded by RFBR, project number 20-01-00670 and by the Interdisciplinary Scientific and Educational School of Moscow University "Brain, Cognitive Systems, Artificial Intelligence". The author is a Scholarship holder of "BASIS" Foundation.

F. Gadducci and T. Kehrer (Eds.): ICGT 2021, LNCS 12741, pp. 102–121, 2021.
https://doi.org/10.1007/978-3-030-78946-6_6

One of important classes of type-logical grammars is Lambek grammars. They are based on the Lambek calculus introduced in [4] (further we denote it as L). This calculus deals with types and sequents. Types are built from primitive types $Pr = \{p_1, p_2, \dots\}$ (which we further denote by small Latin letters $p, q, \dots$) using operations $\backslash, \cdot, /$; for example, $(p \cdot q)/(p\backslash q)$ is a type. A sequent is a structure of the form $A_1, \dots, A_n \to A$ where $n > 0$ and A_i, A are types. The Lambek calculus in the Gentzen style we consider in this paper explains how to derive sequents. There is one axiom and six inference rules of L:

$$\frac{}{A \to A} \; (\text{Ax})$$

$$\frac{\Pi \to A \quad \Gamma, B, \Delta \to C}{\Gamma, \Pi, A\backslash B, \Delta \to C} \; (\backslash \to) \qquad \frac{A, \Pi \to B}{\Pi \to A\backslash B} \; (\to \backslash) \qquad \frac{\Gamma, A, B, \Delta \to C}{\Gamma, A \cdot B, \Delta \to C} \; (\cdot \to)$$

$$\frac{\Pi \to A \quad \Gamma, B, \Delta \to C}{\Gamma, B/A, \Pi, \Delta \to C} \; (/ \to) \qquad \frac{\Pi, A \to B}{\Pi \to B/A} \; (\to /) \qquad \frac{\Pi \to A \quad \Psi \to B}{\Pi, \Psi \to A \cdot B} \; (\to \cdot)$$

Here capital Latin letters denote types, and capital Greek letters denote sequences of types (besides, Π, Ψ are nonempty).

Example 1.1. Below an example of a derivation is presented:

$$\frac{\dfrac{\dfrac{s \to s \quad q \to q}{s/q, q \to p} \; (/ \to) \quad s \to s}{s/q, s, s\backslash q \to s} \; (\backslash \to) \quad q \to q}{s/q, s/q, q, s\backslash q \to s} \; (/ \to)$$

The Lambek calculus has algebraic and logical nature (e.g. it can be considered as a substructural logic of intuitionistic logic). Nicely, it can be used to describe formal and natural languages, which can be done using so-called Lambek grammars. A Lambek grammar consists of an alphabet Σ, of a finite binary relation $\triangleright$ between symbols of Σ and types of the Lambek calculus and of a distinguished type S. The grammar recognizes (synonym for *generates* regarding categorial grammars) the language of all strings $a_1 \dots a_n$, for which there exist types $T_1, \dots, T_n$ such that $a_i \triangleright T_i$, and $T_1, \dots, T_n \to S$ is derivable in L.

Example 1.2. The Lambek grammar over the alphabet $\{a, b\}$ such that $S = s \in Pr$ and $a \triangleright s/q$, $b \triangleright s\backslash q$, $b \triangleright q$ generates the language $\{a^n b^n \mid n > 0\}$. E.g., given the string $aabb$, one transforms it into a sequent $s/q, s/q, q, s\backslash q \to s$ and derives it in L as is shown in Example 1.1.

From 1960-s until nowadays different extensions of L were proposed to capture linguistic phenomena of interest (e.g. parasitic gaps, wh-movements etc.; see [5]). Many mathematical properties of L and of Lambek grammars have been discovered. The one, which is of interest in this work, is the Pentus theorem (see [7]). It states that Lambek grammars recognize exactly the class of context-free languages (without the empty word). Therefore, in the string case Lambek grammars and context-free grammars are equivalent.

Moving away from categorial grammars, we consider another approach: *hyperedge replacement grammar* (HRG in short). These grammars were developed in order to generalize context-free grammars to hypergraphs. Namely, they generate hypergraphs by means of productions: a production allows one to replace a hyperedge of a hypergraph with another hypergraph. Hyperedge replacement grammars (HRGs) have a number of properties in common with context-free grammars such as the pumping lemma, the Parikh theorem, the Greibach normal form etc. An overview of HRGs can be found in [2]. In this work, we follow definitions and notation from this handbook chapter as far as possible.

Comparing two fields of research, we came up with a question: is it possible to generalize type-logical formalisms to hypergraphs in a natural way? A desired extension should satisfy two properties: there must be an embedding of string type-logical grammars in it, and it should be connected to hyperedge replacement grammars, as type-logical grammars are connected to context-free grammars. Moreover, one expects that the Lambek calculus underlying Lambek grammars can also be generalized to hypergraphs resulting in some kind of a graph logic.

Our first attempt was concerned with generalizing basic categorial grammars to hypergraphs. The result called hypergraph basic categorial grammars is introduced in [8]. The idea of such grammars is just inverting a derivation of HRGs and remodeling it from bottom to top. This idea is not new; it is, e.g. closely connected to the concept of abstract categorial grammars (see [3]). In [8], such inverting is done straightforwardly: there are types, which are complex terms made of hypergraphs; using them one labels hyperedges and then checks whether a graph labeled by types can be reduced to a single-edge hypergraph using a uniform reduction law.

In our preprint [9], we do the next step and introduce a generalization of the Lambek calculus to hypergraphs called the hypergraph Lambek calculus (HL in short). This calculus naturally generalizes L to hypergraphs, and, moreover, contains its different variants as fragments. On the other hand, the inference rules of this calculus are defined through hyperedge replacement, and thus HL is related to HRGs. In this paper, we introduce the definition of HL in Sect. 3, establish an embedding of L in HL, and then turn to defining and studying a grammar formalism that can be built on the basis of HL. Such formalism is called *hypergraph Lambek grammar* (HL-grammar for short); it is introduced in Sect. 4. In Sect. 5 we establish three main results: HL-grammars are not weaker than HRGs; the parsing problem for HL-grammars is NP-complete; HL-grammars can recognize finite intersections of hypergraph context-free languages. These three facts enable one to conclude that HL-grammar is an attractive alternative to HRG: both formalisms are NP-complete while HL-grammars are more powerful.

2   Preliminaries

$\mathbb{N}$ includes 0. Σ^* is the set of all strings over the alphabet Σ. $\Sigma^{\circledast}$ is the set of all strings consisting of distinct symbols. The length $|w|$ of the word w is the number of symbols in w. The set of all symbols contained in a word w is denoted by $[w]$. If $f : \Sigma \to \Delta$ is a function from one set to another, then it is naturally extended to a function $f : \Sigma^* \to \Delta^*$ $(f(\sigma_1 \ldots \sigma_k) = f(\sigma_1) \ldots f(\sigma_k))$.

Let C be some fixed set of labels for whom the function $type : C \to \mathbb{N}$ is considered (C is called a ranked alphabet).

Definition 2.1. A hypergraph G over C *is a tuple* $G = \langle V, E, att, lab, ext \rangle$ *where V is a set of nodes, E is a set of hyperedges, $att : E \to V^{\circledast}$ assigns a string (i.e. an ordered set) of attachment nodes to each hyperedge, $lab : E \to C$ labels each hyperedge by some element of C in such a way that $type(lab(e)) = |att(e)|$ whenever $e \in E$, and $ext \in V^{\circledast}$ is a string of external nodes.*

Components of a hypergraph G are denoted by $V_G, E_G, att_G, lab_G, ext_G$ resp.

In the remainder of the paper, hypergraphs are usually called just graphs, and hyperedges are called edges. The set of all graphs with labels from C is denoted by $\mathcal{H}(C)$. Graphs are usually named by letters G and H.

In drawings of graphs, black dots correspond to nodes, labeled squares correspond to edges, att is represented by numbered lines, and external nodes are depicted by numbers in brackets. If an edge has exactly two attachment nodes, it can be depicted by an arrow (which goes from the first attachment node to the second one).

Note that Definition 2.1 implies that attachment nodes of each hyperedge are distinct, and so are external nodes. This restriction can be removed (i.e. we can consider graphs with loops), and all further definitions will be preserved; however, in this paper, we stick to the above definition.

Definition 2.2. *The function $type_G$ (or type, if G is clear) returns the number of nodes attached to an edge in a graph G: $type_G(e) := |att_G(e)|$. If G is a graph, then $type(G) := |ext_G|$.*

Definition 2.3. *A sub-hypergraph (or just subgraph) H of a graph G is a hypergraph such that $V_H \subseteq V_G$, $E_H \subseteq E_G$, and for all $e \in E_H$ $att_H(e) = att_G(e)$, $lab_H(e) = lab_G(e)$.*

Definition 2.4. *If $H = \langle \{v_i\}_{i=1}^n, \{e_0\}, att, lab, v_1 \ldots v_n \rangle$, $att(e_0) = v_1 \ldots v_n$ and $lab(e_0) = a$, then H is called* a handle. *It is denoted by $a^{\bullet}$.*

Definition 2.5. *An isomorphism between graphs G and H is a pair of bijective functions $\mathcal{E} : E_G \to E_H$, $\mathcal{V} : V_G \to V_H$ such that $att_H \circ \mathcal{E} = \mathcal{V} \circ att_G$, $lab_G = lab_H \circ \mathcal{E}$, $\mathcal{V}(ext_G) = ext_H$. In this work, we do not distinguish between isomorphic graphs.*

Strings can be considered as graphs with the string structure. This is formalized in

Definition 2.6. *A string graph induced by a string $w = a_1 \ldots a_n$ is a graph of the form $\langle \{v_i\}_{i=0}^n, \{e_i\}_{i=1}^n, att, lab, v_0 v_n \rangle$ where $att(e_i) = v_{i-1} v_i$, $lab(e_i) = a_i$. In this work, we denote it by $SG(w)$.*

We additionally introduce the following definition (not from [2]):

Definition 2.7. *Let $H \in \mathcal{H}(C)$ be a graph, and let $f : E_H \to C$ be a function. Then $f(H) := \langle V_H, E_H, att_H, lab_{f(H)}, ext_H \rangle$ where $lab_{f(H)}(e) = f(e)$ for all e in E_H. It is required that $type(lab_H(e)) = type(f(e))$ for $e \in E_H$.*

If one wants to relabel only one edge e_0 within H with a label a, then the result is denoted by $H[e_0 := a]$.

2.1 Hyperedge Replacement Grammars

Hyperedge replacement grammars (HRGs in short) are based on the procedure of hyperedge replacement. The replacement of an edge e_0 in G with a graph H can be done if $type(e_0) = type(H)$ as follows:

1. Remove e_0;
2. Insert an isomorphic copy of H (H and G have to consist of disjoint sets of nodes and edges);
3. For each i, fuse the i-th external node of H with the i-th attachement node of e_0.

The result is denoted by $G[e_0/H]$. It is known that if several edges of a graph are replaced by other graphs, then the result does not depend on the order of replacements; moreover the result is not changed, if replacements are done simultaneously (see [2]). The following notation is in use: if $e_1, \ldots, e_k$ are distinct edges of a graph H and they are simultaneously replaced by graphs $H_1, \ldots, H_k$ resp. (this requires $type(H_i) = type(e_i)$), then the result is denoted $H[e_1/H_1, \ldots, e_k/H_k]$.

A *hyperedge replacement grammar (HRG)* is a tuple $HGr = \langle N, \Sigma, P, S \rangle$, where N and Σ are disjoint finite ranked alphabets (of nonterminal and terminal symbols resp.), P is a set of productions, and $S \in N$. Each production is of the form $A \to H$ where $A \in N$, $H \in \mathcal{H}(N \cup \Sigma)$ and $type(A) = type(H)$. A grammar generates the language of all terminal graphs (i.e. graphs with terminal labels only) that can be obtained from $S^\bullet$ by applying productions from P. Such a language is called a hypergraph context-free language (HCFL in short). Two grammars are said to be equivalent if they generate the same language.

3 Hypergraph Lambek Calculus

In this section, the hypergraph Lambek calculus (HL) is introduced. As in the string case, we are going to define types, sequents, an axiom and rules of the calculus; now, however, they are expected to be based on hypergraphs. The intuition of the definitions we aim to introduce comes from the procedure of convertion of a context-free grammar into an equivalent Lambek grammar in the string case. This procedure can be illustrated by the following

Example 3.1. Consider a context-free grammar $Gr = \langle \{S, Q\}, \{a, b\}, P, S \rangle$ with the set P including three productions: $S \to aQ$, $Q \to Sb$, $Q \to b$. This grammar generates the language $\{a^n b^n \mid n > 0\}$. E.g. $S \Rightarrow aQ \Rightarrow aSb \Rightarrow aaQb \Rightarrow aabb$

shows that $aabb$ is generated by Gr. Note that Gr is lexicalized; that is, there is exactly one terminal symbol in the right-hand side of each production. This enables one to convert Gr into an equivalent Lambek grammar as follows:

$$\begin{aligned} S &\to aQ & &\rightsquigarrow & a &\triangleright s/q \\ Q &\to Sb & &\rightsquigarrow & b &\triangleright s\backslash q \\ Q &\to b & &\rightsquigarrow & b &\triangleright q \end{aligned}$$

This procedure is quite intuitive. For example, the production $S \to aQ$ can be read as "a structure of the type S can be obtained, if one concatenates a symbol a and a structure of the type Q (in such order)". In comparison, $a \triangleright s/q$ can be understood as the statement "a is such a symbol that, whenever a structure of the type q appears to its right, they together form a structure of the type s". Clearly, the above two statements are equivalent. In general, a correspondence $a \triangleright A/B$ in a Lambek grammar can be informally understood as follows: a is such a symbol that it waits for a structure (a string) of the type B from the right in order to form a string of the type A together with it. Similarly, one can describe the meaning of a correspondence $a \triangleright B\backslash A$ with the only difference that a symbol requires a structure of the type B from the left.

The lexicalized normal form (or the weak Greibach normal form) for context-free grammars can be generalized to HRGs, which is studied in [10].

Definition 3.1. *An HRG HGr is in the weak Greibach normal form if there is exactly one terminal label in the right-hand side of each production.*

Let $A \to H$ be a production in such a grammar; that is, H is a graph with exactly one terminal label. In order to perform a convertion similar to the one explained above for context-free grammars, one would like to "extract" the only terminal label from H and to associate this label with a type, which would look like A/H. In H, however, we should mark the place, from which we take the only terminal label; let us do this using a special $\$$ label (which would be allowed to label edges of different types[1]). Besides, in order to distinguish between string divisions and a new hypergraph division we shall write $\div$ instead of $/$ or $\backslash$.

Example 3.2. Below we present a production of some HRG and the result of its convertion (the result has no formal sense yet, it is just a game with symbols):

Let us investigate how $\backslash$ and $/$ of the Lambek calculus correlate with $\div$. It is known that context-free grammars can be embedded in HRGs using string graphs. For example, the grammar Gr from Example 3.1 can be transformed into an HRG with the following productions: $S \to \mathrm{SG}(aQ)$, $Q \to \mathrm{SG}(Sb)$, $Q \to \mathrm{SG}(b)$. Now let us apply the above convertion to the first and the second productions:

---

[1] To be consistent with the definition of a hypergraph one may assume that there are different symbols $\$_n, n \geq 0$ instead such that $type(\$_n) = n$.

$$S \rightarrow {}_{(1)} \bullet \xrightarrow{\;a\;} \bullet \xrightarrow{\;\;Q\;\;} \bullet {}_{(2)} \quad \rightsquigarrow \quad a \triangleright s \div \left({}_{(1)} \bullet \xrightarrow{\;\$\;} \bullet \xrightarrow{\;\;q\;\;} \bullet {}_{(2)} \right)$$

$$Q \rightarrow {}_{(1)} \bullet \xrightarrow{\;\;S\;\;} \bullet \xrightarrow{\;b\;} \bullet {}_{(2)} \quad \rightsquigarrow \quad b \triangleright q \div \left({}_{(1)} \bullet \xrightarrow{\;\;s\;\;} \bullet \xrightarrow{\;\$\;} \bullet {}_{(2)} \right)$$

The resulting types in the right-hand side may be considered as graph counterparts of types s/q and $s\backslash q$ resp. Note that the difference between the left and the right divisions is represented in these types by the position of the $\$$ symbol.

For now, this convertion is only juggling symbols, and $\div$ does not have a functional definition yet. It will be presented later. Now, let us discuss the issue of generalizing the multiplication operation. In the string case, $A \cdot B$ can be informally understood as the set of all strings of the form uv where u is of the type A, and v is of the type B; that is, $A \cdot B$ corresponds to the concatenation operation. In the hypergraph case, one needs a "generalized concatenation". Implementing this idea we introduce a unary operation $\times$. If M is a hypergraph labeled by types, then $\times(M)$ represents all substitution instances of M, that is, all hypergraphs that are obtained from M by replacing each edge labeled by some type by a hypergraph of this type. The connection between $\times$ and $\cdot$ is shown in Sect. 3.3.

3.1   Formal Definition of Types and Sequents

In this section, we give formal meanings to the operations introduced above. Let us fix a countable set Pr of primitive types and a function $type : Pr \rightarrow \mathbb{N}$ such that for each $n \in \mathbb{N}$ there are infinitely many $p \in Pr$ for which $type(p) = n$. Types are constructed from primitive types using division $\div$ and multiplication $\times$ operations. Simultaneously, the function $type$ is defined on types: this is obligatory since we are going to label edges by types.

Definition 3.2. *The set $Tp(\mathrm{HL})$ of types is defined inductively as follows:*

1. *$Pr \subseteq Tp(\mathrm{HL})$.*
2. *Let N ("numerator") be in $Tp(\mathrm{HL})$. Let D ("denominator") be a graph such that exactly one of its edges (call it e_0) is labeled by $\$$, and the other edges (possibly, there are none of them) are labeled by elements of $Tp(\mathrm{HL})$; let also $type(N) = type(D)$. Then $T = (N \div D)$ also belongs to $Tp(\mathrm{HL})$, and $type(T) := type_D(e_0)$.*
3. *Let M be a graph such that all its edges are labeled by types from $Tp(\mathrm{HL})$ (possibly, there are no edges at all). Then $T = \times(M)$ belongs to $Tp(\mathrm{HL})$, and $type(T) := type(M)$.*

In types with division, D is usually drawn as a graph in brackets, so instead of $(N \div D)$ (a formal notation) a graphical notation $N \div (D)$ is in use. Sometimes brackets are omitted.

Example 3.3. The following structures are types:

$$- \; A_1 = p \div \left(\boxed{\;\$\;} \xrightarrow{\;1\;} \bullet {}_{(1)} \quad \boxed{\;p\;} \xrightarrow{\;1\;} \bullet \right) ;$$

$$- A_2 = \times \left(\bullet_{(1)} \quad \boxed{A_1}\overset{1}{\relbar\!\!\relbar} \bullet_{(2)} \right);$$

$$- A_3 = \times \left(\bullet \quad \boxed{p}\overset{1}{\relbar\!\!\relbar}\bullet \right) \div \left(\boxed{\$} \quad \boxed{p}\overset{1}{\relbar\!\!\relbar}\bullet \right).$$

Here $type(p) = 1$, $type(A_1) = 1$, $type(A_2) = 2$, $type(A_3) = 0$.

Definition 3.3. A graph sequent *is a structure of the form* $H \to A$ *where* $A \in Tp(\text{HL})$ *is a type,* $H \in \mathcal{H}(Tp(\text{HL}))$ *is a graph labeled by types and* $type(H) = type(A)$. H *is called the antecedent of the sequent, and* A *is called the succedent of the sequent.*

Let $\mathcal{T}$ be a subset of $Tp(\text{HL})$. We say that $H \to A$ is over $\mathcal{T}$ if $G \in \mathcal{H}(\mathcal{T})$ and $A \in \mathcal{T}$.

Example 3.4. The following structure is a graph sequent (where A_2, A_3 are from Example 3.3):

$$\overset{A_2}{\underset{}{\bullet\!\longleftarrow\!\bullet}}\overset{A_2}{\underset{}{\longrightarrow\!\bullet}} \quad \to \quad A_3$$

3.2 Axiom and Rules

The hypergraph Lambek calculus (denoted HL) we introduce here is a logical system that defines what graph sequents are derivable (=provable). HL includes one axiom and four rules, which are introduced below. Each rule is illustrated by an example exploiting string graphs.

The only axiom is the following: $A^\bullet \to A$, $A \in Tp(\text{HL})$.

Rule $(\div \to)$. Let $N \div D$ be a type and let $E_D = \{d_0, d_1, \ldots, d_k\}$ where $lab(d_0) = \$$. Let $H \to A$ be a graph sequent and let $e \in E_H$ be labeled by N. Let finally $H_1, \ldots, H_k$ be graphs labeled by types. Then the rule $(\div \to)$ is the following:

$$\frac{H \to A \quad H_1 \to lab(d_1) \quad \cdots \quad H_k \to lab(d_k)}{H[e/D][d_0 := N \div D][d_1/H_1, \ldots, d_k/H_k] \to A} \; (\div \to)$$

This rule explains how a type with division appears in an antecedent: we replace an edge e by D, put a label $N \div D$ instead of $\$$ and replace the remaining labels of D by corresponding antecedents.

Example 3.5. Consider the following rule application with T_i being some types and with T being equal to $q \div SG(T_2 \$ T_3)$:

$$\frac{SG(pq) \to T_1 \quad SG(rs) \to T_2 \quad SG(tu) \to T_3}{SG(prs\, T\, tu) \to T_1} \; (\div \to)$$

Rule $(\to \div)$. Let $F \to N \div D$ be a graph sequent; let $e_0 \in E_D$ be labeled by $\$$. Then

$$\frac{D[e_0/F] \to N}{F \to N \div D} \; (\to \div)$$

Formally speaking, this rule is improper since it is formulated from bottom to top. It is understood, however, as follows: if there are such graphs D, F and such a type N that in a sequent $H \to N$ the graph H equals $D[F/e_0]$, and $H \to N$ is derivable, then $F \to N \div D$ is also derivable.

Example 3.6. Consider the following rule application where T is some type (here we draw graphs instead of writing $SG(w)$ to visualize the rule application):

$$\frac{(1) \bullet \xrightarrow{p} \bullet \xrightarrow{q} \bullet \xrightarrow{r} \bullet (2) \to T}{(1) \bullet \xrightarrow{p} \bullet \xrightarrow{q} \bullet (2) \to T \div \left((1) \bullet \xrightarrow{\$} \bullet \xrightarrow{r} \bullet (2) \right)} (\to \div)$$

Rule $(\times \to)$. Let $G \to A$ be a graph sequent and let $e \in E_G$ be labeled by $\times(F)$. Then

$$\frac{G[e/F] \to A}{G \to A} (\times \to)$$

This rule again is formulated from bottom to top. Informally speaking, there is a subgraph of an antecedent in the premise, and it is "compressed" into a single $\times(F)$-labeled edge.

Example 3.7. Consider the following rule application where U is some type:

$$\frac{(1) \bullet \xrightarrow{p} \bullet \xrightarrow{q} \bullet \xrightarrow{r} \bullet \xrightarrow{s} \bullet (2) \to U}{(1) \bullet \xrightarrow{p} \bullet \xrightarrow{\times(SG(qr))} \bullet \xrightarrow{s} \bullet (2) \to U} (\times \to)$$

Rule $(\to \times)$. Let $\times(M)$ be a type and let $E_M = \{m_1, \ldots, m_l\}$. Let $H_1, \ldots, H_l$ be graphs over $Tp(\mathrm{HL})$. Then

$$\frac{H_1 \to lab(m_1) \quad \ldots \quad H_l \to lab(m_l)}{M[m_1/H_1, \ldots, m_l/H_l] \to \times(M)} (\to \times)$$

This rule is quite intuitive: several sequents can be combined into a single one via some graph structure M.

Example 3.8. Consider the following rule application with T_i being some types:

$$\frac{SG(pq) \to T_1 \quad SG(rs) \to T_2 \quad SG(tu) \to T_3}{SG(pqrstu) \to \times(SG(T_1 T_2 T_3))} (\to \times)$$

Definition 3.4. *A graph sequent $H \to A$ is derivable in HL (HL $\vdash H \to A$) if it can be obtained from axioms using rules of HL. A corresponding sequence of rule applications is called* a derivation *and its representation as a tree is called* a derivation tree.

Example 3.9. The sequent from Example 3.4 is derivable in HL. Here is its derivation tree:

$$
\cfrac{
\cfrac{
\cfrac{
\cfrac{
\cfrac{p^\bullet \to p \quad p^\bullet \to p}{(1)\;\bullet\!-\!\boxed{A_1}\;\boxed{p}\!-\!\bullet \quad \to \quad p \quad\quad p^\bullet \to p}\;(\div \to)
}{(1)\;\bullet\!-\!\boxed{A_1}\;\boxed{A_1}\!-\!\bullet \quad \boxed{p}\!-\!\bullet \quad \to \quad p}\;(\div \to)
}{\bullet\!-\!\boxed{A_1} \;\bullet\; \boxed{A_1}\!-\!\bullet \quad \boxed{p}\!-\!\bullet \quad \to \quad \times\!\left(\;\bullet\; \boxed{p}\!-\!\bullet\right)}\;(\to \times)
}{\bullet\!-\!\boxed{A_1} \;\bullet\; \boxed{A_1}\!-\!\bullet \quad \to \quad A_3}\;(\to \div)
}{\bullet\!-\!\boxed{A_1} \quad \overset{A_2}{\bullet\!-\!\!\to\!\bullet} \quad \to \quad A_3}\;(\times \to)
}{\overset{A_2}{\bullet\!\leftarrow\!-\!\bullet}\quad\overset{A_2}{\bullet\!-\!\!\to\!\bullet} \quad \to \quad A_3}\;(\times \to)
$$

3.3   Embedding of the Lambek Calculus in HL

As expected, the Lambek calculus can be embedded in HL using string graphs. Consider the following embedding function $tr : Tp(\mathrm{L}) \to Tp(\mathrm{HL})$:

- $tr(p) := p, \quad p \in Pr, type(p) = 2;$
- $tr(A/B) := tr(A) \div \mathrm{SG}(\$\, tr(B));$
- $tr(B\backslash A) := tr(A) \div \mathrm{SG}(tr(B)\, \$);$
- $tr(A \cdot B) := \times(\mathrm{SG}(tr(A)\, tr(B))).$

Example 3.10. The type $r\backslash(p \cdot q)$ is translated into the type

$$
\times\!\left((1)\;\bullet\!\xrightarrow{\;p\;}\!\bullet\!\xrightarrow{\;q\;}\!\bullet\,(2)\right) \div \left((1)\;\bullet\!\xrightarrow{\;r\;}\!\bullet\!\xrightarrow{\;\$\;}\!\bullet\,(2)\right)
$$

A string sequent $\varGamma \to A$ is transformed into a graph sequent as follows: $tr(\varGamma \to A) := \mathrm{SG}(tr(\varGamma)) \to tr(A)$. Let $tr(Tp(\mathrm{L}))$ be the image of tr.

Theorem 3.1

1. *If* $\mathrm{L} \vdash \varGamma \to C$, *then* $\mathrm{HL} \vdash tr(\varGamma \to C)$.
2. *If* $\mathrm{HL} \vdash G \to T$ *is a derivable graph sequent over* $tr(Tp(\mathrm{L}))$, *then for some* $\varGamma$ *and* C *we have* $G \to T = tr(\varGamma \to C)$ *(in particular, G has to be a string graph) and* $\mathrm{L} \vdash \varGamma \to C$.

The proof of this theorem is straightforward; it can be found in [9] along with the further discussion of HL.

4   Hypergraph Lambek Grammars

Now, we are finally ready to introduce a grammar formalism based on the hypergraph Lambek calculus.

Definition 4.1. *A* hypergraph Lambek grammar (HL-grammar) *is a tuple* $HGr = \langle \Sigma, S, \triangleright \rangle$ *where* Σ *is a finite ranked alphabet,* $S \in Tp(\mathrm{HL})$ *is a distinguished type, and* $\triangleright \subseteq \Sigma \times Tp(\mathrm{HL})$ *is a finite binary relation such that* $a \triangleright T$ *implies* $type(a) = type(T)$.

We call the set $dict(HGr) = \{T \in Tp(\mathrm{HL}) : \exists a : a \triangleright T\}$ a dictionary of HGr.

Definition 4.2. The language $L(HGr)$ recognized by a hypergraph Lambek grammar $HGr = \langle \Sigma, S, \triangleright \rangle$ *is the set of all hypergraphs* $G \in \mathcal{H}(\Sigma)$, *for which a function* $f_G : E_G \to Tp(\mathrm{HL})$ *exists such that:*

1. $lab_G(e) \triangleright f_G(e)$ *whenever* $e \in E_G$;
2. $\mathrm{HL} \vdash f_G(G) \to S$.

Example 4.1. Consider the HL-grammar $EGr_1 := \langle \{a, b\}, s, \triangleright_1 \rangle$ where $s, q \in Pr$, $type(s) = type(q) = 2$ and $\triangleright_1$ is defined below:

- $a \triangleright_1 s \div \mathrm{SG}(\$q)$; $- b \triangleright_1 q \div \mathrm{SG}(s\$)$; $- b \triangleright_1 q$;

   This grammar recognizes the language of string graphs $\{\mathrm{SG}(a^n b^n) \mid n > 0\}$. E.g. given the string graph $\mathrm{SG}(aabb)$, one relabels its edges by corresponding types and obtains the graph $\mathrm{SG}\left[(s \div \mathrm{SG}(\$q)) \ (s \div \mathrm{SG}(\$q)) \ q \ (q \div \mathrm{SG}(s\$))\right]$. Now it remains to derive the sequent $\mathrm{SG}\left[(s \div \mathrm{SG}(\$q)) \ (s \div \mathrm{SG}(\$q)) \ q \ (q \div \mathrm{SG}(s\$))\right] \to s$:

$$\dfrac{\dfrac{\dfrac{\mathrm{SG}(s) \to s \quad \mathrm{SG}(q) \to q}{\mathrm{SG}\left[(s \div \mathrm{SG}(\$q)) \ q\right]^\bullet \to s} (\div \to) \quad \mathrm{SG}(s) \to s}{\mathrm{SG}\left[(s \div \mathrm{SG}(\$q)) \ s \ (q \div \mathrm{SG}(s\$))\right] \to s} (\div \to) \quad \mathrm{SG}(q) \to q}{\mathrm{SG}\left[(s \div \mathrm{SG}(\$q)) \ (s \div \mathrm{SG}(\$q)) \ q \ (q \div \mathrm{SG}(s\$))\right] \to s} (\div \to)$$

It is not hard to notice that this derivation resembles that from Example 1.1. The above derivation illustrates Theorem 3.1.

Example 4.2. Consider an HL-grammar $EGr_2 = \langle \{a\}, A_3, \triangleright_2 \rangle$ where $a \triangleright_2 A_2$ (A_2, A_3 are from Example 3.3). Then the graph $\bullet\!\!\xrightarrow{\ a\ }\!\bullet\!\xrightarrow{\ a\ }\!\!\bullet$ belongs to $L(HGr)$. In order to show this, we need to relabel all edges of this graph by types corresponding to current labels via $\triangleright_2$ and then to check derivability of a resulting sequent. In this example, we can only relabel a by A_2. Now it suffices to check derivability of a sequent

$$\bullet\!\xleftarrow{\ A_2\ }\!\bullet\!\xrightarrow{\ A_2\ }\!\bullet \quad \to \quad A_3$$

which is derivable according to Example 3.9.

   Note that, if the graph in the antecedent contained not 2, but arbitrary number (say, n) of A_2-labeled edges outgoing from a single node, then the sequent would be derivable as well: its derivation (from bottom to top) would consist of n applications of $(\times \to)$, of one application of $(\to \div)$ and of $(\to \times)$, and finally of n applications of $(\div \to)$. It can be proved that no more graph sequents with A_2 in the antecedent and A_3 in the succedent are derivable, so HGr recognizes the language of stars.

The definitions and principles of HL-grammars may look sophisticated; however, their main idea is the same as for Lambek grammars: instead of generating graphs by means of productions, an HL-grammar takes the whole graph at first, then tries to relabel its edges via a correspondence ▷ and to derive the resulting sequent (with the succedent S) in HL.

Hypergraph Lambek grammars recognize hypergraph languages; thus they represent an alternative tool to HRGs. As in the string case, the major problem is describing the class of languages recognized by HL-grammars and comparing it with the class of languages generated by HRGs. In the string case the following theorem holds:

Theorem 4.1. *The class of languages recognized by Lambek grammars coincides with the class of context-free languages without the empty word.*

This theorem has two directions. The first one (CFGs ⤳ LGs) was proved by Gaifman in 1960 [1] while the other one (LGs ⤳ CFGs) was proved by Pentus in 1993 [7]. The first part is more simple; its proof is based on the Greibach normal form for context-free grammars (see Example 3.1). The second part appeared to be a hard problem; Pentus proved it using delicate logical and algebraic techniques including the free group interpretation and interpolants.

Summing up, in the string case context-free grammar and Lambek grammar approaches are equivalent. Regarding the graph case our first expectation was that similar things happen: HRGs and HL-grammars are equivalent disregarding, possibly, some nonsubstantive cases. As in the string case, we can introduce the analogue of the Greibach normal form for HRGs and study how to convert HRGs in this normal form into HL-grammars. However, this was not clear at all whether it is possible to perform the convertion of HL-grammars into equivalent HRGs: the proof of Pentus exploits free group interpretation, which is hard to generalize to graphs (we have no idea how to do this). Surprisingly, this convertion cannot be done at all! We figured out that hypergraph Lambek grammars recognize a wider class of languages than hypergraph context-free languages. Moreover, even the pumping lemma and the Parikh theorem do not hold for HL-grammars.

4.1   Properties of HL-grammars

Let us formulate several formal properties of HL and of HL-grammars that will help us later to prove the main theorem.

Theorem 4.2 (Cut Elimination). *If graph sequents $H \to A$ and $G \to B$ are derivable, and $e_0 \in E_G$ is labeled by A, then $G[e_0/H] \to B$ is also derivable.*

The proof of this theorem can be found in [9]. This theorem directly implies

Proposition 4.1 (Reversibility of $(\times \to)$ and $(\to \div)$)

1. *If* $\mathrm{HL} \vdash H \to C$ *and* $e_0 \in E_H$ *is labeled by* $\times(M)$, *then* $\mathrm{HL} \vdash H[e_0/M] \to C$.
2. *If* $\mathrm{HL} \vdash H \to N \div D$ *and* $e_0 \in E_D$ *is labeled by* $\$$, *then* $\mathrm{HL} \vdash D[e_0/H] \to N$.

Definition 4.3. *A type A is called simple if one of the following holds:*

- *A is primitive;*
- *$A = \times(M)$, $E_M = \{m_1, \ldots, m_l\}$ and $lab(m_1), \ldots, lab(m_l)$ are simple;*
- *$A = N \div D$, $E_D = \{d_0, \ldots, d_k\}$, $lab(d_0) = \$$, N is simple, and $lab(d_1), \ldots,$ $lab(d_k)$ are primitive.*

The next result is proved straightforwardly, but it is very useful in investigating what can be recognized by an HL-grammar.

Theorem 4.3. *Let $\mathrm{HL} \vdash H \to P$ where H is labeled by simple types and P is either primitive or is of the form $\times(K)$ where all edge labels in K are primitive. Then there exists a simple derivation of $H \to P$, i.e. such a derivation that*

1. *The rule $(\to \times)$ either does not appear or is applied once to one of the leaves of the derivation tree.*
2. *In each application of $(\div \to)$ all the premises, except for, possibly, the first one, are of the form $q^{\bullet} \to q$, $q \in Pr$.*
3. *If a sequent $H' \to p$ within the derivation tree contains a type of the form $\times(M)$ in the antecedent, then the rule, after which $H' \to p$ appears in the derivation, must be $(\times \to)$.*

5   Power of Hypergraph Lambek Grammars

We start with showing that languages generated by HRGs can be generated by HL-grammars as well, except for some nonsubstantive cases (related to the empty word issue in the string case). In order to do this we use the weak Greibach normal form for HRGs introduced in [10] (Definition 3.1). Let us denote the number of isolated nodes in H by $isize(H)$.

Definition 5.1. *A hypergraph language L is isolated-node bounded if there is a constant $M > 0$ such that for each $H \in L$ $isize(H) < M \cdot |E_H|$.*

Theorem 5.1. *For each HRG generating an isolated-node bounded language there is an equivalent HRG in the weak Greibach normal form.*

This theorem is proved in [10]. Now, we can prove

Theorem 5.2. *For each HRG generating an isolated-node bounded language there is an equivalent hypergraph Lambek grammar.*

Proof. Let an HRG be of the form $HGr = \langle N, \Sigma, P, S \rangle$. Applying Theorem 5.1 we can assume that HGr is in the weak Greibach normal form.

The proof is essentially similar to that in the string case. Consider elements of N as elements of Pr with the same function $type$ defined on them. Since HGr is in the weak Greibach normal form, each production in P is of the form $\pi = X \to G$ where G contains exactly one terminal edge e_0 (say $lab_G(e_0) = a \in \Sigma$). We convert this production into a type $T_\pi := X \div G[e_0 := \$]$. Then we introduce the HL-grammar $HGr' = \langle \Sigma, S, \triangleright \rangle$ where $\triangleright$ is defined as follows: $a \triangleright T_\pi$. An illustration of this transformation is given in Example 3.2. If $G = a^{\bullet}$, then we can simply write $a \triangleright X$ instead.

The main objective is to prove that $L(HGr) = L(HGr')$. Firstly, we are going to prove that $T^{\bullet} \overset{k}{\Rightarrow} H$ for $T \in N$, $H \in \mathcal{H}(\Sigma)$ **only if** HL $\vdash f(H) \rightarrow T$ for some $f : E_H \rightarrow Tp(\text{HL})$ such that $lab_H(e) \triangleright f(e)$ for all $e \in E_H$ (this would imply $L(HGr) \subseteq L(HGr')$). This is done by induction on k.

Basis. If $k = 1$, then $\pi = T \rightarrow H$ belongs to P and $E_H = \{e_0\}$. Then we can derive HL $\vdash H[e_0 := T \div H[e_0 := \$]] \rightarrow T$ in one step using $(\div \rightarrow)$.

Step. Let the first step of the derivation be of the form $T \Rightarrow G$ $(\pi = T \rightarrow G \in P)$ and let $E_G = \{e_0, \ldots, e_n\}$ where $lab_G(e_0) \in \Sigma$ and $lab_G(e_i) \in N$ otherwise. Let $G_i \in \mathcal{H}(\Sigma)$ be a graph that is obtained from $T_i = lab_G(e_i)$ in the derivation process $(i = 1, \ldots, n)$. Note that $H = G[e_1/G_1, \ldots, e_n/G_n]$. By the induction hypothesis, HL $\vdash f_i(G_i) \rightarrow T_i$ for such $f_i : E_{G_i} \rightarrow Tp(\text{HL})$ that $lab_{G_i}(e) \triangleright f_i(e)$. Then f_i can be combined into a single function f as follows: $f(e) := f_i(e)$ whenever $e \in G_i$ and $f(e_0) := T_\pi$. Then we construct the following derivation (recall that $T_\pi = T \div G[e_0 := \$]$):

$$\frac{T^{\bullet} \rightarrow T \quad f_1(G_1) \rightarrow T_1 \quad \cdots \quad f_n(G_n) \rightarrow T_n}{(G[e_0 := \$])[e_0 := T_\pi][e_1/f_1(G_1), \ldots, e_n/f_n(G_n)] \rightarrow T} \; (\div \rightarrow)$$

This completes the proof since $G[e_0 := T_\pi][e_1/f_1(G_1), \ldots, e_n/f_n(G_n)] = f(H)$.

Secondly, we explain why $L(HGr') \subseteq L(HGr)$. Note that types in the dictionary of HGr' are simple; thus for each derivable sequent of the form $H \rightarrow S$ where H is over this dictionary we can apply Theorem 4.3 and obtain a derivation where each premise except for, possibly, the first one is an axiom. Now we can transform a derivation tree in HL into a derivation of the HRG HGr: each application of $(\div \rightarrow)$ such that a type T_π appears after it is transformed into an application of π in HGr. Formally, we have to use induction again. □

This theorem also shows that hypergraph basic categorial grammars introduced in [8] can be embedded in HL-grammars similarly. The rule $(\div)$, which is the main transformation in [8], is closely related to $(\div \rightarrow)$.

Corollary 5.1. *The membership problem for HL-grammars is NP-complete.*

Proof. The problem is in NP since if a graph H belongs to $L(HGr)$ given H and HGr, then this can be justified by listing a function f from edges of H to types from $dict(HGr)$ and by a derivation tree of $f(H) \rightarrow S$ including descriptions of all arising isomorphisms (here S is a distinguished type in HGr). Such a certificate has polynomial size w.r.t. H.

The problem is NP-complete, because there is an NP-complete isolated-node bounded language generated by an HRG (see [2]), and hence by some HL-grammar too. □

5.1   Finite Intersections of HCFLs

It is known that multiplication in L (i.e. an operation $A \cdot B$) may be considered as some kind of conjunction of A and B such that we have both A and B combined in a single type. In the graph case, we can use multiplication (i.e. $\times$) in a more general way than for strings: any graph structure can be put inside $\times$. What if there is a way to imitate behaviour of a real conjunction using $\times$ and thus to model intersections of languages? Below we investigate this idea.

Definition 5.2. *An ersatz conjunction* $\wedge_E(T_1, \ldots, T_k)$ *of types* $T_1, \ldots, T_k \in Tp(\mathrm{HL})$ *(such that* $type(T_1) = \cdots = type(T_k) = m$*) is the type* $\times(H)$ *where*

1. $V_H = \{v_1, \ldots, v_m\}$;
2. $E_H = \{e_1, \ldots, e_k\}$;
3. $att_H(e_i) = v_1 \ldots v_m$;
4. $lab_H(e_i) = T_i$;
5. $ext_H = v_1 \ldots v_m$.

Example 5.1. Let T_1, T_2, T_3 be types with *type* equal to 2. Then their *ersatz* conjunction equals $\wedge_E(T_1, T_2, T_3) = \times \left(\begin{array}{c} T_1 \\ (1) \xleftarrow{} T_2 \xrightarrow{} (2) \\ T_3 \end{array} \right).$

Using this construction we can prove the main result of this paper.

Theorem 5.3. *If* $HGr'_1, \ldots, HGr'_k$ *are HRGs generating isolated-node bounded languages, then there is an HL-grammar* HGr *such that* $L(HGr) = L(HGr'_1) \cap \cdots \cap L(HGr'_k)$.

Proof. Following the proof of Theorem 5.2 we construct an HL-grammar HGr_i for each $i = 1 \ldots, k$ such that $L(HGr'_i) = L(HGr_i)$. We assume without loss of generality that types involved in HGr_i and HGr_j for $i \neq j$ do not have common primitive subtypes (let us denote the set of primitive subtypes of types in $dict(HGr_i)$ as Pr_i). Let us denote $HGr_i = \langle \Sigma, s_i, \rhd_i \rangle$. Note that $type(s_1) = \cdots = type(s_k)$ (otherwise $L(HGr_1) \cap \cdots \cap L(HGr_k) = \emptyset$, and the theorem holds due to trivial reasons). The main idea then is to do the following: given $a \rhd_i T_i$, $i = 1, \ldots, k$ we join $T_1, \ldots, T_k$ using ersatz conjunction. A distinguished type of the new grammar will also be constructed from $s_1, \ldots s_k$ using $\wedge_E$. Then a derivation is expected to split into k independent parts corresponding to derivations in grammars $HGr_1, \ldots, HGr_k$. However, there is a nuance that spoils simplicity of this idea; it is related to the issue of isolated nodes. This nuance leads to a technical trick, which we call "tying balloons".

Let us fix $(k-1)$ new primitive types $b_1, \ldots, b_{k-1}$ ("balloon" labels) such that $type(b_i) = 1$. For $j < k$ we define a function $\varphi_j : dict(HGr_j) \to Tp(\mathrm{HL})$ as follows: $\varphi_j(p) = p$ whenever $p \in Pr$; $\varphi_j(p \div D) = \times(M) \div D'$ where

1. $D' = \langle V_D, E_D, att_D, lab_D, ext_D w \rangle$ where $[w] = V_D \setminus [ext_D]$ (that is, w consists of nodes that are not external in D).
2. Denote $m = |w| = |V_D| - |ext_D|$, and $t = type(p)$. Then $M = \langle \{v_1, \ldots, v_{t+m}\}, \{e_0, e_1, \ldots, e_m\}, att, lab, v_1 \ldots v_{t+m} \rangle$ where $att(e_0) = v_1 \ldots v_t$, $lab(e_0) = p$; $att(e_i) = v_{t+i}$, $lab(e_i) = b_j$ whenever $i = 1, \ldots, m$.

Informally, we make all nodes in the denominator D external, while $\times(M)$ "ties a balloon" labeled b_j to each node corresponding to a nonexternal one in D. Presence of these "balloon edges" is compensated by modified types of the grammar HGr_k. Namely, we define a function $\varphi_k : dict(HGr_k) \to Tp(\mathrm{HL})$ as follows: $\varphi_k(p) = p$ whenever $p \in Pr$; $\varphi_k(p \div D) = p \div D'$ where $D' = \langle V_D, E_D \cup \{e_1, \ldots, e_{(k-1)m}\}, att, lab, ext_D \rangle$ such that:

1. $m = |V_D| - |ext_D|$;
2. $e_1, \ldots, e_{(k-1)m}$ are new edges;
3. $att|_{E_D} = att_D,\ lab|_{E_D} = lab_D$;
4. If $v_1, \ldots, v_m$ are all nonexternal nodes of D, then $att(e_i) = v_{\lceil i/(k-1) \rceil}$ for $i = 1, \ldots, (k-1)m$. In other words, we attach $(k-1)$ new edges to each nonexternal node of D.
5. $lab(e_i) = b_{g(i)},\ i = 1, \ldots, (k-1)m$ where $g(i) = [i \bmod (k-1)]$ if $(k-1) \nmid i$ and $g(i) = k-1$ otherwise. That is, for each $b_i, i = 1, \ldots, (k-1)$ and for each nonexternal node there is a b_i-labeled edge attached to it.

Example 5.2. Let $k = 3$, and let $T = p \div \left(\begin{array}{c} \text{(1)} \xrightarrow{\$\quad q} \text{(2)} \quad \bullet \end{array} \right)$. Then

$$- \varphi_1(T) = \times \left(\begin{array}{c} \overset{p}{\longrightarrow} \\ \text{(1)}\quad\text{(2)} \end{array} \begin{array}{c} b_1 \\ 1 \\ \text{(3)} \end{array} \begin{array}{c} b_1 \\ 1 \\ \text{(4)} \end{array} \right) \div \left(\begin{array}{c} \text{(1)} \xrightarrow{\$\ \text{(3)}\ q} \text{(2)} \quad \text{(4)} \\ \bullet \end{array} \right)$$

$$- \varphi_3(T) = p \div \left(\begin{array}{c} \text{(1)} \xrightarrow{\$\qquad q} \text{(2)} \\ 1\ \ 1 \qquad 1\ \ 1 \\ b_1\quad b_2 \quad b_1 \quad b_2 \end{array} \right)$$

Now we are ready to introduce HGr: $HGr = \langle \Sigma, S, \rhd \rangle$ where

$-\ a \rhd T \Leftrightarrow T = \wedge_E(\varphi_1(T_1), \ldots, \varphi_k(T_k))$ and $\forall i = 1, \ldots, k \quad a \rhd_i T_i$;
$-\ S = \wedge_E(s_1, \ldots, s_k)$.

The proof of $L(HGr) = L(HGr_1) \cap \cdots \cap L(HGr_k)$ is divided into two parts: the $\subseteq$-inclusion proof and the $\supseteq$-inclusion proof.

Proof of the $\supseteq$-inclusion. A hypergraph $H \in \mathcal{H}(\Sigma)$ belongs to $L(HGr_1) \cap \cdots \cap L(HGr_k)$ if and only if there are relabeling functions $f_i : E_H \to Tp(\mathrm{HL})$ such that $lab_H(e) \rhd_i f_i(e)$ for all $e \in E_H$, and $\mathrm{HL} \vdash f_i(H) \to s_i$. Using these relabelings we construct a relabeling $f : E_H \to Tp(\mathrm{HL})$ as follows: if $f_i(e) = T_i$, then $f(e) := \wedge_E(\varphi_1(T_1), \ldots, \varphi_k(T_k))$. It follows directly from the definition that $lab_H(e) \rhd f(e)$. Now we construct a derivation of $f(H) \to \wedge_E(s_1, \ldots, s_k)$ from bottom to top:

1. We apply rules $(\times \to)$ to all ersatz conjunctions in the antecedent. This yields a graph without $\times$-labels, which has k "layers" belonging to grammars $HGr_1, \ldots, HGr_k$.

2. We remodel a derivation of $f_1(H) \to s_1$, which consists of $(\div \to)$-applications only, using types of the form $\varphi_1(f_1(e)), e \in E_H$ that are present in $f(H)$. The only difference now is that nonexternal nodes do not "disappear" (recall that a derivation is considered from bottom to top) but edges labeled by types with $\times$ appear. Every time when $\times$ appears in the left-hand side we immediately apply $(\times \to)$, which results in adding one edge labeled by a primitive type from Pr_1 and in adding balloon edges to all nodes that would disappear in the derivation of $f_1(H) \to s_1$.

The result of this part of a derivation is that now all types corresponding to HGr_1 left the antecedent, except for the only s_1-labeled edge attached to the external nodes of the antecedent in the right order; besides, for each nonexternal node in the antecedent there is now a balloon edge labeled by b_1 attached to it.

3. We perform $(k-2)$ more steps similarly to Step 2. using types of the form $\varphi_i(f_i(e)), 1 < i < k$ and thus remodeling a derivation of the sequent $f_i(H) \to s_i$. Upon completion of all these steps the antecedent contains:
   - Types of the form $\varphi_k(f_k(e)), e \in E_H$;
   - $(k-1)$ edges labeled by $s_1, \ldots, s_{k-1}$ resp. and attached to external nodes of the antecedent;
   - Balloon edges such that for each $j \in \{1, \ldots, k-1\}$ and for each nonexternal node there is a b_j-labeled edge attached to it.

4. We remodel a derivation of $f_k(H) \to s_k$ using types of the form $\varphi_k(f_k(e))$. A situation differs from those at steps 2. and 3. because now nonexternal nodes do disappear, and each time when this happens all balloon edges attached to a nonexternal node disappear as well.

After this step, all balloon edges are removed, and we obtain a graph with $type(s_1)$ nodes such that all of them are external, and with k edges labeled by $s_1, \ldots, s_k$ such that their attachment nodes coincide with external nodes of the graph. This ends the proof since $\wedge_E(s_1, \ldots, s_k)$ is exactly this graph standing under $\times$.

Proof of the $\subseteq$-inclusion. Let H be in $L(HGr)$; then there is a function $\Phi : E_H \to Tp(\text{HL})$ such that $\Phi(e) = \wedge_E(\varphi_1(T_1(e)), , \ldots, \varphi_k(T_k(e)))$ whenever $e \in E_H$, $lab(e) \rhd_i T_i(e)$, and $\Phi(H) \to S$ is derivable in HL. We aim to decompose the derivation of this sequent into k ones in grammars $HGr_1, \ldots, HGr_k$. In order to do this we transform the derivation in stages:

Stage 1. Using Proposition 4.1 we replace each edge in $\Phi(H)$ labeled by a type of the form $\times(M)$ by M. A new sequent (denote it by $H' \to S$) is derivable as well.

Stage 2. The sequent $H' \to S$ fits in Theorem 4.3; hence there exists its simple derivation. Let us fix some simple derivation of $H' \to S$ and call it Δ.

Furthermore we consider all derivations from bottom to top. In particular, if we state "X is after Y" regarding some places X and Y in a derivation, then we mean that X is above Y in the derivation tree (e.g. regarding Example 3.9 we would say that $(\to \times)$ is applied after $(\to \div)$).

Stage 3. Design of types $\varphi_i(T)$ differs in the case $i < k$ and $i = k$. Namely, if $\varphi_i(T)$ for $i < k$ participates in the rule $(\div \to)$ in Δ, this affects only primitive types from Pr_i; on the contrary, participating of $\varphi_k(T)$ in $(\div \rightarrow)$ affects types from Pr_k but also balloon types $b_1, \ldots, b_{k-1}$, which appear after rule applications of $(\div \rightarrow)$ and $(\times \rightarrow)$ to types of the form $\varphi_i(T)$, $i < k$. This allows us to come up with the following conclusion: if an application of the rule $(\div \rightarrow)$ to a type of the form $\varphi_k(T)$ preceeds a rule application of $(\div \rightarrow)$ to a type of the form $\varphi_i(T)$ for $i < k$, then we can change their order (note also that all nodes in the denominator of $\varphi_i(T)$ are external). Thus Δ can be remade in such a way that all rules affecting $\varphi_k(T)$ will occur upper than rules affecting $\varphi_i(T), i < k$ in a derivation (and it will remain simple). Let us call a resulting derivation Δ'.

Stage 4. A denominator of a type $\varphi_i(T)$ for $i < k$ contains edges labeled by elements of Pr_i only. Since Δ' is simple, applications of the rule $(\div \rightarrow)$ to types of the form $\varphi_i(T)$ and $\varphi_j(T')$ for $i \neq j$ are independent, and their order can be changed. This means that we can reorganize Δ' in the following way (from bottom to top):

1. Set $i = 1$.
2. Apply the rule $(\div \rightarrow)$ to a type of the form $\varphi_i(T)$ and right away the rule $(\times \rightarrow)$ to its numerator.
3. If there still are types of the form $\varphi_i(T)$, repeat step 2;
4. If $i = k - 1$, go forward; otherwise, set $i = i + 1$ and go back to step 2.
5. Apply the rule $(\div \rightarrow)$ to types of the form $\varphi_k(T)$.
6. Now, an antecedent of the major sequent (denote this sequent as $G \rightarrow S$) does not include types with $\div$ or $\times$. S is of the form $\times(M_S)$, and Theorem 4.3 provides that the last rule applied has to be $(\rightarrow \times)$; therefore, $G = M_S$ and we reach the sequent $M_S \rightarrow S$. Consequently, $G = M_S$ consists of k edges labeled by $s_1, \ldots, s_k$ resp.

Let us call this derivation Δ_0. Observe that, after steps 1–4 in the above description, balloon edges with labels $b_1, \ldots, b_{k-1}$ may occur in the antecedent of a sequent (denote this sequent, which appears after step 4, as $G' \rightarrow S$). There is only one way for them to disappear: they have to participate in the rule $(\div \rightarrow)$ with a type of the form $\varphi_k(T)$ (since only for such types it is the case that their denominators may contain balloon edges). Note, however, that balloon edges within the denominator of $\varphi_k(T)$ are attached only to nonexternal nodes. Therefore, balloon edges in G' can be attached only to nonexternal nodes as well. Besides, if some balloon edge labeled by b_i is attached to a node $v \in V_{G'} \setminus [ext_{G'}]$, then the set of balloon edges attached to v has to consist of exactly $k - 1$ edges labeled by $b_1, \ldots, b_{k-1}$ (because in the denominator of $\varphi_k(T)$ exactly such edges are attached to each nonexternal node). Finally, note that after step 5 all nonexternal nodes disappear since M_S contains exactly $type(S)$ nodes, all of which are external. This allows us to conclude that balloon edges have to be present on all nonexternal nodes (otherwise, a nonexternal node cannot go away interacting with a type of the form $\varphi_k(T)$). Informally, a balloon edge labeled by b_i indicates that a node was used by a type from the i-th grammar HGr_i, and $\varphi_k(T)$ verifies that each nonexternal node is used by the i-th grammar exactly once.

Summarizing all the above observations, we conclude that, after steps 1–4, there is exactly one balloon edge labeled by b_i on each nonexternal node of G'

for all $i = 1, \ldots, k-1$ (and no balloon edge is attached to some external node of G'). The only way for b_i to appear attached to a node (recall that we consider the derivation from bottom to top) is to participate in the rule $(\times \rightarrow)$ after the application of $(\div \rightarrow)$ to a type of the form $\varphi_i(T)$. Now we are ready to decompose Δ_0 into k ones:

- For $1 \le i < k$ we consider step 2 of Δ_0 with that only difference that we disregard balloon edges and additional external nodes added in the construction of HGr. Then the combination of rules $(\div \rightarrow)$ and $(\times \rightarrow)$ applied to a type $\varphi_i(T)$ turns into an application of the rule $(\div \rightarrow)$ to T in the HGr_i. Take into account that the only type that is built of elements of Pr_i and remains to step 6 is s_i attached to external nodes in the right order. Therefore, if we remove from H' all edges not related to HGr_i and relabel each edge having a label $\varphi_i(T)$ by T (call the resulting graph H'_i), then $H'_i \rightarrow s_i$ is derivable.
- For $i = k$ everything works similarly; however, instead of step 2 we have to look at step 5 and again not to consider balloon edges. Then an application of $(\div \rightarrow)$ to $\varphi_k(T)$ is transformed into a similar application of $(\div \rightarrow)$ to T in HGr_k. After the whole process, only s_k remains, so, if H'_k is a graph obtained from H' by removing edges not related to HGr_k and changing each label of the form $\varphi_k(T)$ by T, then $H'_k \rightarrow s_k$ is derivable.

Finally note that $H'_i = \Phi_i(H)$ where $\Phi_i(e) = T_i(e)$. The requirement $lab(e) \rhd_i T_i(e)$ completes the proof, because thus $H \in L(HGr_i)$ for all $i = 1, \ldots, k$. $\square$

The balloon trick is used in the proof to control that making all nodes in denominators of $\varphi_i(T)$ external ($i < k$) does not lead to using, e.g., a nonexternal isolated node in rules $(\div \rightarrow)$ more than once.

Corollary 5.2. *The language* $\{SG(a^{2n^2}), n > 0\}$ *can be generated by some HL-grammar.*

Proof. The string language $L_1 = (\{a^n b^n \mid n > 0\})^+$ is context-free, and so is $L_2 = \{a^k b^{m_1} a^{m_1} b^{m_2} a^{m_2} \ldots b^{m_{k-1}} a^{m_{k-1}} b^l \mid k, m_i, l > 0\}$. Consequently, languages $SG(L_1) = \{SG(w) \mid w \in L_1\}$ and $SG(L_2)$ are generated by some HRGs. The language $L_3 = L_1 \cap L_2$ equals $L_3 = \{(a^n b^n)^n \mid n > 0\}$, so $SG(L_3)$ is a finite intersection of HCFLs and can be generated by some HL-grammar $\langle \{a, b\}, S, \rhd \rangle$. Finally, note that $HGr := \langle \{a\}, S, \rhd' \rangle$ where $a \rhd' T \Leftrightarrow a \rhd T$ or $b \rhd T$ recognizes the language $L = \{SG(a^{2n^2}), n > 0\}$. $\square$

Corollary 5.3. *The pumping lemma and the Parikh theorem do not hold for languages generated by HL-grammars.*

Proof. The language $\{SG(a^{2n^2}), n > 0\}$ contradicts both theorems. $\square$

6   Conclusion

In the string case, there is a disparity between context-free grammars and Lambek grammars: while generating the same set of languages, the former can be parsed in polynomial time (the CYK algorithm) but the membership problem for the latter is NP-complete (see [6]). Situation changes dramatically in the hypergraph case: hypergraph Lambek grammars introduced in this paper have the same algorithmic complexity as hyperedge replacement grammars but they generate much more languages.

Of a particular note is that HL-grammars recognize more languages than finite intersections of hypergraph context-free languages (due to article limits, we shall not prove this here). Moreover, nontrivial upper bounds restricting the power of HL-grammars (like the pumping lemma for HRGs) are unknown; it is subject to be investigated further. In particular, it would be interesting to answer the question whether languages generated by HL-grammars are closed under intersection (Theorem 5.3 gives us a hope that this could be true). For now there is no clear way for us how to prove this.

To conclude, HL-grammars represent a curious mechanism opposed to HRGs in the underlying concepts. From the point of view of parsing, HL-grammar is not a simple formalism; however, it is more powerful than HRGs and hence is worth consideration. It is also important to notice that the hypergraph Lambek calculus itself is an interesting logic, so we hope that in the future all these formalisms will be studied deeper in their different aspects.

Acknowledgments. I thank my scientific advisor prof. Mati Pentus for fruitful discussions.

References

1. Bar-Hillel, Y., Gaifman, C., Shamir, E.: On categorial and phrase-structure grammars. Bulletin of the Research Council of Israel F(9), 1–16 (1963)
2. Drewes, F., Kreowski, H.-J., Habel, A.: Hyperedge replacement graph grammars (1997)
3. Kanazawa, M.: Second-order abstract categorial grammars as hyperedge replacement grammars. J. Log. Lang. Inf. **19**, 137–161 (2010)
4. Lambek, J.: The mathematics of sentence structure. Am. Math. Monthly **65**(3), 154–170 (1958)
5. Moortgat, M.: Multimodal linguistic inference. J. Log. Lang. Inf. **5**(3/4), 349–385 (1996)
6. Pentus, M.: Lambek calculus is NP-complete. Theor. Comput. Sci. **357**(1–3), 186–201 (2006)
7. Pentus, M.: Lambek grammars are context free. In: Proceedings of the Eighth Annual Symposium on Logic in Computer Science (LICS 1993), Montreal, Canada. IEEE Computer Society (1993)
8. Gadducci, F., Kehrer, T. (eds.): ICGT 2020. LNCS, vol. 12150. Springer, Cham (2020). https://doi.org/10.1007/978-3-030-51372-6
9. Pshenitsyn, T.: Introduction to a Hypergraph Logic Unifying Different Variants of the Lambek Calculus. Preprint at arXiv.org. https://arxiv.org/abs/2103.01199
10. Pshenitsyn, T.: Weak Greibach Normal Form for Hyperedge Replacement Grammars. In: Electronic Proceedings in Theoretical Computer Science, vol. 330, pp. 108–125. Open Publishing Association (2020)

Evaluation Diversity for Graph Conditions

Sven Schneider$^{(\boxtimes)}$ ⓘ and Leen Lambers ⓘ

University of Potsdam, Hasso Plattner Institute, Potsdam, Germany
{sven.schneider,leen.lambers}@hpi.de

Abstract. Graphs are used as a universal data structure in various domains. Sets of graphs (and likewise graph morphisms) can be specified using the graph logic GL of Graph Conditions (GCs). The *evaluation* of a graph against a GC results in a satisfaction judgement on whether the graph is specified by the GC. GL is as expressive as first-order logic on graphs and infinitely many graphs may be evaluated against a given GC. Therefore, a complete compact overview of *how* a given GC may be evaluated for varying graphs can support GC validation, testing, debugging, and repair.

As a main contribution, we generate such an overview for a given GC in the form of a complete finite set of diverse evaluations for varying associated graphs formally given by so called Evaluation Trees (ETs). Each of these ETs concretely describes *how* its associated graph is evaluated against the given GC by presenting each evaluation step. The returned ETs are *complete* since each possible ET subsumes one of the returned ETs and *diverse* by not containing superfluous ETs subsuming smaller ETs. We apply an implementation of our ET generation procedure in the tool AUTOGRAPH to a running example.

Keywords: Graph logic · Graph conditions · Logic coverage · Coverage criteria · Model generation · Validation · Debugging

1 Introduction

Graphs are used for the representation of e.g. UML-based system designs [18], runtime models [9,11], and graph databases [17]. Languages such as OCL [19], Nested Graph Conditions (GCs) [10], and query languages such as CYPHER are used to (a) specify sets of graphs stating e.g. consistency constraints, (b) specify sets of graph morphisms stating e.g. application conditions in graph transformation systems, and (c) obtain, aggregate, and change information contained in the graphs. We focus on GCs as a *formal* specification language with often sufficient expressiveness yet few syntactical constructs, which are evaluated for graphs (or graph morphisms) to determine satisfaction judgements.

By generating a complete compact overview of evaluations, we aim at improving support for the *use cases* of validation, testing, debugging, and repair of GCs and their evaluation. GC validation refers to the process of ensuring that the given GC specifies exactly the intended set of graphs (or graph morphisms).

© Springer Nature Switzerland AG 2021
F. Gadducci and T. Kehrer (Eds.): ICGT 2021, LNCS 12741, pp. 122–141, 2021.
https://doi.org/10.1007/978-3-030-78946-6_7

An implementation of GC satisfaction is correct when it evaluates a given GC against a graph (or a graph morphism) returning the correct satisfaction judgement (which is more difficult to achieve when employing optimizations). Testing attempts to demonstrate non-validity (by providing graphs incorrectly (not) specified by the GC) or in-correctness (by providing pairs of GCs and graphs for which incorrect satisfaction judgements are returned). The goal of debugging is to determine faults resulting in such counterexamples, which are then to be resolved during repair.

We propose a novel *generation procedure* for concrete detailed representations of how a GC is evaluated against a graph in the form of Evaluation Trees (ETs). This generation procedure takes the syntactical representation of the GC at all of its nesting levels into account. The returned ETs can e.g. be inspected during validation or debugging, used as test cases, modified during repair, or communicated to other tools such as theorem provers for further verification steps. For example, inspecting the returned ETs allows (since the syntactical representation of the GC is reflected in the ET) to locate faults in the GC by determining those sub-GCs that are evaluated in the ET in an unintended way. Moreover, our ET generation procedure supports the construction of use-case oriented sets of ETs varying in the generated ETs by using different Evaluation Strategys (ESs) such as the *short circuit left to right* ES.

Our ET generation procedure constructs for a GC a complete finite set of diverse ETs where each ET concretely describes an evaluation for the GC for some corresponding small graph constructed alongside. *Completeness* of the returned ETs means that they also symbolically describe *all* ETs obtained for arbitrary graphs. *Diversity* of the returned ETs means that any two such ETs describe evaluations performing different evaluation steps. The completeness and diversity of the generated ETs are important for the considered use cases to ensure that a full yet minimal overview of possible evaluations is provided. The ET generation procedure has been implemented in AUTOGRAPH [21,23,24].

In Sect. 2, we recall the graph logic GL, in Sect. 3, we define ETs, in Sect. 4, we define ESs capturing how ETs are constructed, in Sect. 5, we present our ET generation procedure, in Sect. 6, we compare with related work, and, in Sect. 7, we conclude and point out future work.

2   Graphs and Graph Logic

We consider finite typed directed graphs (short graphs) following their formalization in [7].[1] For our running example, we use the type graph TG in Fig. 1a. In visualizations of graphs typed over TG (see Fig. 1b for an example), names of nodes indicate their typing (e.g. the node a1 is of type $:A$) and edges are only numbered (the type graph TG does not allow for confusion here since any

---

[1] Type graphs and typed graphs can be understood as formalizations of UML class diagrams and UML object diagrams.

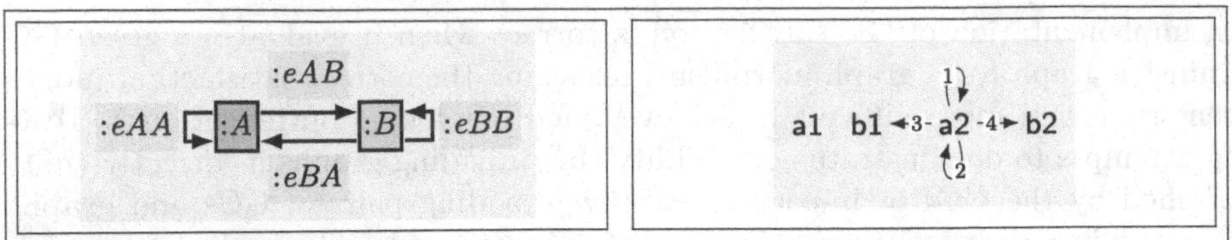

(a) Type graph TG

(b) Graph for running example

$$\phi = \neg\exists(\,a, \exists(\,a \rightleftarrows^{1}, \top) \vee \neg\exists(\,a \text{-}1\blacktriangleright b, \top))$$

(c) Graph Condition (GC) for running example stating that there may not be some node a that either has a self loop or no edge to some node b

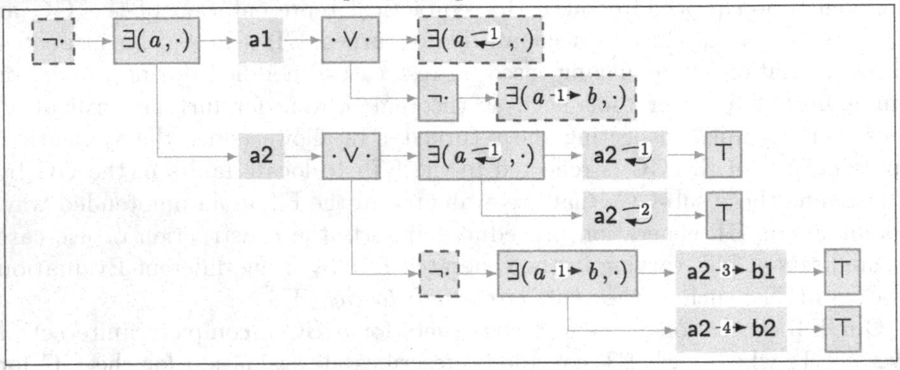

(d) Some ET for the graph from Figure 1b and the GC from Figure 1c

$$\gamma_1 = \neg\exists(\,a, \exists(\,a \rightleftarrows^{1}, \top) \vee \neg\exists(\,a \text{-}1\blacktriangleright b, \top), \{m_1 \mapsto \gamma_2, m_2 \mapsto \gamma_3 \vee \gamma_4\}, \emptyset)$$
$$\gamma_2 = \exists(\,a \rightleftarrows^{1}, \top, \emptyset, \emptyset) \vee \neg\exists(\,a \text{-}1\blacktriangleright b, \top, \emptyset, \emptyset)$$
$$\gamma_3 = \exists(\,a \rightleftarrows^{1}, \top, \{m_3 \mapsto \top, m_4 \mapsto \top\}, \emptyset)$$
$$\gamma_4 = \neg\exists(\,a \text{-}1\blacktriangleright b, \top, \{m_5 \mapsto \top, m_6 \mapsto \top\}, \emptyset)$$

where

$$
\begin{aligned}
m_1(a) &= \text{a1} & m_2(a) &= \text{a2} \\
m_3(a \rightleftarrows^{1}) &= \text{a2} \rightleftarrows^{1} & m_4(a \rightleftarrows^{1}) &= \text{a2} \rightleftarrows^{2} \\
m_5(a \text{-}1\blacktriangleright b) &= \text{a2-}3\blacktriangleright\text{b1} & m_6(a \text{-}1\blacktriangleright b) &= \text{a2-}4\blacktriangleright\text{b2}
\end{aligned}
$$

(e) The mathematical representation of the ET from Figure 1e according to Definition 3

Fig. 1. Type graph, typed graph, GC, and ET for running example

two distinct edges in the type graph differ in source or target). We only employ monomorphisms (short monos) between graphs, which map all nodes and edges injectively.

We now recall the graph logic GL as introduced in [10], which allows for the specification of sets of graphs and monos using the GCs of GL. Intuitively, for a given host graph G, a GC over some subgraph H of G identified by some mono

$m : H \hookrightarrow G$ states the presence (or absence) of certain graph elements. For this purpose, monos of the form $f : H \hookrightarrow H'$ are used to describe how the mono m should (or should not) be extended to a mono $m' : H' \hookrightarrow G$ of a larger subgraph H'. The combination of propositional operators, existential quantification, and nesting results in an expressiveness equivalent to first-order logic on graphs [5]. Note that we use the abbreviation $[m_1, m_2] = \{i \in \mathbf{N} \mid m_1 \leq i \leq m_2\}$ in the remainder for the set of all natural numbers from m_1 to m_2.

Definition 1 (Graph Conditions (GCs)). *If H is a graph, then ϕ is a graph condition (GC) over H, written $\phi \in \mathsf{GC}(H)$, if one of the following items applies.*

- $\phi = \neg\phi'$ *and* $\phi' \in \mathsf{GC}(H)$.
- $\phi = \vee(\phi_1, \ldots, \phi_n)$ *and* $\forall i \in [1, n]. \; \phi_i \in \mathsf{GC}(H)$.
- $\phi = \exists(f : H \hookrightarrow H', \phi')$ *and* $\phi' \in \mathsf{GC}(H')$.

Note that empty disjunction is defined using $[1, 0] = \emptyset$ as the base case and that further GC operators such as $\top$, $\bot$, $\wedge$, and $\forall$ can be derived but are omitted to ease presentation. Moreover, in all subsequent definitions, we implicitly assume analogous definitions for such derived operators as well.

We now define the two satisfaction relations of GL capturing (a) when a mono $m : H \hookrightarrow G$ into a host graph G satisfies a GC over H and (b) when a graph G satisfies a GC over the empty graph $\emptyset$.

Firstly, a mono satisfies negations and disjunctions as expected. Moreover, a GC of the form $\exists(f : H \hookrightarrow H', \phi')$ is satisfied by $m : H \hookrightarrow G$ when m can be extended to a match $m' : H' \hookrightarrow G$ that is consistent with f by letting the triangle $m' \circ f = m$ commute such that the extended mono m' then satisfies the sub-GC ϕ'. In the remainder, we use the abbreviation $T(f, m) = \{m' : H' \hookrightarrow G \mid m' \circ f = m\}$ for the set of all such monos m' that may need to be checked for satisfaction of ϕ'.

Secondly, a graph G satisfies a GC over the empty graph $\emptyset$ when the initial morphism $i(G) : \emptyset \hookrightarrow G$ satisfies the given GC.

Definition 2 (Satisfaction of GCs). *A mono $m : H \hookrightarrow G$ satisfies a GC ϕ over H, written $m \models \phi$ or $m \in [\![\phi]\!]$, if one of the following items applies.*

- $\phi = \neg\phi'$ *and* $\neg(m \models \phi')$.
- $\phi = \vee(\phi_1, \ldots, \phi_n)$ *and* $\exists i \in [1, n]. \; m \models \phi_i$.
- $\phi = \exists(f : H \hookrightarrow H', \phi')$ *and* $\exists m' \in T(f, m). \; m' \models \phi'$.

A graph G satisfies a GC ϕ over the empty graph $\emptyset$, written $G \models \phi$, if the initial morphism $i(G) : \emptyset \hookrightarrow G$ satisfies ϕ.

For example, the GC from Fig. 1c is not satisfied by the graph from Fig. 1b because (Reason 1) the $:A$ node a1 has no exiting edge to some $:B$ node. Another explanation is that (Reason 2) the $:A$ node a2 has the self-loop 1 or that it has the self-loop 2. While our subsequently presented approach also covers the case of satisfaction of a GC by some mono, we focus on the satisfaction of a GC by some graph to ease our presentation.

Later on, we employ the operation divGraphs introduced in [23,24] and implemented in the tool AUTOGRAPH, which generates a complete finite set of diverse *graphs* satisfying a given GC defined over the empty graph.[2]

Fact 1 (Operation divGraphs **[23,24]).** *If ϕ is a GC over the empty graph $\emptyset$, then* divGraphs$(\phi) = S$ *is a set of graphs satisfying the following items.*

- SOUNDNESS*: every graph G in S satisfies ϕ.*
- COMPLETENESS*: every graph G satisfying ϕ has a subgraph G' in S.*
- DIVERSITY*: every graph G in S has no strict subgraph G' in S.*

The operation divGraphs *may not always terminate returning a finite set S but it (a) terminates when the GC is not satisfiable and (b) gradually computes the set S allowing to obtain a partial result even when being aborted prematurely.*

For example, divGraphs returns for the GC $\exists(a \text{-1} \rightarrow b, \top) \wedge \exists(a \rightleftharpoons^1, \top)$ the two graphs $\widehat{1} \widehat{} a1 \text{ -2} \rightarrow b1$ and $\widehat{1} \widehat{} a1 \quad a2 \text{ -2} \rightarrow b1$ where the two a nodes are (are not) overlapped.[3] However, divGraphs returns for the GC from Fig. 1c only the empty graph, which is contained in all graphs satisfying the considered GC. In general, this result is obtained for all GCs of the form $\neg\exists(f : \emptyset \hookrightarrow H, \phi')$ when $H \neq \emptyset$ for any sub-GC ϕ'. Hence, witnesses as generated by divGraphs do not always provide insights into the entire GC as required for the use cases of validation, testing, debugging, and repair discussed in the introduction.

3   Evaluation Trees

We now introduce Evaluation Trees (ETs) to systematically capture all viable proofs for the (non-)satisfaction of some GC by some graph. From a tool-based perspective, implementations of the satisfaction relation of GL from Definition 2 return for a graph and a GC a boolean result indicating whether the graph satisfies the GC. However, such boolean results do not explain why (or why not) the graph satisfies the GC.

ETs record the steps performed in an evaluation of a GC for some graph. In particular, all monos considered during the evaluation are stored with the ETs constructed for them recursively. We now define ETs as a simple data structure

---

[2] The operation divGraphs used here is a simplification of the operation presented in [23,24], which returned symbolic models representing all graphs satisfying a given GC. Technically, we apply the operation from [23,24], remove additional information with which the returned graphs are equipped and remove all graphs from that set for which the set contains a subgraph already.

[3] Note that the runtime of divGraphs is usually insignificant when it is applied offline (i.e., at design time) to small hand-written GCs only (despite the need to solve the NP-hard subgraph isomorphism problem when computing the possibly exponential number of graph overlappings). If one or both of these assumptions are not satisfied in an application context, further evaluations are required to ensure a practical runtime in that setting and the property (b) from Fact 1 on partial results may become relevant.

only, to be able to consider various so called evaluation strategies later on, which result in different ETs for a pair of a GC and a graph.

The structure of an ET for some GC corresponds to the structure of this GC, which means that ETs are also constructed using the same three operators for negation, disjunction, and existential quantification. Moreover, to denote that a sub-GC has not been considered during an evaluation, we use the additional operator *unevaluated* (written U). Besides the propositional operators, for the case of existential quantification, the ET for a GC $\exists(f, \phi')$ is of the form $\exists(f, \phi', m_T, m_F)$ where m_T and m_F are partial maps. These maps record a considered mono $m' : H' \hookrightarrow G$ from $T(f, m)$ by mapping it to an ET constructed for m' and the sub-GC ϕ'. We use support(m_T) and support(m_F) to denote the sets of monos mapped by m_T and m_F. The map m_T maps those considered monos that satisfy the GC and m_F maps those considered monos that do not satisfy the GC. While GCs are defined over their context graph H, ETs are defined over the monos $m : H \hookrightarrow G$ of these context graphs into the host graph, which are also used in the satisfaction relation of GCs for monos.

Definition 3 (Evaluation Trees (ETs)). *If $m : H \hookrightarrow G$ is a mono, and ϕ is a GC over H, then γ is an evaluation tree (ET) over m and ϕ, written $\gamma \in$ ET(m, ϕ) , if one of the following items applies.*

- *$\phi = \neg\phi'$, $\gamma = \neg\gamma'$ and $\gamma' \in$ ET(m, ϕ').*
- *$\phi = \vee(\phi_1, \ldots, \phi_n)$, $\gamma = \vee(\gamma_1, \ldots, \gamma_n)$, and $\forall i \in [1, n]. \gamma_i \in$ ET(m, ϕ_i).*
- *$\phi = \exists(f : H \hookrightarrow H', \phi')$, $\gamma = \exists(f : H \hookrightarrow H', \phi', m_T, m_F)$, m_T and m_F are finite partial maps of monos $m' \in T(f, m)$ to ETs $\gamma' \in$ ET(m', ϕ'), support$(m_T) \subseteq \llbracket\phi\rrbracket$, and support$(m_F) \subseteq \llbracket\neg\phi\rrbracket$,*
- *$\gamma = $ U(ϕ).*

Moreover, γ is an ET over a graph G and a GC ϕ over the empty graph $\emptyset$, if γ is an ET over the initial morphism $i(G) : \emptyset \hookrightarrow G$ and ϕ.
Lastly, γ is compact[4], written compact(γ), if the union of all support sets of the m_T and m_F components of all sub-ETs $\exists(f, \phi', m_T, m_F)$ of γ is jointly epimorphic.

ETs are similar to satisfaction trees from [22] but support the additional *unevaluated* operator U to identify sub-GCs that have not been evaluated.

For our running example, consider the compact ET from Fig. 1d, in which the nesting of the ET and the mappings of m_T and m_F components are depicted using nodes and arrows. The same ET is given in Fig. 1e as γ_1 in its formal syntax where all m_F components are empty. The depicted ET records the two matches a1 and a2 of the a node and records for a2 two matches of the loop on the a node and two matches for edges exiting the a node to the b nodes b1 and b2.

---

[4] Intuitively, compactness means that all graph elements of the host graph G have been matched by some considered mono during the evaluation.

Subsequently, we use the standard three-valued propositional Kleene logic using the base set $\mathbf{T} = \{F, U, T\}$ where, in addition to the rules of two-valued logic, the rules $\neg U = U$, $U \vee F = U$, $U \vee U = U$, and $U \vee T = T$ are used.[5]

To be able to define correct evaluation strategies constructing ETs in the next section, we now define the satisfaction judgement given by an ET.[6] In fact, this satisfaction judgement is obtained by interpreting the ET as a three-valued logic formula where (a) each top-level sub-ET of the form $\exists(f, \phi', m_T, m_F)$ is an atomic proposition that is satisfied iff[7] m_T contains at least one mapping of a mono m' to some ET representing a proof for the sub-GC ϕ' and (b) each top-level sub-ET of the form $U(\phi')$ is an atomic proposition with value U.

Definition 4 (Operation sat). *If γ is an ET, then $\tau \in \mathbf{T}$ is the satisfaction judgement of γ, written $\mathsf{sat}(\gamma) = \tau$, if one of the following items applies.*

- $\gamma = \neg\gamma'$ *and* $\tau = \neg\mathsf{sat}(\gamma')$.
- $\gamma = \vee(\gamma_1, \ldots, \gamma_n)$ *and* $\tau = \vee\{\mathsf{sat}(\gamma_i) \mid i \in [1, n]\}$.
- $\gamma = \exists(f, \phi, m_T, m_F)$ *and* $\tau = (m_T \neq \emptyset)$.
- $\gamma = U(\phi)$ *and* $\tau = U$.

In visualizations of ETs as in Fig. 1d, we depict the satisfaction judgement for each sub-ET by using a green solid border, a red dashed border, and an orange dotted border (used in ETs in later figures) for the cases of T (satisfaction), F (non-satisfaction), and U (unknown satisfaction).

To be able to define diversity later on for a set of ETs, we define that an ET γ_1 is contained in an ET γ_2 with the same satisfaction judgement, when (a) all evaluation steps showing satisfaction at some level in γ_1 have been performed correspondingly when deriving γ_2 and (b) all evaluation steps showing non-satisfaction at some level in γ_2 have been performed correspondingly when deriving γ_1. The two ETs may be constructed for different host graphs G_1 and G_2 but are constructed for the same GC ϕ.

Definition 5 (Containment of ETs). *If ϕ is a GC over H, γ_1 is an ET over $m_1 : H \hookrightarrow G_1$ and ϕ, γ_2 is an ET over $m_2 : H \hookrightarrow G_2$ and ϕ, and $\mathsf{sat}(\gamma_1) = \mathsf{sat}(\gamma_2)$, then γ_1 is contained in γ_2, written $(m_1, \gamma_1) \leq (m_2, \gamma_2)$ or simply $\gamma_1 \leq \gamma_2$, if one of the following items applies.*

- $\forall i \in \{1, 2\}$. $\gamma_i = \neg\gamma_i'$ *and* $(m_1, \gamma_1') \leq (m_2, \gamma_2')$.
- $\forall i \in \{1, 2\}$. $\gamma_i = \vee(\gamma_1^i, \ldots, \gamma_n^i)$ *and* $\forall j \in [1, n]$. $(m_1, \gamma_j^1) \leq (m_2, \gamma_j^2)$.
- $\forall i \in \{1, 2\}$. $\gamma_i = \exists(f, \phi', m_T^i, m_F^i)$, *there is an injective function* $F_T : m_T^1 \hookrightarrow m_T^2$ *such that* $\forall(m, \gamma) \in m_T^1$. $(m, \gamma) \leq F_T(m, \gamma)$, *and there is an injective function* $F_F : m_F^2 \hookrightarrow m_F^1$ *such that* $\forall(m, \gamma) \in m_F^2$. $F_F(m, \gamma) \leq (m, \gamma)$.

---

[5] Intuitively, the operations of this three-valued logic can be computed based on numerical values using the bijective homomorphism $\langle \cdot \rangle$, which maps the elements of the base set $\mathbf{T}$ as follows: $\langle F \rangle = -1$, $\langle U \rangle = 0$, and $\langle T \rangle = 1$. Then, negation is translated into the multiplication with -1 and disjunction is translated into taking the maximum of its arguments. E.g. $\neg(U \vee T) = F$ because $-1 \times \max(0, 1) = -1$.

[6] The operation sat is an adaptation of the predicate $\models_{ST}$ from [22] accommodating for the additional *unevaluated* operator U.

[7] The returned satisfaction judgement for $\exists(f, \phi, m_T, m_F)$ is either T or F.

- $\gamma_1 = \gamma_2 = U(\phi')$.

Also, $\gamma_1 < \gamma_2$ if $\gamma_1 \leq \gamma_2$ but not $\gamma_2 \leq \gamma_1$.

For our running example, the ET from Fig. 1d contains for example an ET where the match of a2 has not been recorded since already the match of a1 was sufficient to proof satisfaction of the existential quantification $\exists(a, \cdot)$.

We may want to construct for some ET some minimal contained ET (with the same satisfaction judgement) as a minimal representation. However, firstly, such a minimal representation would not be unique due to e.g. different monos mapped by m_T components of sub-ETs. Secondly, allowing empty m_F components of sub-ETs allows for ETs with little value since (for $H \neq \emptyset$) $\neg\exists(f : \emptyset \hookrightarrow H, \phi, \emptyset, \emptyset)$ could be a minimal such ET for any mono $m : \emptyset \hookrightarrow G$. Both of these points are discussed in more detail when we define evaluation strategies and our synthesis problem.

4   Evaluation Strategies

We now define Evaluation Stratigies (ESs) as operations that generate (for a given graph and GC) a single ET (possibly) nondeterministically or, as formalized below, a possibly non-singleton set of ETs. ESs ensure (for soundness) that the correct satisfaction judgement can be obtained from each generated ET using the operation sat and (for completeness) that non-satisfaction is demonstrated by considering all possible matches. As these requirements are stated for all inputs, they are enforced for all nesting levels of resulting ETs.

Definition 6 (Evaluation Strategy (ES)). *An evaluation strategy (ES) $\mathfrak{es}$ determines for a mono $m : H \hookrightarrow G$ and a GC ϕ over H a non-empty set T of ETs γ over m and ϕ. Formally, $\mathfrak{es}$ is a function containing elements from $\{(m, \phi) \mapsto T \mid m : H \hookrightarrow G, \phi \in GC(H), T \neq \emptyset, T \subseteq ET(m, \phi)\}$. Moreover, each ES $\mathfrak{es}$ must satisfy the following two properties.*

- SOUNDNESS: *Returned ETs must provide the correct satisfaction judgement using sat w.r.t. GC satisfaction. Formally, $\gamma \in \mathfrak{es}(m, \phi)$ implies $(m \models \phi$ iff[8] $\mathsf{sat}(\gamma))$.*
- COMPLETENESS: *For a non-satisfied $\exists(f, \phi)$ all possible monos must be checked. Formally, $\exists(f, \phi, \emptyset, m_F) \in \mathfrak{es}(m, \exists(f, \phi))$ implies $\mathsf{support}(m_F) = T(f, m)$.*

Lastly, an ES $\mathfrak{es}$ is deterministic, if $\mathfrak{es}(m, \phi) = T$ is a singleton for each input (m, ϕ).

We now introduce three different ESs, which are relevant for different reasons. These three ESs differ in their handling of disjunction and in the extent to which matches are recorded for existential quantification.

The *all circuit ES*[9] evaluates (a) a GC $\vee(\phi_1, \ldots, \phi_n)$ by constructing an ET for each of the sub-GCs ϕ_i and (b) a GC $\exists(f, \phi)$ for some match m by

---

[8] Since $m \models \phi$ is either T or F, this also has to hold for $\mathsf{sat}(\gamma)$.
[9] This ES is given by the operation cst in [22].

constructing an ET γ for each match $m' \in \mathcal{T}(f, m)$ recording each pair (m', γ) in m_T or m_F depending on $\mathsf{sat}(\gamma)$. For example, this ES returns the ET from Fig. 1d. Also, this ES was used for graph repair in [22] where the obtained ET represents how (or how not) a graph G satisfies a consistency constraint. Lastly, e.g. for early debugging steps, this ES can be more appropriate by returning a single ET for a given graph in contrast to the subsequent non-deterministic ESs.

The *short circuit left to right ES* is relevant as it corresponds to implementations of the GC satisfaction relation in tools such as AUTOGRAPH avoiding unnecessary evaluation steps. It evaluates (a) a GC $\vee(\phi_1, \ldots, \phi_n)$ by constructing an ET for the sub-GCs in the order of their appearance from left to right and stopping when for some GC ϕ_i an ET γ with $\mathsf{sat}(\gamma) = \mathsf{T}$ is constructed using the ET $\mathsf{U}(\phi_j)$ for each subsequent sub-GC ϕ_j and (b) a GC $\exists(f, \phi)$ for a match m by constructing an ET γ for the matches $m' \in \mathcal{T}(f, m)$ recording each pair (m', γ) in m_T or m_F depending on $\mathsf{sat}(\gamma)$ until either all matches have been considered or some pair has been added to m_T.

Lastly, the *shortest circuit ES* generates no superfluous evaluation steps for disjunctions or superfluous pairs (m', γ) in m_T or m_F components. It evaluates (a) a GC $\vee(\phi_1, \ldots, \phi_n)$ by constructing an ET for each sub-GC when these are all not satisfied or only constructs an ET for one of the satisfied sub-GCs using the ET $\mathsf{U}(\phi_j)$ for all other sub-GCs and (b) a GC $\exists(f, \phi)$ for a match m by constructing an ET γ for all matches $m' \in \mathcal{T}(f, m)$ when the GC is not satisfied and constructing a single ET for a single match from $\mathcal{T}(f, m)$ when the GC is satisfied. This ES is of particular relevance since the resulting ETs can be understood as minimal proofs for (non-)satisfaction, which are therefore suitable for manual inspection in the context of e.g. validation.

The described behavior of these three ESs is specified by means of Evaluation Modes (EMs) given by regular expressions over $\mathbf{T} = \{\mathsf{F}, \mathsf{U}, \mathsf{T}\}$. A word w generated by such an EM then specifies that the ith ET constructed for some disjunction or existential quantification must have the satisfaction judgement given by the ith element of w.

Definition 7 (Evaluation Mode (EM)). *An evaluation mode (EM)* em *is a regular expression over* $\mathbf{T} = \{\mathsf{F}, \mathsf{U}, \mathsf{T}\}$*. Moreover, we define the EMs* AC, SCLR, *and* SC.

- *The* all circuit *evaluation mode* $\mathsf{AC} = (\mathsf{T} + \mathsf{F})^*$ *requires the execution of all checks.*
- *The* short circuit left to right *evaluation mode* $\mathsf{SCLR} = \mathsf{F}^* + \mathsf{F}^*\mathsf{TU}^*$ *requires that checks are executed from left to right stopping when the first* T *result is obtained not executing subsequent checks.*
- *The* shortest circuit *evaluation mode* $\mathsf{SC} = \mathsf{F}^* + \mathsf{U}^*\mathsf{TU}^*$ *requires that all checks are executed to show non-satisfaction or to execute a single check to show satisfaction.*

Finally, $\mathcal{S}(\mathsf{em}, n) = \{(w, \tau) \mid w \in \mathsf{em} \wedge \mathsf{length}(w) = n \wedge \tau = (\exists i.\ w_i = \mathsf{T})\}$ *is the set of words from* em *of length* n *equipped with their satisfaction judgement* τ.

We now define the three ESs for the three EMs using the operation eval.[10][11]

Definition 8 (Operation eval). *If* em $\in$ {AC, SCLR, SC} *is an evaluation mode,* $m : H \hookrightarrow G$ *is a mono,* ϕ *is a GC over* H, $\tau \neq \mathsf{U}$, *and* $\phi \xrightarrow{\text{em},\tau,m} \gamma$ *via the rules from Fig. 2, then* γ *is an ET for* em, m, *and* ϕ, *written* $\gamma \in$ eval(em)(m, ϕ).

We derive the following basic properties as a characterization for eval(AC), eval(SCLR), and eval(SC).

Theorem 1 (Characterization). eval(AC), eval(SCLR), *and* eval(SC) *are ESs and return only ETs with boolean satisfaction judgement. Of these three ESs, only* eval(AC) *is deterministic.*

Proof.

- We show that eval(AC), eval(SCLR), and eval(SC) are ESs (essentially non-emptiness of the resulting set of ETs) by induction on the structure of the given GC showing that there is always at least one ET constructed. Moreover, soundness is also shown by induction along the same lines and completeness follows directly from the definitions of each of the ESs.
- We show that eval(AC) is deterministic also by induction on the structure of the given GC showing that there is always a unique ET constructed.
- We show that eval(SCLR) and eval(SC) are not deterministic since they generate more than one ET for $\exists(a, \top)$ and a1 a2 .
- We show that every ET obtained using eval(AC), eval(SCLR), and eval(SC) has a boolean satisfaction judgement by induction on the structure of the given GC showing that U can only be obtained for a direct sub-GC of a disjunction but that the disjunction then results again in a boolean satisfaction judgement.

While we focus on the three described ESs, there are further ESs. For example, disjunctions may be evaluated in random order or from right to left, existential quantification is evaluated *bottom up* in e.g. [3], and matches are determined based on the given graph and profiling information in e.g. [2].

5   Evaluation Generation

As explained before, we want to construct a set of ETs for a given GC. We now introduce the notion of evaluation diversity to capture sets of ETs containing no superfluous elements that are covered by other ETs of that set.

Definition 9 (Evaluation Diversity). *A set of ETs S is diverse, written* diverse(S), *if there are no distinct ETs γ_1 and γ_2 in S satisfying $\gamma_1 \leq \gamma_2$.*

---

[10] To simplify our presentation, we also employ the regular expression based specification via $(w, \tau) \in \mathcal{S}(\text{em}, n)$ for existential quantification. We thereby restrict ourselves to finite sets of monos $\mathcal{T}(f, m)$, which is guaranteed e.g. for finite host graphs. However, the used rules can easily be relaxed to handle also infinite match sets $\mathcal{T}(f, m)$.

[11] Steps in Fig. 2 are i.a. labeled with the satisfaction judgement τ for the derived ET.

$$\text{NOT} \quad \frac{\phi \xrightarrow{em,\tau,m} \gamma}{\neg\phi \xrightarrow{em,\neg\tau,m} \neg\gamma} \, b \neq \mathsf{U} \qquad\qquad \text{UNEVALUATED} \quad \frac{}{\phi \xrightarrow{em,\mathsf{U},m} \mathsf{U}(\phi)}$$

$$\text{OR} \quad \frac{\forall i \in [1,n].\ \phi_i \xrightarrow{em,w_i,m} \gamma_i}{\vee(\phi_1,\ldots,\phi_n) \xrightarrow{em,\tau,m} \vee(\gamma_1,\ldots,\gamma_n)} \, (w,\tau) \in \mathcal{S}(em,n)$$

$$\text{EXISTS} \quad \frac{\forall i \in \{i \mid w_i = \mathsf{T} \vee w_i = \mathsf{F}\}.\ \phi \xrightarrow{em,w_i,m_i} \gamma_i}{\exists(f,\phi) \xrightarrow{em,\tau,m} \exists(f,\phi,m_\mathsf{T},m_\mathsf{F})} \quad \begin{array}{l} \mathcal{T}(f,m) = \{m_1,\ldots,m_n\} \\ (w,\tau) \in \mathcal{S}(em,n) \\ m_\mathsf{T} = \{m_i \mapsto \gamma_i \mid w_i = \mathsf{T}\} \\ m_\mathsf{F} = \{m_i \mapsto \gamma_i \mid w_i = \mathsf{F}\} \end{array}$$

Fig. 2. Rules for operation eval from Definition 8 generating an ET for a match. The rules are given in structural operational semantics (SOS) notation. The name, the optional side-conditions, the optional premises (statements on the defined relation), and the conclusion (a statement on the defined relation) are given left, right, above, and below the horizontal line.

$$\text{NOT} \quad \frac{\phi \xrightarrow{em,\tau} \xi}{\neg\phi \xrightarrow{em,\neg\tau} \neg\xi} \, \tau \neq \mathsf{U} \qquad\qquad \text{UNEVALUATED} \quad \frac{}{\phi \xrightarrow{em,\mathsf{U}} \mathsf{U}(\phi)}$$

$$\text{OR} \quad \frac{\forall i \in [1,n].\ \phi_i \xrightarrow{em,w_i} \xi_i}{\vee(\phi_1,\ldots,\phi_n) \xrightarrow{em,\tau} \vee(\xi_1,\ldots,\xi_n)} \, (w,\tau) \in \mathcal{S}(em,n) \qquad \text{EXISTS0} \quad \frac{}{\exists(f,\phi) \xrightarrow{em,\mathsf{F}} \exists_0(f,\phi)}$$

$$\text{EXISTSF} \quad \frac{\phi \xrightarrow{em,\mathsf{F}} \xi}{\exists(f,\phi) \xrightarrow{em,\mathsf{F}} \exists_\mathsf{F}(f,\phi,\xi)} \qquad\qquad \text{EXISTST} \quad \frac{\phi \xrightarrow{em,\mathsf{T}} \xi}{\exists(f,\phi) \xrightarrow{em,\mathsf{T}} \exists_\mathsf{T}(f,\phi,\xi)}$$

Fig. 3. Rules for operation gc2ep from Definition 12 generating an Evaluation Pattern (EP) for a GC

$$\text{NOT} \quad \frac{\xi \xrightarrow{\tau,m} \gamma}{\neg\xi \xrightarrow{\neg\tau,m} \neg\gamma} \qquad\qquad \text{UNEVALUATED} \quad \frac{}{\mathsf{U}(\phi) \xrightarrow{\mathsf{U},m} \mathsf{U}(\phi)}$$

$$\text{OR} \quad \frac{\forall i \in [1,n].\ \xi_i \xrightarrow{\tau_i,m} \xi_i}{\vee(\xi_1,\ldots,\xi_n) \xrightarrow{\tau,m} \vee(\gamma_1,\ldots,\gamma_n)} \, \tau = (\exists i.\ \tau_i = \mathsf{T})$$

$$\text{EXISTS0} \quad \frac{}{\exists_0(f,\phi) \xrightarrow{\mathsf{F},m} \exists(f,\phi,\emptyset,\emptyset)} \, \mathcal{T}(f,m) = \emptyset$$

$$\text{EXISTSF} \quad \frac{\xi' \xrightarrow{\mathsf{F},m_k} \gamma_k \text{ and } \nexists i \in [1,n] - \{k\}.\ m_i \models \phi}{\exists_\mathsf{F}(f,\phi,\xi') \xrightarrow{\mathsf{F},m} \exists(f,\phi,\emptyset,\{(m_k,\gamma_k)\})} \, \begin{array}{l} \mathcal{T}(f,m) = \{m_1,\ldots,m_n\} \\ k \in [1,n] \end{array}$$

$$\text{EXISTST} \quad \frac{\xi' \xrightarrow{\mathsf{T},m_k} \gamma_k}{\exists_\mathsf{T}(f,\phi,\xi') \xrightarrow{\mathsf{T},m} \exists(f,\phi,\{(m_k,\gamma_k)\},\emptyset)} \, \begin{array}{l} \mathcal{T}(f,m) = \{m_1,\ldots,m_n\} \\ k \in [1,n] \end{array}$$

Fig. 4. Rules for operation ep2et from Definition 14 generating an ET for an EP

Note that the ESs eval(SCLR) and eval(SC) do not always return diverse sets of ETs already (e.g. $\exists(a, \top)$ leads to a non-diverse set of two ETs for **a1 a2**). Moreover, in the use cases discussed in the introduction, no concrete graph G is given as required for a construction of ETs using one of these ESs.

We now state our main synthesis problem to be solved subsequently. In particular, we rely on diversity to rule out superfluous ETs, compactness to rule out unnecessarily large underlying graphs, and completeness to ensure that all possible evaluations are covered (thereby covering evaluations demonstrating satisfaction as well as non-satisfaction).

Definition 10 (Synthesis Problem). *Given an ES* es *and a GC* ϕ, *construct for* ϕ *some set* S *of ETs (for possibly varying graphs) such that* S *is finite,* S *is diverse,* S *contains only compact ETs, and* S *is complete in the sense that each ET derivable using* es *contains some ET from* S.

Note that any solution S to this problem is unique.

Lemma 1 (Uniqueness of Solutions). *Any solution* S *to the synthesis problem from Definition 10 is unique up to isomorphism of the underlying graphs.*

Proof. Let S' be another solution and let w.l.o.g. γ_1 be an ET from S for which no ET $\gamma_2 \in S$ has an isomorphic underlying graph. By correctness of S' there is some $\gamma_3 \in S'$ with $\gamma_3 \leq \gamma_1$. Then either $\gamma_1 \leq \gamma_3$ or $\gamma_3 < \gamma_1$. If $\gamma_1 \leq \gamma_3$, then both underlying graphs are isomorphic since γ_1 and γ_3 are compact implying a contradiction. If $\gamma_3 < \gamma_1$, then there is some $\gamma_4 \in S$ with $\gamma_4 \leq \gamma_3$ implying $\gamma_4 \leq \gamma_1$, which contradicts diversity of S.

We now present our approach to solve the presented synthesis problem.

As a first step, we define EPs, which describe abstractly evaluations for a given GC. For disjunctions, it contains for each sub-GC an EP, which is U(ϕ) for the case of an unevaluated sub-GC ϕ. For existential quantifications, it may state that during evaluation for some match m (a) no match $m' \in \mathcal{T}(f, m)$ can be found using the $\exists_0$ operator, (b) that the GC is satisfied since some match $m' \in \mathcal{T}(f, m)$ can be found satisfying the sub-GC along the lines of the sub-EP using the $\exists_\top$ operator, or (c) that the GC is not satisfied and that some match $m' \in \mathcal{T}(f, m)$ can be found not-satisfying the sub-GC along the lines of the sub-EP using the $\exists_F$ operator.[12]

Definition 11 (Evaluation Patterns (EPs)). *If* H *is a graph, then* ξ *is an evaluation pattern (EP) over* H, *written* $\xi \in \mathsf{EP}(H)$, *if one of the following items applies.*

- $\xi = \neg\xi'$ *and* $\xi' \in \mathsf{EP}(H)$.
- $\xi = \vee(\xi_1, \ldots, \xi_n)$ *and* $\forall i \in [1, n]. \ \xi_i \in \mathsf{EP}(H) \cup \{\mathsf{U}(\phi) \mid \phi \in \mathsf{GC}(H)\}$.
- $\xi = \exists_0(f : H \hookrightarrow H', \phi)$ *and* $\phi \in \mathsf{GC}(H')$.

---

[12] Note that for (c), there may be in general multiple ways as to why a sub-GC is not satisfied and fixing one particular reason for all matches would not allow for a set of ETs that is complete in the sense of Definition 10.

- $\xi = \exists_\mathsf{T}(f : H \hookrightarrow H', \phi, \xi')$, $\phi \in \mathsf{GC}(H')$, and $\xi' \in \mathsf{EP}(H')$.
- $\xi = \exists_\mathsf{F}(f : H \hookrightarrow H', \phi, \xi')$, $\phi \in \mathsf{GC}(H')$, and $\xi' \in \mathsf{EP}(H')$.

We now define the operation gc2ep for constructing a set of EPs for a given EM and GC. The parameter τ captures the satisfaction judgement that would result in evaluations following the generated EP and is required for the rules EXISTSF and EXISTST to construct suitable sub-EPs. Note that this operation also depends on the given EM for the case of disjunction.

Definition 12 (Operation gc2ep). *If* $\mathfrak{em} \in \{\mathsf{AC}, \mathsf{SCLR}, \mathsf{SC}\}$ *is an evaluation mode,* ϕ *is a GC over* H*,* ξ *is an EP over* H*,* $\tau \neq \mathsf{U}$*, and* $\phi \xrightarrow{\mathfrak{em},\tau} \xi$ *via the rules from Fig. 3, then* ξ *is an evaluation pattern for* $\mathfrak{em}$ *and* ϕ*, written* $\xi \in \mathsf{gc2ep}(\mathfrak{em}, \phi)$*.*

For our running example, gc2ep returns the EPs given in Table 1a for the GC from Fig. 1c where, for improved readability, sub-GCs are replaced by the symbol $-$. For instance, (a) the EP ξ_4 states that a satisfying match for a must be found such that matches for $a \rightleftharpoons^1$ and $a^{-1} \rightarrow b$ can be found and (b) the EP ξ_7 states that only non-satisfying matches for a are found where at least one of them cannot be extended to a match of $a \rightleftharpoons^1$ but to a match of $a^{-1} \rightarrow b$.

As a next step, we determine using the operation induced a GC for a given EP. For each nesting level of existential quantification, we basically obtain a valuation for the next-level existential quantifications as to whether a match can be found for them. Such a valuation is given as a conjunction of positive/negative literals of the form $\exists(f, \phi)/\neg\exists(f, \phi)$. Also, EPs of the form $\mathsf{U}(\phi')$ inside a disjunction do not contribute to the valuation (technically, they are replaced by $\top$ disappearing in the resulting conjunction).

Definition 13 (Operation induced). *If* ξ *is an EP over a graph* H*, then* ϕ *is the GC over* H *induced by* ξ*, written* $\mathsf{induced}(\xi) = \phi$*, if one of the following items applies.*

- $\xi = \neg\xi'$ *and* $\phi = \mathsf{induced}(\xi')$.
- $\xi = \vee(\xi_1, \ldots, \xi_n)$, $\phi = \wedge(\phi_1, \ldots, \phi_n)$, *and* $\forall i \in [1, n]. \, \phi_i = \mathsf{induced}(\xi_i)$.
- $\xi = \exists_0(f, \phi')$ *and* $\phi = \neg\exists(f, \top)$.
- $\xi = \exists_\mathsf{T}(f, \phi', \xi')$ *and* $\phi = \exists(f, \mathsf{induced}(\xi'))$.
- $\xi = \exists_\mathsf{F}(f, \phi', \xi')$ *and* $\phi = \exists(f, \mathsf{induced}(\xi'))$.
- $\xi = \mathsf{U}(\phi')$ *and* $\phi = \top$.

For our running example, induced returns the GCs given in Table 1b for the EPs from Table 1a where ϕ'_i is obtained from ξ_i. While induced does not depend on an EM, it is applied later on only for the EPs constructed for the EM at hand.

As a next step, we resolve the induced GCs representing nested valuations by applying divGraphs to obtain for each induced GC a sound, complete, and diverse set of graphs according to Fact 1. Note that EPs may not be realizable in the sense that their induced GC is not satisfiable e.g. the GC $\exists(a, \top) \wedge \exists(a, \top)$ results in the EP $\exists_\mathsf{T}(a, \top, \top) \wedge \exists_0(a, \top)$, which induces the GC $\exists(a, \top) \wedge \neg\exists(a, \top)$

Table 1. Intermediate and final results for running example

Evaluation Pattern	Evaluation Modes		
	AC	SCLR	SC
$\xi_1 = \neg\exists_0(a, -)$	✔	✔	✔
$\xi_2 = \neg\exists_T(a, -, \exists_T(a\overset{1}{\leftarrow}, -, T) \vee \neg\exists_0(a^{-1}\!\!\rightarrow b, -))$	✔	✘	✘
$\xi_3 = \neg\exists_T(a, -, \exists_T(a\overset{1}{\leftarrow}, -, T) \vee U(\neg\exists(a^{-1}\!\!\rightarrow b, T)))$	✘	✔	✔
$\xi_4 = \neg\exists_T(a, -, \exists_T(a\overset{1}{\leftarrow}, -, T) \vee \neg\exists_T(a^{-1}\!\!\rightarrow b, -, T))$	✔	✘	✘
$\xi_5 = \neg\exists_T(a, -, \exists_0(a\overset{1}{\leftarrow}, -) \vee \neg\exists_0(a^{-1}\!\!\rightarrow b, -))$	✔	✔	✘
$\xi_6 = \neg\exists_T(a, -, U(\exists(a\overset{1}{\leftarrow}, T)) \vee \neg\exists_0(a^{-1}\!\!\rightarrow b, -))$	✘	✘	✔
$\xi_7 = \neg\exists_F(a, -, \exists_0(a\overset{1}{\leftarrow}, -) \vee \neg\exists_T(a^{-1}\!\!\rightarrow b, -, T))$	✔	✔	✔

(a) EPs generated using gc2ep from Def. 12 for running example

$$\phi'_1 = \neg\exists(a, T)$$
$$\phi'_2 = \exists(a, \exists(a\overset{1}{\leftarrow}, T) \wedge \neg\exists(a^{-1}\!\!\rightarrow b, T))$$
$$\phi'_3 = \exists(a, \exists(a\overset{1}{\leftarrow}, T))$$
$$\phi'_4 = \exists(a, \exists(a\overset{1}{\leftarrow}, T) \wedge \exists(a^{-1}\!\!\rightarrow b, T))$$
$$\phi'_5 = \exists(a, \neg\exists(a\overset{1}{\leftarrow}, T) \wedge \neg\exists(a^{-1}\!\!\rightarrow b, T))$$
$$\phi'_6 = \exists(a, \neg\exists(a^{-1}\!\!\rightarrow b, T))$$
$$\phi'_7 = \exists(a, \neg\exists(a\overset{1}{\leftarrow}, T) \wedge \exists(a^{-1}\!\!\rightarrow b, T))$$

(b) GCs generated using induced from Def. 13 for running example

Graphs	Evaluation Trees
G_1 ∅	
G_2 $a\overset{1}{\leftarrow}$	
G_3 $a\overset{1}{\leftarrow}$	
G_4 $1\!\!\rightarrow a\text{-}2\!\!\rightarrow b$	
G_5 a	
G_6 a	
G_7 $a^{-1}\!\!\rightarrow b$	

(c) Pairs of graphs and ETs generated using ep2et from Def. 14 for running example for which divGraphs returns the empty set. For our running example, divGraphs returns only singleton sets (because only for ϕ'_4 two graphs are overlapped even resulting in this case in a single overlapping) of graphs given in Table 1c (left column) where G_i is the single graph obtained for ϕ'_i.

As a last construction step, we derive ETs γ using the operation ep2et for an EP ξ and one of the graphs G obtained for it. This operation is similar to an ES since it constructs an ET for a mono (the initial morphism i(G)) and a GC (implicitly contained in ξ). However, it is no ES since it depends on the given EP restricting the ETs to be constructed and does not depend on an EM for that purpose. Also note that the constructed ETs do not satisfy the completeness property of ESs since for a sub-GC that is not satisfied as specified using an EP $\exists_\mathsf{F}(f, \phi', \xi')$ only a single mono is recorded in the m_F component.

Definition 14 (Operation ep2et). *If* $m : H \hookrightarrow G$ *is a mono,* ξ *is an EP over* H, $\tau \neq \mathsf{U}$, *and* $\xi \xrightarrow{\tau,m} \gamma$ *via the rules from Fig. 4, then* γ *is an ET for* m *and* ξ, *written* $\gamma \in$ ep2et(m, ξ).

For our running example, ep2et returns the ETs given in Table 1c (right column) where γ_i is obtained from ξ_i and G_i. Below of Definition 2, Reason 1 for non-satisfaction covered the ETs γ_6 and Reason 2 for non-satisfaction covered the ETs γ_3. Finally, note that all seven ETs are not mutually contained in each other and that the ETs γ_4 and γ_5 are contained in the ET from Fig. 1d (matching i.a. a to a2 and a1, respectively).

The operation divGraphs does not explain why a certain graph is returned in terms of connecting the graph elements in such a returned graph with the graph elements occurring in graphs from the given GC. As a consequence, there may be an ambiguity when constructing an ET for such a returned graph using ep2et resulting in sets of ETs that are not diverse. For example, the GC $\exists(a1 \ a2, \top)$ results in the EP $\exists_\top(a1 \ a2, \top, \top)$, which induces the GC $\exists(a1 \ a2, \top)$ for which divGraphs only returns the graph a1 a2. For this graph, ep2et then constructs two ETs γ_1 and γ_2 in which $a1$ and $a2$ are mapped to a1 and a2 in γ_1 and to a2 and a1 in γ_2. Hence, due to this simple swapping, these two ETs are mutually contained in each other via $\leq$. To solve this problem, we incrementally remove ETs violating diversity once the ETs have been constructed.

Definition 15 (Generation of ETs). *If* em $\in \{\mathsf{AC}, \mathsf{SCLR}, \mathsf{SC}\}$ *is an EM, and* ϕ *is a GC over* $\emptyset$, *then* genEvaluations(em)(ϕ) *is some largest diverse set of ETs contained in* $\{\gamma \mid \xi \in$ gc2ep$(\mathsf{em}, \phi), \phi' =$ induced$(\xi), G \in$ divGraphs(ϕ'), $\gamma \in$ ep2et$(\mathsf{i}(G), \xi)\}$.

Note that genEvaluations(em)(ϕ) in fact defines a unique result due to Lemma 1.

To summarize for our running example, genEvaluations constructs the ETs from Table 1c (right column) as follows: AC leads to $\{\gamma_1, \gamma_2, \gamma_4, \gamma_5, \gamma_7\}$, SCLR leads to $\{\gamma_1, \gamma_3, \gamma_5, \gamma_7\}$, and SC leads to $\{\gamma_1, \gamma_3, \gamma_6, \gamma_7\}$.

Finally, we state that the presented operation genEvaluations(em) correctly constructs the desired set of diverse ETs.

Theorem 2 (Solution to Synthesis Problem). *If* em $\in \{AC, SCLR, SC\}$ *is an EM, then* genEvaluations(em) *solves the synthesis problem from Definition 10 for* eval(em).

Proof.

- genEvaluations(em)(ϕ) is a set of ETs by construction.
- FINITENESS OF genEvaluations(em)(ϕ): each of the computation steps produces at most a finite number of results.
- DIVERSITY OF genEvaluations(em)(ϕ): due to the last step of constructing the maximal diverse subset.
- COMPACTNESS OF ETS IN genEvaluations(em)(ϕ): fix some ET γ over some $m : H \hookrightarrow G$. Each graph element in the host graph G is created to find a certain mono during the construction of γ. If the mono actually constructed later on using ep2et matches a different subgraph, then the host graph G would not have been minimal contradicting diversity of the result of divGraphs.
- COMPLETENESS OF genEvaluations(em)(ϕ): fix em $\in \{AC, SCLR, SC\}$. Fix $\gamma \in$ eval(em) over some $m : \emptyset \hookrightarrow G$. Construct the EP ξ from γ by replacing $\exists(f, \phi, m_T, m_F)$ by (a) $\exists_0(f, \phi)$ when $m_T = m_F = \emptyset$, (b) $\exists_T(f, \phi, \xi')$ when $(m, \gamma') \in m_T$ and ξ' is obtained by γ', and (c) $\exists_F(f, \phi, \xi')$ when $m_T = \emptyset$ and $(m, \gamma') \in m_F$ and ξ' is obtained by γ'. Note that there may be different EPs here, which is to be expected since multiple generated ETs may be contained in some given ET obtained using eval(em). The constructed EP ξ would be constructed as well using gc2ep(em, ϕ). The resulting induced GC $\phi' =$ induced(ξ) for which divGraphs(ϕ') returns the graph G' containing all elements matched by γ. Then, ep2et(m, ξ) constructs an ET γ' that matches elements in G' as γ matches elements in G. Obviously, γ' is compact since G' was constructed to contain only the matched elements. Lastly, even if some other ET γ'' would be contained in γ' resulting in its removal, γ'' could be used as the witness to be determined instead of γ'.

However, divGraphs may not terminate when applying genEvaluations, which only allows to generate ETs gradually as graphs are gradually generated by divGraphs in this case. Also, a set of two ETs constructed for different EPs is always diverse. Hence, even when divGraphs does not terminate for all induced GCs, the ETs constructed for induced GCs where divGraphs terminates are all ETs to be actually returned by genEvaluations.

6   Related Work

We are unaware of other approaches attempting to derive complete sets of diverse evaluations. However, there is a plethora of existing *graph* generation approaches such as [16, 26, 28]. In these approaches, quantitative diversity measures [8, 25, 27] (capturing diversity of a set of graphs using a numerical value) are employed as a proxy of quality but the best measure to be used is difficult to determine in advance. In [16], the user must provide graph transformation rules, which are then used to generate graphs. Similarly, in [26] the user provides a graph in which possible graph transformations are encoded by annotating elements suitably. In [28], graphs representing graph database instances are generated by using distributions describing the likelihood of nodes and edges of certain types. Moreover, in [13,14], models are generated using logic-based (for relational models) and algebraic techniques (for specific graphs). Similarly, the ALLOY tool [12] generates models of a relational specification using a bounded model size assumption, which can be understood as graphs. A common challenge for these approaches is to minimize the number of computations not leading to outputs. For example (a) when the set of model candidates is much larger than the set of actual models satisfying all constraints in concrete settings or (b) when a diverse subset is to be obtained by generating a large number of graphs and then selecting a subset of it that is optimal w.r.t. diversity. In contrast, our own model generation procedure from [23] generates a complete finite set of diverse graphs for a given GC. On the one hand, the graphs generated by this procedure are not as detailed as the ETs computed here and the graph generation procedure focuses only on the GC semantics returning the same graphs for equivalent GCs and does not always take the entire GC into account possibly returning the same graphs when sub-GCs are replaced. On the other hand, as for the generated ETs in our present approach, the GC itself characterizes qualitative diversity as a proxy.

In white-box software testing, code coverage is considered. In particular, propositional logic-based coverage criteria such as MCDC/ACC [1,4] are used e.g. in avionics and automotive to describe specific sets of evaluations of propositional logic formulas. That is, they describe tests that are to be constructed for such a formula in the form of valuations of its atomic propositions. For example, predicate coverage requires a satisfying and a non-satisfying valuation, combinatorial coverage requires all combinations of valuations of atomic propositions, while MCDC/ACC criteria require that each atomic proposition affects the outcome once. In our approach, we extend ACC to the case of nesting in GCs but restrict valuations (given by induced GCs) of atomic propositions (given by existential quantifications) to those actually evaluated, covering therefore also short or shortest circuit evaluations. Similar to propositional logic coverage, tool support for generating tests covering nested control flow graphs have been discussed in [6] but are limited due to the expressiveness of considered programming languages.

7   Conclusion and Future Work

We have introduced a generation procedure, as implemented in the tool AUTO-GRAPH, taking a Graph Condition (GC) as input and generating a complete finite set of diverse evaluations given by Evaluation Trees (ETs). The returned ETs provide a suitable finite overview of the semantics of the GC capturing the evaluation steps of e.g. manual proofs or of standard implementations using different Evaluation Strategies (ESs). The generated ETs are complete by symbolically capturing all possible evaluations of the given GC against any possible graph while also providing concrete evaluations for small graphs constructed along with them. Moreover, the returned ETs are diverse by describing sufficiently different evaluations. Our ET generation procedure hence improves support for the use cases of validation, testing, debugging, and repair of GCs and their evaluation.

Besides applying our proposed approach to continue our work on model based testing of graph databases from [15] and on graph repair w.r.t. consistency constraints from [22], we are working on techniques (a) to determine whether an ES is able to execute a particular ET, (b) to generate even more fine-grained evaluations incorporating the process of pattern matching when required, and (c) to systematically generate larger ETs for a given GC. Moreover, we want to extend our approach for the use cases of validation, testing, debugging, and repair of graph transformation rules [7] and their application. Lastly, our approach may be integrated with e.g. OCL using existing translations [20] between OCL and graph conditions.

References

1. Ammann, P., Offutt, A.J., Huang, H.: Coverage criteria for logical expressions. In: 14th International Symposium on Software Reliability Engineering (ISSRE 2003), Denver, CO, USA, 17–20 November 2003, pp. 99–107. IEEE Computer Society (2003). https://doi.org/10.1109/ISSRE.2003.1251034
2. Barkowsky, M., Giese, H.: Hybrid search plan generation for generalized graph pattern matching. In: Guerra, E., Orejas, F. (eds.) ICGT 2019. LNCS, vol. 11629, pp. 212–229. Springer, Cham (2019). https://doi.org/10.1007/978-3-030-23611-3_13
3. Beyhl, T., Blouin, D., Giese, H., Lambers, L.: On the operationalization of graph queries with generalized discrimination networks. In: Echahed, R., Minas, M. (eds.) ICGT 2016. LNCS, vol. 9761, pp. 170–186. Springer, Cham (2016). https://doi.org/10.1007/978-3-319-40530-8_11
4. Chilenski, J.J., Miller, S.P.: Applicability of modified condition/decision coverage to software testing. Softw. Eng. J. **9**(5), 193–200 (1994). https://doi.org/10.1049/sej.1994.0025
5. Courcelle, B.: The expression of graph properties and graph transformations in monadic second-order logic. In: Rozenberg, G. (ed.) Handbook of Graph Grammars and Computing by Graph Transformations, Volume 1: Foundations, pp. 313–400. World Scientific (1997). ISBN 9810228848
6. Cseppento, L., Micskei, Z.: Evaluating code-based test input generator tools. Softw. Test. Verification Reliab. **27**(6) (2017). https://doi.org/10.1002/stvr.1627
7. Ehrig, H., Ehrig, K., Prange, U., Taentzer, G.: Fundamentals of Algebraic Graph Transformation. MTCSAES. Springer, Heidelberg (2006). https://doi.org/10.1007/3-540-31188-2
8. Feldt, R., Poulding, S.M., Clark, D., Yoo, S.: Test set diameter: quantifying the diversity of sets of test cases. In: 2016 IEEE International Conference on Software Testing, Verification and Validation, ICST 2016, Chicago, IL, USA, 11–15 April 2016, pp. 223–233. IEEE Computer Society (2016). ICST.2016.33

9. Ghahremani, S., Giese, H., Vogel, T.: Efficient utility-driven self-healing employing adaptation rules for large dynamic architectures. In: Wang, X., Stewart, C., Lei, H. (eds.) 2017 IEEE International Conference on Autonomic Computing, ICAC 2017, Columbus, OH, USA, 17–21 July 2017, pp. 59–68. IEEE Computer Society (2017). https://doi.org/10.1109/ICAC.2017.35. ISBN 978-1-5386-1762-5

10. Habel, A., Pennemann, K.: Correctness of high-level transformation systems relative to nested conditions. Math. Struct. Comput. Sci. **19**(2), 245–296 (2009). https://doi.org/10.1017/S0960129508007202

11. Hochgeschwender, N., Schneider, S., Voos, H., Bruyninckx, H., Kraetzschmar, G.K.: Graph-based software knowledge: storage and semantic querying of domain models for run-time adaptation. In: 2016 IEEE International Conference on Simulation, Modeling, and Programming for Autonomous Robots, SIMPAR 2016, San Francisco, CA, USA, 13–16 December 2016, pp. 83–90. IEEE (2016). https://doi.org/10.1109/SIMPAR.2016.7862379

12. Jackson, D., Milicevic, A., Torlak, E., Kang, E., Near, J.: Alloy: a language & tool for relational models (2017). https://alloytools.org/. Accessed 05 July 2020

13. Kakita, S., Watanabe, Y., Densmore, D., Davare, A., Sangiovanni-Vincentelli, A.L.: Functional model exploration for multimedia applications via algebraic operators. In: Sixth International Conference on Application of Concurrency to System Design (ACSD 2006), 28–30 June 2006, Turku, Finland, pp. 229–238. IEEE Computer Society (2006). https://doi.org/10.1109/ACSD.2006.8

14. Kang, E., Jackson, E.K., Schulte, W.: An approach for effective design space exploration. In: Calinescu, R., Jackson, E. (eds.) Monterey Workshop 2010. LNCS, vol. 6662, pp. 33–54. Springer, Heidelberg (2011). https://doi.org/10.1007/978-3-642-21292-5_3

15. Lambers, L., Schneider, S., Weisgut, M.: Model-based testing of read only graph queries. In: 13th IEEE International Conference on Software Testing, Verification and Validation Workshops, ICSTW 2020, Porto, Portugal, 24–28 October 2020, pp. 24–34. IEEE (2020). https://doi.org/10.1109/ICSTW50294.2020.00022

16. Nassar, N., Kosiol, J., Kehrer, T., Taentzer, G.: Generating large EMF models efficiently. FASE 2020. LNCS, vol. 12076, pp. 224–244. Springer, Cham (2020). https://doi.org/10.1007/978-3-030-45234-6_11

17. Neo4J Team. Neo4J. http://neo4j.com/

18. Object Management Group. Unified Modeling Language. https://www.uml.org/

19. OMG. Object Constraint Language (2014). http://www.omg.org/spec/OCL/

20. Radke, H., Arendt, T., Becker, J.S., Habel, A., Taentzer, G.: Translating essential OCL invariants to nested graph constraints focusing on set operations. In: Parisi-Presicce, F., Westfechtel, B. (eds.) ICGT 2015. LNCS, vol. 9151, pp. 155–170. Springer, Cham (2015). https://doi.org/10.1007/978-3-319-21145-9_10

21. Schneider, S.: AutoGraph. https://github.com/schneider-sven/AutoGraph

22. Schneider, S., Lambers, L., Orejas, F.: A logic-based incremental approach to graph repair featuring delta preservation. Int. J. Softw. Tools Technol. Transf. (2020, accepted)

23. Schneider, S., Lambers, L., Orejas, F.: Automated reasoning for attributed graph properties. Int. J. Softw. Tools Technol. Transf. **20**(6), 705–737 (2018). https://doi.org/10.1007/s10009-018-0496-3

24. Schneider, S., Lambers, L., Orejas, F.: Symbolic model generation for graph properties. In: Huisman, M., Rubin, J. (eds.) FASE 2017. LNCS, vol. 10202, pp. 226–243. Springer, Heidelberg (2017). https://doi.org/10.1007/978-3-662-54494-5_13

25. Semeráth, O., Farkas, R., Bergmann, G., Varró, D.: Diversity of graph models and graph generators in mutation testing. Int. J. Softw. Tools Technol. Transf. **22**(1), 57–78 (2020). https://doi.org/10.1007/s10009-019-00530-6

26. Semeráth, O., Nagy, A.S., Varró, D.: A graph solver for the automated generation of consistent domain-specific models. In: Chaudron, M., Crnkovic, I., Chechik, M., Harman, M. (eds.) Proceedings of the 40th International Conference on Software Engineering, ICSE 2018, Gothenburg, Sweden, 27 May–03 June 2018, pp. 969–980. ACM (2018). https://doi.org/10.1145/3180155.3180186

27. Syriani, E., Bill, R., Wimmer, M.: Domain-specific model distance measures. J. Object Technol. **18**(3), 3:1–3:19 (2019). https://doi.org/10.5381/jot.2019.18.3.a3

28. The Linked Data Benchmark Council (LDBC). Datagen. Accessed 04 May 2021

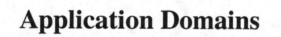

Application Domains

Host-Graph-Sensitive RETE Nets for Incremental Graph Pattern Matching

Matthias Barkowsky[✉] and Holger Giese

Hasso-Plattner Institute at the University of Potsdam,
Prof.-Dr.-Helmert-Str. 2-3, 14482 Potsdam, Germany
{matthias.barkowsky,holger.giese}@hpi.de

Abstract. Efficient querying of large graph structures is a problem at
the heart of several application domains such as social networks and
model driven engineering. In particular in the context of model driven
engineering, where the same query is executed frequently over an evolving graph structure, incremental techniques based on discrimination networks such as RETE nets are a popular solution. However, the construction of adequate RETE nets for a specific problem instance is a challenge
in and of itself. In this paper, we propose an approach to RETE net construction for queries in the form of simple graph patterns that considers
not only the structure of the query, but also the structure of the graph the
query is being executed over in order to improve the net's performance
with respect to execution time and memory consumption. Furthermore,
we suggest a technique for adapting the net structure to changing characteristics of the underlying graph. We evaluate the presented concepts
empirically based on queries and data from two independent benchmarks.

1 Introduction

Efficient querying of large graph structures is a challenge at the heart of several
application domains such as social networks and model driven engineering [3].
As it relates to the subgraph homomorphism problem, which is NP-complete
for arbitrary inputs [14], this problem is mostly tackled by heuristic approaches
such as local search based on heuristically computed search plans [5,11,22]. In
the context of model driven engineering, model-sensitive techniques for search
plan generation have been shown to outperform approaches that do not consider
information about the queried model for typical application scenarios [22].

However, model driven engineering often deals with evolving models over
which the same query needs to be executed frequently [3]. Local-search-based
approaches are not suitable for handling such situations, since they have to
recompute query results from scratch every time the model is queried, even
though the (potentially large) model undergoes only minor, incremental changes

This work was developed mainly in the course of the project modular and incremental Global Model Management (project number 336677879), which is funded by the
Deutsche Forschungsgemeinschaft.

F. Gadducci and T. Kehrer (Eds.): ICGT 2021, LNCS 12741, pp. 145–163, 2021.
https://doi.org/10.1007/978-3-030-78946-6_8

in between queries. Generalized Discrimination Networks [13] and in particular RETE nets [10], which maintain data structures that store intermediate results and can be updated incrementally, present a more appropriate solution to this problem. Similar to search plans employed for local search, the topology of the employed RETE nets has a significant performance impact.

During its evolution, a queried model's characteristics may change significantly, leading to deteriorating performance of a previously adequate discrimination network. This is particularly pronounced in the case where the evolution starts with an empty model, which does not yet expose any meaningful structural information that could be used for computing a suitable network structure. In contrast to techniques based on local search, current approaches for incremental graph querying therefore construct the employed discrimination network at design time of the query [20,21]. Hence, the construction process does not incorporate information about the queried model.

In this paper, we introduce a heuristic, model-sensitive technique for RETE net construction for graph queries in the form of simple graph patterns. Our technique is based on a cost function, which takes information about the queried model into account. We then propose an approach for dynamically adapting the employed net's topology as the model evolves, which allows tailoring the RETE net's structure to potentially changing model characteristics. We evaluate our approach using queries and generated data from the LDBC Social Network Benchmark [8] and the Train Benchmark [18].

The remainder of the paper is structured as follows: We first reiterate the notion of graphs, graph queries and RETE nets in Sect. 2. We then introduce our technique for constructing model-sensitive RETE nets in Sect. 3 and discuss an approach for reacting to host graph evolution in Sect. 4. In Sect. 5, we evaluate the effectiveness of our technique and compare our implementation to a state-of-the-art tool for incremental graph pattern matching. An overview of related work is given in Sect. 6. Finally, we present our conclusion and planned future work in Sect. 7.

2   Prerequisites

Graphs and Graph Queries. A *graph* G is a tuple $G = (V^G, E^G, s^G, t^G)$, where V^G is the set of vertices, E^G is the set of edges and $s^G : E^G \to V^G$ and $t^G : E^G \to V^G$ are functions assigning each edge its source vertex and target vertex, respectively [7]. A *graph morphism* $f : G \to H$ from G into another graph $H = (V^H, E^H, s^H, t^H)$ is a pair of mappings $f^V : V^G \to V^H$ and $f^E : E^G \to E^H$ such that $f^V \circ s^G = s^H \circ f^E$ and $f^V \circ t^G = t^H \circ f^E$. We call f^V the *vertex morphism* and f^E the *edge morphism*.

The graph G can be typed over a *type graph* $TG = (V^{TG}, E^{TG}, s^{TG}, t^{TG})$ by means of a graph morphism $type^G$, which assigns a type from TG to each element in G, resulting in the *typed graph* $G^T = (G, type^G)$. A *typed graph morphism* $f^T : G^T \to H^T$ between G^T and another typed graph $H^T = (H, type^H)$ comprises a graph morphism $f : G \to H$ such that $type^H \circ f = type^G$.

A simple graph query as considered in this paper is characterized by a typed *query graph Q* and can be executed over a *host graph H* typed over the same type graph. The execution yields a set of typed graph morphisms from Q into H called *matches* and is called *graph pattern matching*. Note that in this paper, we focus on the basic case without advanced features such as attribute constraints.

An example graph query from a social network domain and the corresponding type graph are displayed in Fig. 1 in object and class diagram notation, respectively. The query searches for *Persons* that are interested in a *Tag* attached to a *Post* created by a friend of a friend of that person.

Typed graphs are often encoded as object graphs [1], where vertices are represented by objects, whereas edges are represented by references. Moreover, since given a valid vertex morphism, the edge mappings only need to be enumerated, many approaches to graph pattern matching only compute explicit mappings for vertices and map edges implicitly [2,5,20]. This means that they only compute vertex morphisms such that there exists at least one corresponding edge morphism, which is the approach we adopt in this paper.

Graph pattern matching corresponds to the subgraph homomorphism problem, which is known to be NP-complete [14]. Nonetheless, several techniques exist that achieve acceptable performance in practice. Most of these solutions fall into one of two categories: local search or incremental approaches based on discrimination networks. Techniques based on local search iteratively map the vertices in the query graph to vertices in the host graph. The order in which vertices are being mapped has a substantial impact on the performance of the approach and is determined by a search plan [11,22]. Local-search-based techniques do not store any information about the graph query's execution and repeated execution hence requires a computation of the query's matches from scratch. They are therefore unsuitable in situations where the query is executed frequently over the same host graph with only minor changes between executions.

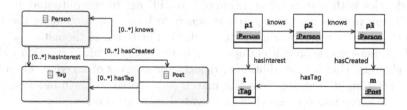

Fig. 1. Example type graph (left) and associated graph query (right)

RETE Nets. These scenarios are tackled more effectively by techniques that decompose the original query into a set of simpler subqueries, which are organized in a directed graph called discrimination network. In the following, we only consider the case where this graph is also acyclic, which is sufficient for graph pattern matching as presented in this section. In a discrimination network, each node is responsible for maintaining the set of matches for one of the subqueries.

Therefore, nodes can make use of the results computed by other nodes of the network, composing their matches into increasingly complex intermediate results, with these dependencies represented as edges in the discrimination network. The intermediate results can be retained after the query's initial execution and thus enable a more efficient search in subsequent executions, where the existing data structures only have to be updated according to the changes to the host graph. While these data structures allow for better query performance with respect to execution time, the storage of intermediate results also results in an overhead in memory consumption compared to querying techniques such as local search.

RETE nets [10] are a special kind of discrimination network, where the query is ultimately decomposed into primitive building blocks, that is, single vertices and edges, and each node in the network has at most two dependencies to other nodes. Common types of nodes used in RETE nets include *input nodes*, *filter nodes*, and *join nodes*. The task of graph pattern matching as presented in Sect. 2 only requires input and join nodes, which is why we focus on these two kinds of nodes in the remainder of the paper.

Input nodes correspond to primitive query subgraphs consisting of a single vertex or edge and are responsible for extracting elements with corresponding type from the host graph. Join nodes combine matches computed by two other nodes representing smaller query subgraphs into matches for a bigger subgraph.

Dependencies between nodes are realized by *indexers*, which store matches produced by the required node in the form of tuples of host graph vertices and provide an index that facilitates efficient access by the dependent node. Another indexer stores the query's result, that is, all matches computed by the node corresponding to the complete query graph. In the following, we will refer to the number of tuples in an indexer as the *size* of that indexer. The size of a RETE net is given by the number of tuples in all indexers of the net.

Using matches stored in indexers corresponding to incoming dependency relations, the execution of a node populates indexers corresponding to outgoing dependencies with matches. Execution of a RETE net to compute matches of a query graph in a specific host graph is hence performed by executing the net's nodes in an order corresponding to a topological ordering of the net.

RETE nets enable an efficient updating of the query result in reaction to host graph changes, that is, the addition or deletion of a vertex or an edge. Therefore, the affected input nodes update the match sets stored in their indexers. The changes to these indexers are then propagated through the net.

Similar to search plans for graph pattern matching based on local search, numerous different RETE net structures exist for the same query graph. Like the search plan in the case of local search, the chosen structure has a significant impact on the RETE net's performance, making it desirable to find an adequate structure for the problem instance at hand.

An example RETE net for the example query in Fig. 1 is displayed in Fig. 2. The net effectively computes all paths $(p1, p2, p3, m, t)$ via successive joins (labelled $\bowtie$) of inputs (labelled $u \rightarrow v$) and finally checks the presence of an edge between $p1$ and t via another join.

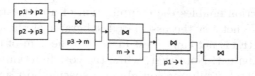

Fig. 2. Example RETE net for the query in Fig. 1

3 Host-Graph-Sensitive RETE Nets

In this section, we introduce our approach to constructing host-graph-sensitive RETE nets. Our technique is based on a heuristic partitioning algorithm, which decomposes a query graph into a RETE net using a cost function based on statistical data extracted from the host graph the query is being executed over.

3.1 Host-Graph-Sensitive Cost Estimation for RETE Nets

The computational effort required for graph pattern matching using a RETE net is mostly constituted of the effort required for propagating intermediate results through the net. This effort heavily depends on the number of such intermediate results, and thus on the number of tuples stored in the net's indexers, which also determines the memory consumption of the RETE net. Hence, the structure of the RETE net should be chosen such that the number of such tuples, which is heavily influenced by the host graph the net is executed over, is minimized.

However, the total number of indexer entries in a RETE net depends on the exact structure of the host graph and hence cannot be determined in advance without effectively executing the net. We therefore employ a cost-estimation function, which takes statistical information about the host graph into account, to estimate the number of tuples stored in an indexer. Because each tuple in an indexer represents a match of the corresponding subgraph of the graph query, the size of an indexer is given by the number of matches of the associated subgraph.

The set of matches $M_H^{Q_1}$ of a subgraph $Q_1 = ((V^{Q_1}, E^{Q_1}, s^{Q_1}, t^{Q_1}), type^{Q_1})$ in a host graph $H = ((V^H, E^H, s^H, t^H), type^H)$ is a subset of the possible vertex morphisms, which corresponds to a cartesian product of (correctly typed) host graph vertices. $dom(Q_1, H) = \prod_{v \in V^{Q_1}} dom^V(v, H)$ hence is an upper bound for $|M_H^{Q_1}|$, where $dom^V(v, H) = |\{w | w \in V^H \wedge type^{Q_1}(v) = type^H(w)\}|$.

Each edge $e \in E^{Q_1}$ with source and target $s, t \in V^{Q_1}$ effectively acts as a filter for vertex morphisms, excluding vertex morphisms from $M_H^{Q_1}$ for which no edge corresponding to e is present in the host graph. The portion of vertex morphisms with a corresponding edge in H is given by $flt^E(e, H) = \frac{dom^E(e,H)}{dom^V(s,H) \cdot dom^V(t,H)}$, where we define dom^E analogously to dom^V.

We compute an estimate for the combined filtering effect of edges in E^{Q_1} by $flt(Q_1, H) = \prod_{e \in E^{Q_1}} flt^E(e, H)$, which allows us to estimate $|M_H^{Q_1}|$ by

$$cost(Q_1, H) := dom(Q_1, H) \cdot flt(Q_1, H). \tag{1}$$

The above function enables the computation of a cost estimate for a single indexer in a RETE net via the associated subgraph of the query graph. The cost of the entire RETE net can be estimated by the sum of the cost estimates of all indexers in the net. Note that, while the cost function is model-sensitive in the sense that it takes information about a specific host graph into account, the estimate does not consider the host graph's exact structure and hence has varying accuracy for different host graphs. In particular, it does not provide any formal guarantees regarding the size of $|M_H^{Q_1}|$. However, it is similar to heuristics employed for local search [11, 22], which have proven useful in practice.

For a host graph conforming to the type graph in Fig. 1 with characteristics displayed on the left in Fig. 3, the number of possible vertex morphisms for the query in Fig. 1 is 2 000 000. The filtering rates of *knows*, *hasCreated*, *hasTag*, and *hasInterest* edges are computed as 0.5, 0.1, 0.1, and 0.25, respectively. The function in Eq. 1 thus estimates the number of matches by 1250.

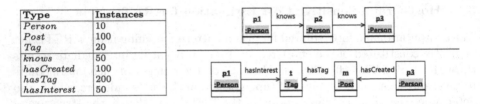

Type	Instances
Person	10
Post	100
Tag	20
knows	50
hasCreated	100
hasTag	200
hasInterest	50

Fig. 3. Sample host graph characteristics for the type graph and example partitioning of the query from Fig. 1

3.2   Heuristic Construction of Host-Graph-Sensitive RETE Nets

All non-trivial intermediate results in a RETE net, that is, all intermediate results other than single edges or vertices, are created by joining two smaller intermediate results. The structure of the RETE net is hence determined by the selection and ordering of joins. The join ordering problem is well known from the domain of relational databases and has been proven to be NP-hard [15]. Computing an optimal structure with respect to the cost function from Sect. 3.1 therefore quickly becomes unfeasible with increasing size of the query graph.

Since all intermediate results, including trivial ones, correspond to matches for a subgraph of the query graph, each join in the RETE net essentially composes two subgraphs of the query graph, with the top-most join producing matches for the entire query graph. Thus, the net corresponds to a recursive, binary partitioning of the query graph. Algorithm 1 outlines a simple partitioning algorithm, which takes a query graph (or subgraph thereof) Q and host graph H as inputs and recursively divides Q into smaller partitions. The returned binary partitioning corresponds to one possible RETE net structure for Q.

In each step of the recursion, Q is partitioned into two proper subgraphs, where each edge is only contained in one of the subgraphs but partitions may share vertices. We employ a greedy strategy based on the cost function from Sect. 3.1, which tries to minimize the combined cost of the two subgraphs in each step.

The algorithm first initializes the two partitions P_1 and P_2, where $P_1 = Q$ and P_2 is the empty graph, and divides Q into weakly connected components. If Q only consists of a single component, it is checked whether Q is already trivial, that is, it only consists of a single edge or vertex. In this case, Q represents a trivial intermediate result and does not require further partitioning.

Otherwise, the procedure `BalanceEdges` shown in Algorithm 2 is called, which tries to improve the partitioning by moving edges from P_1 to P_2. First, the current combined cost of P_1 and P_2 is stored as $cost_{old}$. For each edge e in P_1, the procedure then computes the resulting combined cost estimate if e was moved to P_2 using the cost function from Sect. 3.1. Note that moving e may include (i) removing its source or target from P_1 if they do not have any other adjacent edge in P_1 and (ii) adding its source or target to P_2 if they are not yet contained in P_2. The edge whose movement would result in the lowest cost and the associated cost estimate are stored as e_{min} and $cost_{min}$, respectively. Then, if $cost_{min} < cost_{old}$, e_{min} is moved to P_2 and the process is repeated if at least two edges are left in P_1, ensuring that P_1 is never empty.

The execution of `BalanceEdges` may result in no edges being moved if for none of the edges initially in P_1, a movement improves the current cost estimate. In this case, to ensure P_1 being a proper subgraph of Q, one edge from P_1 is moved using `ForceEdgeMove`. Therefore, we compute a pair of vertices with the largest distance in P_1 and move an edge adjacent to one of these vertices. The intuition behind this heuristic is that the creation of matches for subgraphs containing both these vertices would involve a multitude of joins and is therefore likely to lead to a large number of intermediate results. Moving an edge adjacent to one such vertex facilitates the distribution of the two vertices into different partitions in subsequent steps. Thereby, such subgraphs are avoided as early intermediate results in a derived RETE net. After the call to `ForceEdgeMove`, the partitioning is refined by another call to `BalanceEdges`.

If Q consists of multiple components, the partitioning is instead performed by `BalanceComponents`, which works in a similar way to `BalanceEdges` but moves components instead of edges. `ForceComponentMove` simply moves the individual component with the highest cost. Finally, the resulting subgraphs P_1 and P_2 are further partitioned by recursive calls to `Partition`.

The presented algorithm always terminates, since P_1 and P_2 are always proper subgraphs of Q and hence their size decreases strictly monotonously in each recursive step. The resulting recursion tree, where each node represents the input graph Q of the associated recursive call, corresponds to a correct join structure for Q in the sense that (i) the graph corresponding to an inner node of the tree is always the union of its child nodes' graphs (ii) the topmost node of the tree corresponds to the complete query graph and (iii) each leaf corresponds

Procedure Partition(Q, H)

    Input   : Q: The query graph

                  H: The host graph

    Output : A recursive, binary partitioning of Q

```
1   P₁ ← Q;
2   P₂ ← ∅;
3   C ← GetComponents(Q);
4   if |C| = 1 then
5       if |E^Q| ≤ 1 then
6           return Q;
7       else
8           BalanceEdges(P₁, P₂, H);
9           if P₂ = ∅ then
10              ForceEdgeMove(P₁, P₂, H);
11              BalanceEdges(P₁, P₂, H);
12      end
13  else
14      BalanceComponents(P₁, P₂, H);
15      if P₂ = ∅ then
16          ForceComponentMove(P₁, P₂, H);
17          BalanceComponents(P₁, P₂, H);
18  end
19  return (Partition(P₁, H), Partition(P₂, H));
```

Algorithm 1: Host-graph-sensitive partitioning algorithm for graph queries

Procedure BalanceEdges(P_1, P_2, H)

    Input   : P_1, P_2: The initial partitions

                  H: The host graph

```
1   do
2       cost_old ← cost(P₁, H) + cost(P₂, H);
3       e_min ← null;
4       cost_min ← ∞;
5       foreach e ∈ E^P₁ do
6           cost_e ← cost(P₁ - e, H) + cost(P₂ + e, H);
7           if cost_e < cost_min then
8               e_min ← e;
9               cost_min ← cost_e;
10      end
11      if cost_min < cost_old then
12          P₁ ← P₁ - e_min;
13          P₂ ← P₂ + e_min;
14  while cost_min < cost_old ∧ |E^P₁| > 1;
```

Algorithm 2: Algorithm for partitioning a single weakly connected component using the cost function from Sect. 3.1

to a unique edge or vertex in Q. The algorithm foregoes an optimal solution with respect to the cost function, but achieves polynomial runtime complexity.

Theorem 1. *For query graph $Q = ((V^Q, E^Q, s^Q, t^Q), type^Q)$ and host graph $H = ((V^H, E^H, s^H, t^H), type^H)$, the runtime complexity of Algorithm 1 is in $O(|V^H| + |E^H| + (|V^Q| + |E^Q|) \cdot (|V^Q|^3 + |E^Q|^2))$.*

Proof (Idea). The cost function from Sect. 3.1 for individual query graph elements can be computed by a single scan of H. Weakly connected components in the query can be computed in $O(|V^Q| + |E^Q|)$ and the runtime complexity of BalanceEdges and BalanceComponents are in $O(|E^Q|^2)$ and $O(|V^Q|^2 + |E^Q|)$, respectively. Using the Floyd-Warshall algorithm [9], ForceEdgeMove and ForceComponentMove can be performed in $O(|V^Q|^3 + |E^Q|)$ and $O(|V^Q| + |E^Q|)$, respectively. Because the recursion tree is a full binary tree, it consists of at most $|V^Q| + |E^Q| - 1$ recursive calls. This yields the stated complexity.   □

As the example query from Fig. 1 consists of a single weakly connected component, its partitioning via Algorithm 1 is effectively performed by calling BalanceEdges. The first call to BalanceEdges does not move any elements, as the movement of no individual edge improves the total estimated cost. Assuming that the subsequent execution of ForceEdgeMove moves the *hasCreated* edge, since $(p1, m)$ is a pair of vertices with maximum distance in the query graph, the second call moves the *hasTag* and then the *hasInterest* edge before the greedy strategy achieves no further improvement. Thus, the partitions displayed on the right in Fig. 3 are created and further decomposed recursively.

4 Reacting to Host Graph Evolution

The evolution of a host graph may significantly alter the graph's characteristics, especially when starting with an empty graph, which does not yet exhibit any of the graph's later characteristics. An initially constructed host-graph-sensitive RETE net can hence become inappropriate. Instead of employing the same RETE net structure for the graph's entire lifetime, it therefore makes sense to consider switching to a structure better suited for handling the current graph. However, this requires the computation of a new RETE net structure. Moreover, the new, initially empty RETE net has to be populated with correct intermediate results.

Recomputation and repopulation cause an overhead in execution time, but may improve the net's quality with respect to memory consumption and future runtime. Thus, a strategy for deciding when to trigger recomputation is required. While a fixed periodic strategy is an option, any such approach runs the risk of an inadequate frequency either causing a substantial computational overhead or too long reaction times. The decision can instead be made based on the processing of changes in the RETE net, which is a suitable approach if the overhead should be limited to a constant factor. This can be achieved by tracking the number of tuples added to and removed from the RETE net's indexers since the last recomputation n and the size of the RETE net at that point d, and only triggering recomputation and repopulation when $n > d$.

For a fixed query and given incremental maintenance of the required host graph statistics, recomputation of the RETE net by Algorithm 1 only takes a constant amount of time. Also, the number of nodes in the computed net has a constant upper bound, the length of tuples in the net's indexers has a constant upper bound, and the net consists of only join and input nodes. Thus, given appropriate data structures, the effort for populating the newly computed net is linear in the net's size. If recomputation is only triggered when $n > d$ and repopulation is aborted as soon as the new net's size exceeds the size of the old net, which is at most $d + n \leq 2 \cdot n$, only a linear overhead in n is incurred on the processing of changes between two recomputations of the RETE net.

Aborting repopulation as described also ensures that recomputation and repopulation never immediately increases the net's size. But although a currently smaller net serves as an indicator, there is no guarantee that recomputation will improve the RETE net long-term. There is hence no guarantee regarding the overall time required for processing a sequence of changes while adapting the net structure compared to the time required using a static net. However, the overhead of recomputation and repopulation is limited to a constant factor.

Still, repopulation in particular remains an expensive operation, as in contrast to recomputation, its execution time depends on the size of the RETE net and thus the host graph. To avoid unnecessary repopulation, we employ the cost function from Sect. 3.1 to assess the quality of the new RETE net before deciding whether to trigger repopulation. Therefore, we compare the total estimated cost of all indexers in the new net to the cost estimate for the old net. If the new RETE net is not estimated to represent an improvement, this suggests that switching the RETE net comes at no benefit and should hence be avoided. While the exact size of the old net is available and could be used for comparison, this comparison would be unfair due to the inaccuracy of the cost function.

We potentially reduce the effort for repopulation further by reusing indexers from the old RETE net, if some portion of the net remains unchanged after recomputation. This is indicated by indexers in the old and new net corresponding to isomorphic query subgraphs. To be reuseable, the tuples stored in the indexers have to be indexed by the same key in both RETE nets. If this is the case, we insert the old indexer into the appropriate position in the new RETE net so that it does not have to be populated again. Furthermore, if it is the only indexer succeeding some input or join node, that node does not have to be executed during repopulation. Note that these conditions are always true for the overall query result. Thus, the topmost indexer storing matches for the entire query is never repopulated and the topmost node is never executed during repopulation.

Algorithm 3 outlines the described approach for controlling RETE net recomputation and repopulation. It processes a sequence of incoming host graph changes in two nested loops and maintains a RETE net that internally tracks the number of changes to its indexers in a dedicated counter variable *changes*.

At the start of the outer loop, the current size of the employed RETE net is stored and the RETE net's change counter is reset. Then, the inner loop processes changes by calls to ProcessChange until the number of changes to the RETE net's indexers exceeds the stored initial size, at which point the loop is left and a new RETE net structure is computed by RecomputeNetStructure.

If the estimated cost of the newly computed structure is lower than the cost of the old structure, the old RETE net is examined for indexers that can be reused in the new net. Afterwards, the new RETE net's indexers are populated via a call to RepopulateToMaxSize, reusing old indexers where possible. However, this process is aborted as soon as the new RETE net's size exceeds the size of the old net. The call to RepopulateToMaxSize returns false if population was aborted and true otherwise. If population was successful, the old RETE net is replaced by the new net. Finally, the next iteration of the outer loop is started.

Procedure ProcessChanges(C, R)

Input : C: Changes to process
         R: Current RETE net

```
1   while true do
2       d ← R.size;
3       R.changes ← 0;
4       while R.changes ≤ d do
5           ProcessChange(R, GetNextChange(C));
6       end
7       R′ ← RecomputeNetStructure();
8       if cost(R′) < cost(R) then
9           I ← FindReusableIndexers(R, R′);
10          valid ← RepopulateToMaxSize(R′, I, R.size);
11          if valid then
12              R ← R′;
13          end
14      end
15  end
```

Algorithm 3: Size-based algorithm for controlling RETE net recomputation

5   Evaluation

We attempt to address the following research questions: (**RQ1**) Can considering information about the host graph during RETE net construction lead to better performing nets? (**RQ2**) Is the overhead incurred by adapting the net to changing host graph characteristics acceptable? (**RQ3**) Can RETE net recomputation improve performance if host graph characteristics change significantly?

Therefore, we experiment[1] with datasets and queries provided by two independent benchmarks, the LDBC Social Network Benchmark (Interactive Workload) [8] and the Train Benchmark [18]. We compare the following strategies:

- **VIATRA:** (host-graph-insensitive) RETE net constructed by VIATRA [20].
- **EMULATE:** RETE net with the same structure as VIATRA.
- **STATIC:** RETE net constructed using the approach described in Sect. 3 over the final host graph
- **DYNAMIC:** RETE net constructed using the approach from Sect. 3 over the initial host graph, with dynamic adaptation as described in Sect. 4

To execute the RETE nets, we employ the VIATRA tool in case of the VIATRA strategy, and our own EMF-based [1] implementation for the other strategies.

5.1 LDBC Social Network Benchmark

We adapt the complex reading queries provided as part of the so-called Interactive workload of the LDBC Social Network Benchmark to match the definition of simple graph pattern matching in Sect. 2 as follows: We remove attribute constraints as well as negative application conditions and ignore two of the 14 queries that contain paths of arbitrary length. We generate a social network containing 1000 persons (and about 300 000 vertices and 1 200 000 edges in total) using the benchmark's data generator, which we transform into a sequence of element creations according to the generated associated timestamps.

Construction Scenario. We replay the obtained creation sequence and measure the time required to process subsequences of 1000 changes for each query.

Figure 4 shows the execution times for queries where the time required by the worst performing strategy was higher than that of the best performing strategy by at least factor 1.25 (queries 1, 5, 10, and 12). For all these queries, STATIC and DYNAMIC perform better than EMULATE. While VIATRA appears to perform better than EMULATE for some queries, its execution times are still higher than those of the host-graph-sensitive strategies. This indicates that fitting the structure of the RETE net to the host graph can lead to better performing nets (**RQ1**). Figure 4 also displays the execution times for the two queries where DYNAMIC performed worst compared to VIATRA and EMULATE (queries 7 and 8). However, the execution time of DYNAMIC was never higher than the execution time of VIATRA or EMULATE by more than factor 1.11.

Even though STATIC represents an idealized variant of our approach that is only applicable if the relevant host graph characteristics are known upfront, DYNAMIC never performs worse than STATIC by more than factor 1.2. Dynamic adaptation as described in Sect. 4 can thus yield adequate RETE nets,

---

[1] All experiments were executed on a Linux SMP Debian 4.19.67-2 machine with Intel Xeon E5-2630 CPU (2.3 GHz clock rate) and 386 GB system memory running OpenJDK version 1.8.0_242.

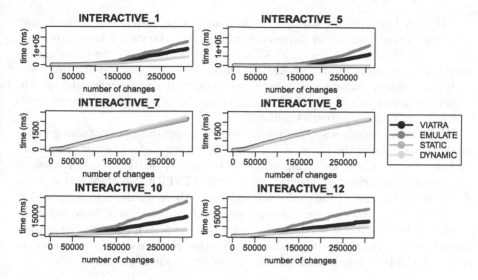

Fig. 4. Runtime measurements for the construction scenario (linear axes)

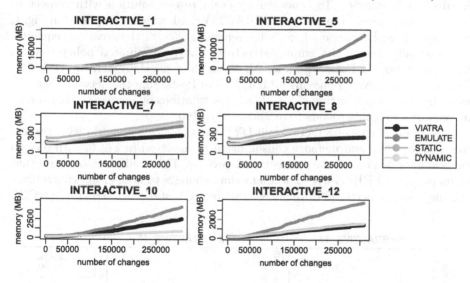

Fig. 5. Memory measurements for the construction scenario (linear axes)

with an acceptable overhead for recomputation and repopulation in our examples (**RQ2**). This is also a result of the host graph exhibiting relevant characteristics early, thus mostly avoiding repopulation of large nets. DYNAMIC performed between 18 and 27 recomputes and between 0 and 16 repopulates for each query.

VIATRA has a better baseline performance with respect to memory consumption, outperforming our implementation by up to factor 2.7 for nine queries. However, as displayed in Fig. 5, VIATRA still performs worse for three queries compared to STATIC and DYNAMIC. Thus, the conceptual observations also apply for memory consumption. Note, however, that during repopulation, the memory consumption of DYNAMIC may temporarily increase as the old net is kept in memory for a potential rollback.

We executed analogous experiments using a larger dataset containing about 10 000 persons. Over the larger social network, all strategies ran out of memory for four to six queries, indicating a limited scalability of RETE nets for simple graph pattern matching in general. Since with VIATRA, we employed a state-of-the-art tool as one of the evaluated strategies, these results also suggest that the dataset with 1000 persons already represents a reasonably large input for simple graph pattern matching. However, the observations for the smaller dataset generally still apply to the larger dataset, as either relative execution times were similar or the host-graph-insensitive techniques ran out of memory earlier.

Evolution Scenario. To conceptually evaluate our solution with respect to **RQ3**, we also execute the query INTERACTIVE_10, which is displayed in Fig. 1, in a hypothetical scenario where, after replaying 25% of the creation sequence, we drastically change the graph's structure. First, we mimic a policy change, where from now on, Posts with a length of less than 225 characters are viewed as Comments, effectively removing all but 0.5% of Posts from the network. Second, we emulate the introduction of friend recommendations by increasing the average number of a Person's friends from about 20 to 100.

The performance of STATIC and DYNAMIC for this scenario with respect to execution time and memory consumption is displayed in Fig. 6, with dashed vertical lines delimiting the restructuring process. The results demonstrate that recomputing the RETE net to accommodate changes to host graph characteristics may lead to better performance compared to a static net structure (**RQ3**).

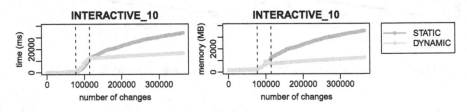

Fig. 6. Performance of STATIC and DYNAMIC for the evolution scenario (linear axes)

5.2 Train Benchmark

The Train Benchmark is specifically tailored to the evaluation of incremental querying techniques. The benchmark consists of a collection of queries that are incrementally executed over a synthetic railway topology. Therefore, execution alternates between checking phases, where query results are retrieved, and transformation phases, where the evolution of the topology is simulated by rule-based transformations using results from the checking phases. The benchmark includes several scenarios, which contain different sets of queries and can be parametrized to either perform a constant number of changes or an amount of changes proportional to the number of query results in each transformation phase.

In order to evaluate our approach using the Train Benchmark, we adapt the benchmark's queries analogously to the adaptations of the LDBC queries. We then execute a benchmark scenario including all of the benchmark's queries, where after an initial query execution, a number of changes equal to 5% of the number of results of special queries is performed repeatedly.

Figure 7 shows the overall execution times of the initial query execution (Read and Check) and the incremental processing of changes (Transformation and Recheck) for a sequence of topologies of increasing size. The results show that the host-graph sensitive techniques require a similar amount of time as the other strategies for the initial execution, but perform slightly better when processing the incremental updates for large models (**RQ1**). Similarly to the experiments with the LDBC Social Network Benchmark, RETE net recomputation does not incur an observable overhead on the execution time of DYNAMIC compared to STATIC (**RQ2**), as the host graph's characteristics do not change significantly after the initial query execution. These results indicate that in typical situations where our technique does not provide a significant benefit, it does not have a serious drawback, either.

5.3 Threats to Validity

Threats to internal validity include unexpected JVM behavior. To minimize this effect, we executed multiple runs of all experiments. Furthermore, we profiled the JVM's garbage collection times and subtracted them from the measured execution times. To mitigate the impact of the implementation, we used the same code to execute RETE nets of all construction strategies and also compared our implementation to an independent tool.

To address threats to external validity, we used queries and data provided as part of independent benchmarks [8,18]. We adapted the queries to fit our definition of graph pattern matching and otherwise used the provided methods of parametrization. The evolution scenario in Sect. 5.1 was not part of the original benchmark. We hence do not make claims regarding the relevance of such a scenario in practice, but demonstrate the potential conceptual benefit.

We further remark that our results are not necessarily generalizable to other application domains and make no quantitative claims regarding performance.

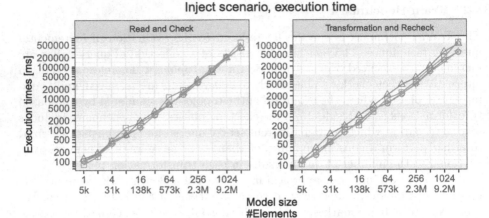

Fig. 7. Runtime measurements for the Train Benchmark and different topologies (logarithmic axes)

6   Related Work

Techniques based on local search form one family of approaches to graph pattern matching. Since the chosen search plan has a substantial performance impact, several techniques for generating an efficient search plan for graph queries exist [5,11,12,22], most of which are host-graph-sensitive. However, most techniques maintain limited information in between query executions, if any, and consequently perform poorly in incremental scenarios. Moreover, while the problem of generating search plans for local search is related to RETE net construction, in most cases the generated search plans only represent a specific subset of the possible RETE net structures, where the RETE net consists of a sequence of nodes that join a primitive pattern element to a growing main result [11,22]. Some approaches consider precomputing and storing tree-like intermediate results, thus representing a hybrid approach between local search and discrimination networks [5,12]. Still, in some cases lifting the imposed restrictions on the RETE net's structure enables the construction of nets with better performance.

RETE nets have been introduced by Forgy in the context of pattern matching for production systems [10], where the proposed construction approach creates nets that correspond to search plans employed for local search. VIATRA [20] is a mature tool for incremental graph pattern matching based on RETE nets. It supports various advanced features, such as attribute constraints and negative application conditions, and thus transcends the notion of graph pattern matching from Sect. 2. However, to the best of our knowledge VIATRA currently does not take host graph characteristics into account for constructing RETE nets.

Varró et al. propose an algorithm for RETE net construction that heuristically optimizes the RETE net with respect to a cost function [21]. While this allows for host-graph-sensitive construction of RETE nets via a cost function like the one introduced in Sect. 3.1, the cost function proposed in [21] is not host-graph-sensitive and tries to minimize the number of required indexers (rather than the number of indexer entries). Consequently, updating the RETE net as the host graph evolves is not considered. In [19], an evaluation of optimization strategies for RETE nets is presented, stating the number of indexer entries as a possible optimization goal and listing cost-based approaches and heuristic reordering of operations as optimization techniques. However, no concrete approach for cost-based optimization is presented.

The task of RETE net construction corresponds to the join ordering problem in relational databases, which has been subject to extensive research [16]. Among others, solutions include greedy heuristics similar to the approach presented in Sect. 3. These techniques often employ cost functions based on so-called join selectivity, which are closely related to the cost function from Sect. 3.1. Join ordering techniques could also be employed in the context of graph pattern matching, but require a translation of the encodings of query and data.

Tree decomposition [17], which constructs a tree representation of a graph where each node in the tree corresponds to a set of the graph's nodes called piece, also relates to RETE net construction. The number of graph nodes in the largest piece is also called tree-width. Due to certain coherence conditions over the decomposition, a tree decomposition of a given query graph could be used as a basis for the construction of discrimination networks and thus RETE nets. While algorithms exist that construct decompositions with low tree-width for certain graphs [6], which would contribute to minimizing the size of a related RETE net, this decomposition does not consider the structure of the host graph.

A more general form of discrimination networks than RETE nets, called Gator network, is presented by Hanson et al. [13]. In contrast to RETE nets, Gator networks may contain nodes with arbitrarily many dependencies, allowing a selection of which intermediate results are to be stored in memory. Gator networks therefore enable improving performance with respect to memory consumption at the cost of execution time. In [13], recomputation to accomodate changing data is not considered. Gator networks have also been employed for incremental graph pattern matching by Beyhl [4], where the employed discrimination network structure is defined manually by a user.

7   Conclusion

In this paper, we presented a host-graph-sensitive approach to RETE net construction for graph pattern matching. Our technique is based on a cost function for estimating the number of entries in the RETE net's indexers based on statistical information about the host graph. The cost function is employed to compute a RETE net structure that is appropriate for the given host graph, using a greedy partitioning algorithm for graph queries. We proposed an approach for adapting the net structure to changing host graph characteristics, which aims to limit the overhead required for recomputation. The presented concepts were evaluated empirically based on two independent benchmarks, demonstrating the feasibility of the approach in realistic application scenarios.

As future work, we plan to generalize our technique to accomodate advanced features for query specification such as attribute constraints and negative application conditions. Furthermore, we want to explore how to apply the developed concepts in the context of more general discrimination networks and extend our evaluation to different application domains.

References

1. EMF: Eclipse Modeling Framework. https://www.eclipse.org/modeling/emf/. Accessed 7 May 2021
2. Arendt, T., Biermann, E., Jurack, S., Krause, C., Taentzer, G.: Henshin: advanced concepts and tools for in-place EMF model transformations. In: Petriu, D.C., Rouquette, N., Haugen, Ø. (eds.) MODELS 2010. LNCS, vol. 6394, pp. 121–135. Springer, Heidelberg (2010). https://doi.org/10.1007/978-3-642-16145-2_9
3. Bergmann, G., et al.: Incremental evaluation of model queries over EMF models. In: Petriu, D.C., Rouquette, N., Haugen, Ø. (eds.) MODELS 2010. LNCS, vol. 6394, pp. 76–90. Springer, Heidelberg (2010). https://doi.org/10.1007/978-3-642-16145-2_6
4. Beyhl, T.: A framework for incremental view graph maintenance. Ph.D. thesis, Hasso Plattner Institute at the University of Potsdam (2018)
5. Bi, F., Chang, L., Lin, X., Qin, L., Zhang, W.: Efficient subgraph matching by postponing cartesian products. In: Proceedings of the 2016 International Conference on Management of Data, pp. 1199–1214. ACM (2016)
6. Bodlaender, H.L.: A linear-time algorithm for finding tree-decompositions of small treewidth. SIAM J. Comput. **25**(6), 1305–1317 (1996)
7. Ehrig, H., Ehrig, K., Prange, U., Taentzer, G.: Fundamentals of Algebraic Graph Transformation. Springer, Heidelberg (2006). https://doi.org/10.1007/3-540-31188-2
8. Erling, O., et al.: The LDBC social network benchmark: interactive workload. In: Proceedings of the 2015 ACM SIGMOD International Conference on Management of Data, pp. 619–630. ACM (2015)
9. Floyd, R.W.: Algorithm 97: shortest path. Commun. ACM **5**(6), 345 (1962)
10. Forgy, C.L.: Rete: A fast algorithm for the many pattern/many object pattern match problem. In: Readings in Artificial Intelligence and Databases, pp. 547–559. Elsevier (1989)
11. Geiß, R., Batz, G.V., Grund, D., Hack, S., Szalkowski, A.: GrGen: a fast SPO-based graph rewriting tool. In: Corradini, A., Ehrig, H., Montanari, U., Ribeiro, L., Rozenberg, G. (eds.) ICGT 2006. LNCS, vol. 4178, pp. 383–397. Springer, Heidelberg (2006). https://doi.org/10.1007/11841883_27
12. Han, W.S., Lee, J., Lee, J.H.: Turboiso: towards ultrafast and robust subgraph isomorphism search in large graph databases. In: Proceedings of the 2013 ACM SIGMOD International Conference on Management of Data, pp. 337–348 (2013)
13. Hanson, E.N., Bodagala, S., Chadaga, U.: Trigger condition testing and view maintenance using optimized discrimination networks. IEEE Trans. Knowl. Data Eng. **14**(2), 261–280 (2002)

14. Hell, P., Nešetřil, J.: On the complexity of H-coloring. J. Comb. Theory. Ser. B **48**(1), 92–110 (1990)
15. Ibaraki, T., Kameda, T.: On the optimal nesting order for computing n-relational joins. ACM Trans. Database Syst. (TODS) **9**(3), 482–502 (1984)
16. Leis, V., Gubichev, A., Mirchev, A., Boncz, P., Kemper, A., Neumann, T.: How good are query optimizers, really? Proc. VLDB Endow. **9**(3), 204–215 (2015)
17. Robertson, N., Seymour, P.D.: Graph minors. III. Planar tree-width. J. Comb. Theory. Ser. B **36**(1), 49–64 (1984)
18. Szárnyas, G., Izsó, B., Ráth, I., Varró, D.: The train benchmark: cross-technology performance evaluation of continuous model queries. Softw. Syst. Model. **17**(4), 1365–1393 (2017). https://doi.org/10.1007/s10270-016-0571-8
19. Szárnyas, G., Maginecz, J., Varró, D.: Evaluation of optimization strategies for incremental graph queries. Periodica Polytechnica Electr. Eng. Comput. Sci. **61**(2), 175–192 (2017)
20. Varró, D., Bergmann, G., Hegedüs, Á., Horváth, Á., Ráth, I., Ujhelyi, Z.: Road to a reactive and incremental model transformation platform: three generations of the VIATRA framework. Softw. Syst. Model. **15**(3), 609–629 (2016). https://doi.org/10.1007/s10270-016-0530-4
21. Varró, G., Deckwerth, F.: A rete network construction algorithm for incremental pattern matching. In: Duddy, K., Kappel, G. (eds.) ICMT 2013. LNCS, vol. 7909, pp. 125–140. Springer, Heidelberg (2013). https://doi.org/10.1007/978-3-642-38883-5_13
22. Varró, G., Deckwerth, F., Wieber, M., Schürr, A.: An algorithm for generating model-sensitive search plans for EMF models. In: Hu, Z., de Lara, J. (eds.) ICMT 2012. LNCS, vol. 7307, pp. 224–239. Springer, Heidelberg (2012). https://doi.org/10.1007/978-3-642-30476-7_15

Rule-Based Top-Down Parsing for Acyclic Contextual Hyperedge Replacement Grammars

Frank Drewes[1] [iD], Berthold Hoffmann[2]([envelope]) [iD], and Mark Minas[3] [iD]

[1] Umeå Universitet, Umeå, Sweden
drewes@cs.umu.se
[2] Universität Bremen, Bremen, Germany
hof@uni-bremen.de
[3] Universität der Bundeswehr München, Neubiberg, Germany
mark.minas@unibw.de

Abstract. Contextual hyperedge replacement (CHR) strengthens the generative power of hyperedge replacement (HR) significantly, thus increasing its usefulness for practical modeling. We define top-down parsing for CHR grammars by graph transformation, and prove that it is correct as long as the generation and use of context nodes in productions does not create cyclic dependencies. An efficient predictive version of this algorithm can be obtained as in the case of HR grammars.

Keywords: Graph transformation · Hyperedge replacement · Contextual hyperedge replacement · Parsing · Correctness

1 Introduction

Contextual hyperedge replacement (CHR, [4,5]) strengthens the generative power of hyperedge replacement (HR, [13]) significantly, by productions with context nodes that refer to nodes which are not connected to the edge being replaced. Unfortunately, both HR and CHR grammars can generate NP-complete graph languages [16]. The authors have therefore devised efficient parsers for subclasses of HR and CHR grammars, implementing so-called predictive top-down (PTD) parsing. Although the concepts and implementation of these parsers have been described at depth in [6], their correctness has only recently been formally confirmed, based on the specification of parsers by means of graph transformation rules, and only for HR grammars [8]. Here we extend the parsers and their correctness proof to CHR grammars. It turns out that a CHR grammar Γ can be turned into a HR grammar generating graphs where the nodes that were context nodes in Γ are *borrowed*, i.e., generated like ordinary nodes. From this graph, the one generated by Γ can be obtained by *contraction*, i.e., merging borrowed nodes with other nodes. We show that this is correct provided that the generation and use of context nodes in CHR productions does

© Springer Nature Switzerland AG 2021
F. Gadducci and T. Kehrer (Eds.): ICGT 2021, LNCS 12741, pp. 164–184, 2021.
https://doi.org/10.1007/978-3-030-78946-6_9

not lead to cyclic dependencies. In this paper, we concentrate on describing a non-deterministic parser, and sketch only briefly how this parser can be made predictive (and efficient), since the latter is completely analoguous to the HR case [8].

The remainder of this paper is structured as follows. After recalling some basic concepts of graph transformation (Sect. 2), we define CHR grammars and their corresponding borrowing HR grammars, and we discuss the requirement of acyclicity (Sect. 3). In Sect. 4, we define a top-down parser for acyclic CHR grammars that processes edges in a linear order (corresponding to leftmost derivations in string grammars), and prove it correct. In the conclusions (Sect. 5), we discuss related and future work.

2 Preliminaries

The set of non-negative integers is denoted by $\mathbb{N}$, and $[n]$ denotes $\{1, \ldots, n\}$ for all $n \in \mathbb{N}$. A^* denotes the set of all finite sequences over a set A; the empty sequence is denoted by ε, and the length of a sequence α by $|\alpha|$. As usual, $\to^+$ and $\to^*$ denote the transitive and the transitive reflexive closure of a binary relation $\to$. For a function $f \colon A \to B$, its extension $f^* \colon A^* \to B^*$ to sequences is defined by $f^*(a_1 \cdots a_n) = f(a_1) \cdots f(a_n)$, for all $n \in \mathbb{N}$ and $a_1, \ldots, a_n \in A$.

2.1 Graphs

The graphs considered in this paper have labeled nodes and edges that may have parallel edges carrying the same label. We also generalize edges to hyperedges, which may connect any number of nodes, not just two.

Throughout the paper, let $\mathbb{L}$ be a global set of labels which is partitioned into two infinite subsets $\dot{\mathbb{L}}$ and $\bar{\mathbb{L}}$, and let $arity \colon \bar{\mathbb{L}} \to \mathbb{N}$ be a function that associates an $arity$ with every label in $\bar{\mathbb{L}}$. Elements of $\dot{\mathbb{L}}$ and $\bar{\mathbb{L}}$ will be used to label nodes and hyperedges, respectively. A finite set $\Sigma \subseteq \mathbb{L}$ is an $alphabet$, and we let $\dot{\Sigma} = \Sigma \cap \dot{\mathbb{L}}$ and $\bar{\Sigma} = \Sigma \cap \bar{\mathbb{L}}$.

Definition 1 (Hypergraph). A $hypergraph$ over an alphabet Σ is a tuple $G = (\dot{G}, \bar{G}, att, lab)$, where $\dot{G}$ and $\bar{G}$ are disjoint finite sets of $nodes$ and $hyperedges$, respectively, the function $att \colon \bar{G} \to \dot{G}^*$ attaches hyperedges to sequences of nodes, and the function $lab \colon \dot{G} \cup \bar{G} \to \Sigma$ maps $\dot{G}$ to $\dot{\Sigma}$ and $\bar{G}$ to $\bar{\Sigma}$ in such a way that $|att(e)| = arity(lab(e))$ for every edge $e \in \bar{G}$.

For brevity, we omit the prefix "hyper" in the sequel. A node is $isolated$ if no edge is attached to it. An edge carrying a label $\sigma \in \Sigma$ is a σ-$edge$, and the Σ'-$edges$ of a graph are those labeled with symbols from $\Sigma' \subseteq \bar{\Sigma}$. G° denotes the discrete subgraph of a graph G, which is obtained by removing all edges. We sometimes write $X(G)$ to denote the set $\dot{G}$ of nodes of G, and instead of "$x \in \dot{G}$ or $x \in \bar{G}$", we may write "$x \in G$". We denote the third and fourth component of a graph G by att_G and lab_G. $\mathcal{G}_\Sigma$ denotes the class of graphs over Σ; a graph

$G \in \mathcal{G}_\Sigma$ is called a *handle* (over Σ) if G has a single edge e and each node of G is attached to e. We denote the set of all handles over Σ by $\mathcal{H}_\Sigma$.

Graphs with unlabeled nodes or edges are a special case obtained by letting $\dot\Sigma$ contain the "invisible" label $\sqcup$, or letting $\bar\Sigma$ contain an invisible label $\sqcup_i$ per arity i. We call a graph *unlabeled* if both nodes and edges are unlabeled. In this case, we omit the labeling in the definition and drawing of the graph.

A set of edges $E \subseteq \bar{G}$ *induces* the subgraph consisting of these edges and their attached nodes. Given graphs $G_1, G_2 \in \mathcal{G}_\Sigma$ with disjoint edge sets, a graph $G = G_1 \cup G_2$ is called the *union* of G_1 and G_2 if G_1 and G_2 are subgraphs of G, $\dot{G} = \dot{G}_1 \cup \dot{G}_2$, and $\bar{G} = \bar{G}_1 \cup \bar{G}_2$. Note that $G_1 \cup G_2$ exists only if common nodes are consistently labeled, i.e., $lab_{G_1}(v) = lab_{G_2}(v)$ for $v \in \dot{G}_1 \cap \dot{G}_2$.

Definition 2 (Graph morphism). Given graphs G and H, a *morphism* $m\colon G \to H$ is a pair $m = (\dot{m}, \bar{m})$ of functions $\dot{m}\colon \dot{G} \to \dot{H}$ and $\bar{m}\colon \bar{G} \to \bar{H}$ that preserve attachments and labels, i.e., $att_H(\bar{m}(v)) = \dot{m}^*(att_G(v))$, $lab_H(\dot{m}(v)) = lab_G(v)$, and $lab_H(\bar{m}(e)) = lab_G(e)$ for all $v \in \dot{G}$ and $e \in \bar{G}$.

The morphism is *injective* or *surjective* if both $\dot{m}$ and $\bar{m}$ are, and a *subgraph inclusion* of G in H if $m(x) = x$ for every $x \in G$; then we write $G \subseteq H$. If m is surjective and injective, it is called an *isomorphism*, and G and H are called *isomorphic*, written as $G \cong H$.

2.2 Graph Transformation

For transforming graphs, we use the classical double-pushout approach of [9], with injective occurrences of rules in graphs.

Definition 3 (Rule). A *graph transformation rule* $r = (P \supseteq I \subseteq R)$ consists of a *pattern graph* P, a *replacement graph* R, and an *interface graph* I included in both P and R. We briefly call r a *rule*, denote it as $r\colon P \rightarrowtail R$, and refer to its graphs by P_r, R_r, and I_r if they are not explicitly named.

An injective morphism $m\colon P \to G$ into a graph G defines an *occurrence* with respect to r if it satisfies the following *dangling condition*: if the occurrence $m(v)$ of a node $v \in P \setminus I$ is attached to some edge $e \in G$, then e is also in $m(P)$.

A rule r *transforms* a graph G at an occurrence m to a graph H by (1) removing $m(x)$ from G for every $x \in P \setminus I$, to obtain a graph K, and (2) constructing H from the disjoint union of K and R by merging $m(x)$ with every $x \in I$. Then we write $G \Rightarrow_r^m H$, but may omit m if it is irrelevant, and write $G \Rightarrow_\mathcal{R} H$ if $\mathcal{R}$ is a set of rules such that $G \Rightarrow_r H$ for some $r \in \mathcal{R}$.

Since the interface of a rule is included in its replacement graph, a transformation step can be constructed in such a way that K is included in H.

2.3 Application Conditions

Sometimes it is necessary to restrict the applicability of a rule by requiring the existence or non-existence of certain subgraphs in the context of its occurrence. Our definition of application conditions is based on [14], but omits nesting as it will not be needed here.

Definition 4 (Conditional rule). For a graph P, the set of *conditions over P* is defined inductively as follows: (i) an inclusion $P \subseteq C$ defines a *basic condition* over P, denoted by $\exists C$. (ii) if c and c' are conditions over P, then $\neg c$, $(c \wedge c')$, and $(c \vee c')$ are conditions over P.

An injective morphism $m \colon P \to G$ *satisfies* a basic condition $\exists C$ if there is an injective morphism $m' \colon C \to G$ whose restriction to P coincides with m. The semantics of Boolean combinations of application conditions is defined in the obvious way; $m \vDash c$ expresses that m satisfies condition c.

A *conditional rule* r' consists of a rule $r = P \rightarrowtail R$ and a condition c over P, and is denoted as $r' \colon c [\!] P \rightarrowtail R$. We let $G \Rightarrow_{r'}^m H$ or simply $G \Rightarrow_{r'} H$ if $m \vDash c$ and $G \Rightarrow_r^m H$. Note that rules without conditions can also be seen as conditional rules with the neutral condition $c = \exists P$. For a set $\mathcal{C}$ of conditional rules, $\Rightarrow_{\mathcal{C}} = \bigcup_{r \in \mathcal{C}} \Rightarrow_r$.

Examples of graphs and rules, with and without conditions, will be shown in the following sections.

3   Contextual Hyperedge Replacement

We recall graph grammars based on contextual hyperedge replacement [4,5], which include hyperedge replacement grammars [13] as a special case.

Definition 5 (Contextual hyperedge replacement). Let Σ be an alphabet and $\mathcal{N} \subseteq \bar{\Sigma}$ a set of *nonterminal edge labels* (*nonterminals*, for short). The *terminal edge labels* (*terminals*) are those in $\mathcal{T} = \bar{\Sigma} \setminus \mathcal{N}$. Accordingly, edges with labels in $\mathcal{N}$ and $\mathcal{T}$ are *nonterminal* and *terminal edges*, respectively.

A rule $p \colon P \rightarrowtail R$ is a *hyperedge replacement production* over Σ (*production*, for short) if the pattern P contains a single edge, which is labeled with a nonterminal, and the interface graph I_p is the discrete subgraph P° consisting of all nodes of P. Isolated nodes in the pattern of p are called *context nodes*; p is called *contextual* if such context nodes exist, and *context-free* otherwise.

A *contextual hyperedge replacement grammar* $\Gamma = \langle \Sigma, \mathcal{N}, \mathcal{P}, Z \rangle$ (*CHR grammar* for short) consists of alphabets Σ and $\mathcal{N} \subseteq \bar{\Sigma}$ as above, a finite set $\mathcal{P}$ of productions over Σ, and a start graph $Z \in \mathcal{G}_\Sigma$. Γ is a (context-free) *hyperedge replacement grammar* (*HR grammar*) if all productions in $\mathcal{P}$ are context-free.

The *language* generated by Γ is given as $\mathcal{L}(\Gamma) = \{G \in \mathcal{G}_{\Sigma \setminus \mathcal{N}} \mid Z \Rightarrow_{\mathcal{P}}^* G\}$.

We use a simple but non-context-free running example because illustrations of parsers would otherwise become too big and complex.

Example 1 (Linked trees). Figure 1 shows our running example, and introduces our conventions for drawing graphs and productions. Nodes are circles, non-terminal edges are rectangular boxes containing the corresponding labels, and terminal edges are shapes like $\blacklozenge, \triangleright, \blacktriangleright$. (In this example, all nodes are unlabeled.) Edges are connected to their attached nodes by lines, called tentacles, which are ordered counter-clockwise around the edge, starting at noon. For productions

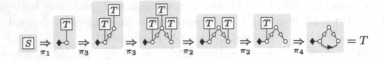

Fig. 1. Productions for linked trees

$$\boxed{S} \Rightarrow_{\pi_1} \quad \Rightarrow_{\pi_3} \quad \Rightarrow_{\pi_3} \quad \Rightarrow_{\pi_2} \quad \Rightarrow_{\pi_2} \quad \Rightarrow_{\pi_4} \quad = T$$

Fig. 2. A derivation of a linked tree T

(and in other rules), we just draw their pattern P and their replacement graph R, and specify the inclusion of the interface nodes by ascribing the same identifier to them in P and R, like x and y in Fig. 1.

Figure 1 defines the productions π_1, π_2, π_3, and π_4 of the CHR grammar Δ. S is the nullary symbol labeling the start graph, and T is a unary nonterminal. A unary ♦-edge is attached to the root of a tree, binary ▷-edges connect nodes to their children, and binary ▶-edges (drawn with curly tentacles) represent links between nodes. Δ generates trees where every node may have a link to any other node, see Fig. 2.

Assumption 1 (CHR grammar). In the sequel, we assume that CHR grammars $\Gamma = \langle \Sigma, \mathcal{N}, \mathcal{P}, Z \rangle$ satisfy the following conditions:

1. The node sequences attached to nonterminal edges are free of repetitions.
2. The start graph Z consists of a single nonterminal edge of arity 0. This nonterminal symbol does not occur in right-hand sides of productions.
3. Γ is reduced, i.e., every production occurs in a derivation of a graph in $\mathcal{L}(\Gamma)$.
4. $\mathcal{L}(\Gamma)$ does not contain graphs with isolated nodes.

These assumptions are made without loss of generality: in [13, Sect. I 4], it is described how HR grammars can be transformed to satisfy Assumptions 1.1–1.2; these results can directly be lifted to CHR grammars. How to attain Assumption 1.3 for CHR grammars is shown in [4, Sect. 3.4]. Assumption 1.4 is made to simplify the technicalities of parsing. To ensure it, unary *virtual* edges can be attached to isolated nodes in the productions and in the graphs generated by the grammar. In Example 1, e.g., the ♦-edge avoids that the grammar generates a single isolated node.

We now recall the well-known notion of derivation trees, which reflect the context-freeness of HR grammars [3, Definition 3.3]. Here we use a slightly modified version that represents derivations of concrete graphs:

Definition 6 (Derivation tree). Let $\Gamma = \langle \Sigma, \mathcal{N}, \mathcal{P}, Z \rangle$ be a HR grammar. The set $\mathbb{T}_\Gamma$ of *derivation trees* over Γ and the mappings $root \colon \mathbb{T}_\Gamma \to \mathcal{H}_\Sigma$ as well as $result \colon \mathbb{T}_\Gamma \to \mathcal{G}_\Sigma$ are inductively defined as follows:

- Each handle G is in $\mathbb{T}_\Gamma$, and $root(G) = result(G) = G$.
- A triple $t = \langle G, p, c \rangle$ consisting of a nonterminal handle G, a production $p \in \mathcal{P}$, and a sequence $c = t_1 t_2 \cdots t_n \in \mathbb{T}_\Gamma^*$ is in $\mathbb{T}_\Gamma$ if the union graphs $G' = G^\circ \cup \bigcup_{i=1}^n root(t_i)$ and $G'' = G^\circ \cup \bigcup_{i=1}^n result(t_i)$ exist, $G \Rightarrow_p G'$, and $X(result(t_j)) \cap X(result(t_j)) = X(root(t_i)) \cap X(root(t_j))$ for all distinct $i, j \in [n]$. Furthermore, we let $root(t) = G$ and $result(t) = G''$.

An example of a derivation and its derivation tree is shown in Fig. 5.

The ordering of derivation trees in $c = t_1 t_2 \cdots t_n$ within a derivation tree $t = \langle G, p, c \rangle$ will become relevant when edges in right-hand sides of rules are ordered, which will be the case in Sect. 4. Then we require that $root(t_1)\, root(t_2) \cdots root(t_n)$ corresponds to the edge ordering in production p.

Let t, t' be any derivation trees. We call t' a child tree of t, written $t' \prec t$, if $t = \langle G, p, t_1 t_2 \cdots t_n \rangle$ and $t' = t_i$ for some i, and we call t' a subtree of t if $t' = t$ or $t = \langle G, p, t_1 t_2 \cdots t_n \rangle$ and t' is a subtree of t_i for some i. A derivation tree t introduces a node u (at its root) if $t = \langle G, p, t_1 t_2 \cdots t_n \rangle$ and $u \in X(root(t_i)) \setminus \dot{G}$ for some i. The set of all these nodes is denoted by $intro(t)$. We define the pre-order traversal $pre(t) \in \mathbb{T}_\Gamma^*$ of a derivation tree t recursively by $pre(t) = t$ if $t \in \mathcal{H}_\Sigma$ and $pre(t) = t\, pre(t_1)\, pre(t_2) \cdots pre(t_n)$ if $t = \langle G, p, t_1 t_2 \cdots t_n \rangle$.

The following theorem is equivalent to Theorem 3.4 in [3]:

Theorem 1. *Let $\Gamma = \langle \Sigma, \mathcal{N}, \mathcal{P}, Z \rangle$ be a HR grammar, $H \in \mathcal{H}_\Sigma$ a handle and $G \in \mathcal{G}_\Sigma$ a graph. There is a derivation tree $t \in \mathbb{T}_{\Gamma}$ with $root(t) = H$ and $result(t) = G$ iff $H \Rightarrow_\mathcal{P}^* G$.*

Note that derivation trees are defined only for HR grammars as in the contextual case, any properly labeled node can be used as a context node as long as it has been created earlier in a derivation. This fact produces dependencies between derivation steps which do not exist in HR derivations. It will turn out in the following that there is a close relationship between a CHR grammar Γ and its so-called borrowing HR grammar $\hat{\Gamma}$: every graph $H \in \mathcal{L}(\Gamma)$ is a "contraction" of a graph $G \in \mathcal{L}(\hat{\Gamma})$. Moreover, the converse is also true as long as Γ is acyclic, a notion to be defined later.

In the following, we assume that $\mathcal{T}$ contains two auxiliary edge labels that are not used elsewhere in Γ: edges carrying the unary label ⊙ will mark borrowed nodes, and binary edges labeled $\neq$ will connect borrowed nodes to all nodes that should be kept separate to them, i.e., not be contracted later on.

Definition 7 (Borrowing grammar). Let $\Gamma = \langle \Sigma, \mathcal{N}, \mathcal{P}, Z \rangle$ be a CHR grammar. For $(p\colon P \multimap R) \in \mathcal{P}$, its *borrowing production* $\hat{p}\colon \hat{P} \to \hat{R}$ is obtained by (a) removing every context node from $\hat{P}$ and $I_{\hat{p}}$ and (b) constructing $\hat{R}$ from R as follows: for every context node v of p, attach a new ⊙-edge to v, and add $\neq$-edges from v to every other node with the label $lab_P(v)$. The *borrowing grammar* $\hat{\Gamma} = \langle \Sigma, \mathcal{N}, \hat{\mathcal{P}}, Z \rangle$ of Γ is given by $\hat{\mathcal{P}} = \{\hat{p} \mid p \in \mathcal{P}\}$.

Note that $\hat{p} = p$ if p is context-free.

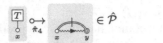

Fig. 3. Borrowing link production   **Fig. 4.** A linked tree with a detached link

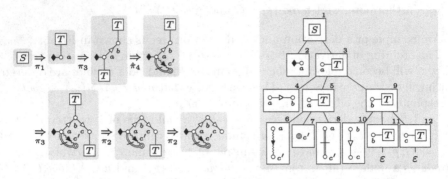

Fig. 5. A derivation of $\hat{T}$ (cf. Fig. 4) and its derivation tree

Definition 8 (Contraction). For a graph G let

$$\dot{G}_{\odot} = \{v \in \dot{G} \mid v = att_G(e) \text{ for a } \odot\text{-edge } e \in \bar{G}\} \text{ and}$$
$$\neq_G = \{(u,v) \in \dot{G} \times \dot{G} \mid uv = att_G(e) \text{ for a } \neq\text{-edge } e \in \bar{G}\}.$$

A morphism $\mu\colon G \to H$ is called a *joining morphism* for G if $\dot{H} = \dot{G} \setminus \dot{G}_{\odot}$, $\bar{H} = \bar{G}$, $\bar{\mu}$ and the restriction of $\dot{\mu}$ to $\dot{G} \setminus \dot{G}_{\odot}$ are inclusions, and $(v, \dot{\mu}(v)) \notin \neq_G$ for every $v \in \dot{G}_{\odot}$. The graph $core(H)$ obtained from H by removing all edges with labels $\odot$ and $\neq$ is called the μ-*contraction* of G or just *a contraction* of G.

Example 2 (Trees with detached links). Figure 3 shows the borrowing link production $\hat{\pi}_4$, and Fig. 4 shows a tree with a detached link that can be generated by a borrowing grammar $\hat{\Delta}$ with productions $\{\pi_1, \pi_2, \pi_3, \hat{\pi}_4\}$. (Here, $\odot$-edges are depicted by drawing the attached node accordingly.) Obviously, the linked tree $\hat{T}$ in Fig. 4 can be generated by the derivation in Fig. 5, and contracted to the linked tree T generated in Fig. 2. Figure 5 also shows the derivation tree of $\hat{T}$. Only the root handles of trees and subtrees are shown, and ε is used to indicate an empty sequence of child trees, thus distinguishing this case from the one where a subtree is a single handle. The numbers on top of the handles illustrate the ordering of the corresponding subtrees in a pre-order traversal of the entire tree.

Definition 9 (Borrowing version of a derivation). Let $\Gamma = \langle \Sigma, \mathcal{N}, \mathcal{P}, Z \rangle$ be a CHR grammar and $\hat{\Gamma}$ its borrowing HR grammar. A derivation

$$Z \Rightarrow_{\hat{p}_1}^{\hat{m}_1} H_1 \Rightarrow_{\hat{p}_2}^{\hat{m}_2} H_2 \Rightarrow_{\hat{p}_3}^{\hat{m}_3} \cdots \Rightarrow_{\hat{p}_n}^{\hat{m}_n} H_n$$

in $\hat{\Gamma}$ is a *borrowing version* of a derivation

$$Z \Rightarrow_{p_1}^{m_1} G_1 \Rightarrow_{p_2}^{m_2} G_2 \Rightarrow_{p_3}^{m_3} \cdots \Rightarrow_{p_n}^{m_n} G_n$$

in Γ if the following hold, for $i = 1, 2, \ldots, n$ and $p_i \colon P \circ\!\!\rightarrow R$:

1. $\hat{p}_i$ is the borrowing production of p_i,
2. if $\bar{P} = \{e\}$ then $\hat{m}_i(e) = m_i(e)$, and
3. for every $x \in \bar{R} \cup (\dot{R} \setminus \dot{P})$, the images of x in G_i and H_i are the same.

By a straightforward induction, it follows that every derivation in Γ has a borrowing version in $\hat{\Gamma}$, and G_i is the μ_i-contraction of H_i for $i \in [n]$, where the joining morphism μ_i is uniquely determined by $\bar{\mu}_i(e) = e$ for all $e \in \bar{H}_i$.

Theorem 2 will show that the converse is also true, i.e., that every contraction of a graph in $\mathcal{L}(\hat{\Gamma})$ can also be derived in Γ, provided that Γ is *acyclic* (Definition 10). Informally, Γ is cyclic if there is a derivation of a graph G in $\hat{\Gamma}$ and a contraction H of G so that there is a cyclic dependency between derivation steps that create nodes and derivation steps that use them as context nodes. These cyclic dependencies then result in derivations of graphs G in $\hat{\Gamma}$ having a contraction H that cannot be derived in Γ because there is no reordering of the derivation steps that yields a valid derivation in Γ.

Cyclic dependencies caused by a joining morphism μ for a derivation tree $t \in \mathbb{T}_{\hat{\Gamma}}$ are formalized using the relation $\sqsubset_\mu$ on subtrees of t. Informally, $t' \sqsubset_\mu t''$ means that t' describes a derivation step (the topmost one that transforms the root handle of t'), which creates a node used as a context node in the corresponding topmost contextual derivation step described by t''. This, together with the definition of acyclic CHR grammars, is defined next.

Definition 10 (Acyclic CHR grammar). Let Γ be a CHR grammar.

For any two subtrees t', t'' of a derivation tree $t \in \mathbb{T}_{\hat{\Gamma}}$, we let $t' \sqsubset_\mu t''$ iff there is a node $u \in intro(t'')$ so that $\mu(u) \neq u$ and $\mu(u) \in intro(t')$.

Γ is *acyclic* if $(\succ \cup \sqsubset_\mu)^+$ is irreflexive for all derivation trees $t \in \mathbb{T}_{\hat{\Gamma}}$ over $\hat{\Gamma}$ and all joining morphisms μ for $result(t)$. Otherwise, Γ is *cyclic*.

Theorem 2. *Let Γ be a CHR grammar and $\hat{\Gamma}$ its borrowing HR grammar. For every graph $H \in \mathcal{L}(\Gamma)$, there is a graph $G \in \mathcal{L}(\hat{\Gamma})$ so that H is a contraction of G. Moreover, every contraction of a graph in $\mathcal{L}(\hat{\Gamma})$ is in $\mathcal{L}(\Gamma)$ if Γ is acyclic.*

Proof. Let Γ be a CHR grammar and $\hat{\Gamma}$ its borrowing HR grammar. The first part of the theorem is a direct consequence of the observation made after Definition 9. To prove the second part of the theorem, let Γ be acyclic and H the μ-contraction of a graph $G \in \mathcal{L}(\hat{\Gamma})$, i.e., $G = result(t)$ for some derivation tree

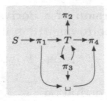

Fig. 6. Grammar graph of Δ

t over $\hat{\Gamma}$. Since $(\succ \cup \sqsubset_\mu)^+$ is irreflexive, one can order the subtrees of t topologically according to $\succ \cup \sqsubset_\mu$, obtaining a sequence $t_1, \ldots, t_n$ of derivation trees so that $1 \leq i < j \leq n$ implies that neither $t_i \prec t_j$ nor $t_j \sqsubset_\mu t_i$. This sequence (when considering just those subtrees that are of the form $\langle G, p, c \rangle$) determines a derivation of G in $\hat{\Gamma}$, and every node u with $\dot{\mu}(u) \neq u$ is created in a later derivation step than $\dot{\mu}(u)$. The derivation is therefore a borrowing version of a derivation of H in Γ. In particular, no node is used as a context node before it has been created. □

Definition 10 does not provide effective means to check whether a CHR grammar is acyclic. However, we can make use of the fact that joining morphisms may only map a borrowed node to a node with the same label. A CHR grammar cannot be cyclic if we can make sure that no nodes with any label can ever be part of a cyclic dependency. The grammar graph of a CHR grammar allows for such reasoning. It describes which rules can create nodes with which labels or use them as context nodes, and it relates nonterminal labels with productions in the sense that productions are applied to nonterminal edges, producing new nonterminal edges. (The creation of terminal edges is irrelevant here.)

Definition 11 (Grammar graph). The *grammar graph* of a CHR grammar $\Gamma = \langle \Sigma, \mathcal{N}, \mathcal{P}, Z \rangle$ is the unlabeled graph G_Γ such that $\dot{G}_\Gamma = \mathcal{P} \cup \mathcal{N} \cup \dot{\Sigma}$ and for each rule $p \colon P \multimap R$, say with $\bar{P} = \{l\}$, $\bar{G}$ contains binary edges from the node $lab(l)$ to p and from p to each node in $\{lab(x) \mid x \in R\} \setminus \mathcal{T}$, as well as an edge from $lab(u)$ to p for every context node u of p.

Example 3 (Acyclicity of the linked tree grammar). Figure 6 shows the grammar graph for Δ. (Recall that "$\sqcup$" is the otherwise omitted invisible node label.) Lemma 1 will reveal that Δ has only harmless dependencies since the cycle $T \to \pi_3 \to T$ in its grammar graph does not contain $\sqcup$.

The next lemma provides a sufficient criterion to check whether a CHR grammar is acyclic, and therefore, whether Theorem 2 can be exploited:

Lemma 1. *The grammar graph of every cyclic CHR grammar has a cycle that contains a node in $\dot{\Sigma}$.*

Proof. Let $\Gamma = \langle \Sigma, \mathcal{N}, \mathcal{P}, Z \rangle$ be a cyclic CHR grammar, $\hat{\Gamma}$ its borrowing HR grammar, G_Γ the grammar graph of Γ, $t \in \mathbb{T}_{\hat{\Gamma}}$ a derivation tree over $\hat{\Gamma}$, and μ

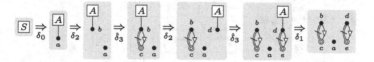

Fig. 7. Productions for generating dags, with borrowing production $\hat{\delta}_3$

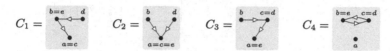

Fig. 8. A derivation with $\hat{\Lambda}$

$$C_1 = \qquad C_2 = \qquad C_3 = \qquad C_4 =$$

Fig. 9. The four contractions of the graph derived in Fig. 8

a joining morphism for $result(t)$ so that there is a sequence $t_1, \ldots, t_n$ of $n > 2$ subtrees of t so that $t_1 = t_n$ and $t_i \, (\succ \cup \sqsubset_\mu) \, t_{i+1}$ for $1 \le i < n$. By the definition of $\succ$ and $\sqsubset_\mu$, t_i is of the form $\langle G_i, p_i, c_i \rangle$ for every i. If $t_i \succ t_{i+1}$ and $\bar{G}_{i+1} = \{e\}$, G_Γ contains edges from p_i to $lab(e)$ and from $lab(e)$ to p_{i+1}. If $t_i \sqsubset_\mu t_{i+1}$, there is a node $u \in intro(t_{i+1})$ so that $\mu(u) \ne u$ and $\mu(u) \in intro(t_i)$, which means that u is the image of a context node of p_{i+1} and the image of a created node of p_i, i.e., G_Γ contains edges from p_i to $lab(u)$ and from $lab(u)$ to p_{i+1}. Hence, G_Γ contains a cycle. Moreover, there must be at least one i so that $t_i \sqsubset_\mu t_{i+1}$ because $\succ$ is irreflexive, proving that the cycle contains a node in $\dot{\Sigma}$. $\square$

The following example demonstrates that directed acyclic graphs (dags) can be generated by a CHR grammar. However, this grammar is cyclic and thus has a grammar graph with a cycle that contains a node in $\dot{\Sigma}$.

Example 4 (CHR grammar for dags). Consider nonterminals S (of arity 0), A (of arity 1), and terminals $\bullet$ (of arity 1) and $\triangleright$ (of arity 2). Figure 7 shows productions δ_0 to δ_3 over these symbols, where nodes attached to a $\bullet$-edge are just drawn as $\bullet$. (These edges are introduced to meet Assumption 1.4 that generated graphs do not contain isolated nodes.) The CHR grammar Λ with these productions and with an S-edge as a start graph generates all unlabeled dags with at least one node. In its borrowing HR grammar $\hat{\Lambda}$, the contextual production δ_3 is replaced by the context-free production $\hat{\delta}_3$ shown on the right of Fig. 7.

A derivation with $\hat{\Lambda}$ is shown in Fig. 8; the resulting terminal graph can be contracted in four possible ways, to the graphs $C_1, \ldots, C_4$ shown in Fig. 9. The contraction C_4 is cyclic, and is the only one that cannot be generated with the productions of the CHR grammar Λ.

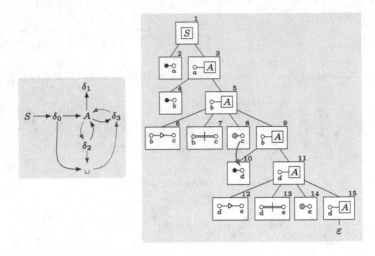

Fig. 10. Grammar graph (left) and derivation tree (right) of grammar $\hat{\Lambda}$

Figure 10 shows the cyclic grammar graph for Λ on the left, and the derivation graph of the derivation in Fig. 8. Here the illegal contraction leading to the cyclic graph C_4 is indicated by the thick bent arrow between nodes 8 and 10.

The fact that the contextual production δ_3 can be applied only after its context node has been generated with production δ_0 or δ_2 makes sure that no cyclic graphs can be generated. This indicates that cyclic CHR grammars are strictly more powerful than acyclic ones.

While the absence of a cycle containing a node in $\dot{\Sigma}$ in the grammar graph implies that a CHR grammar is acyclic, the converse is unfortunately not true: there are acyclic grammars whose grammar graphs have such cycles. A criterion to characterize acyclic CHR grammars needs to be determined in future work.

4   Top-Down Parsing for Acyclic CHR Grammars

We define top-down parsers for acyclic CHR grammars by parsing the corresponding borrowing HR grammars with stack automata that take the necessary merging of borrowed nodes with other nodes into account. The stack automata perform transitions of states that are called configurations. Configurations are represented as graphs, and transitions are described by graph transformation rules. This definition is more precise than the original definition of top-down parsing in [6], but avoids technical complications occurring in [7], where graphs are represented textually as sequences of literals transformed by parser actions. In particular, no explicit substitution and renaming operations on node identifiers are required. Further, this approach simplifies handling the borrowing and merging required for CHR parsing.

For ease of presentation, we consider an arbitrary but fixed CHR grammar $\Gamma = \langle \Sigma, \mathcal{N}, \mathcal{P}, Z \rangle$ throughout the rest of this paper, and let $\hat{\Gamma} = \langle \Sigma, \mathcal{N}, \hat{\mathcal{P}}, Z \rangle$ be its borrowing HR grammar according to Definition 7.

Top-down parsers attempt to construct a derivation of a graph that matches a given input graph. Our top-down parser processes the edges of an input graph G in a nondeterministically chosen linear order when it attempts to construct a derivation for G. Technically, this order is represented by equipping derived edges with two additional tentacles by which they are connected to nodes labeled with a fresh symbol $\bullet$ to form a linear thread.

The definition of the top-down parser for borrowing HR grammars differs from that of HR grammars as follows: in [8], the binding of stack to input nodes and edges was represented by identifying them; here we connect matched nodes with binding edges, and remove matched edges. This results in a more elegant construction, yields more intuitive parses, and simplifies the correctness proof.

Definition 12 (Threaded graph). The *threaded alphabet* $\Sigma^{\bullet}$ of Σ is given by $\dot{\Sigma}^{\bullet} = \dot{\Sigma} \cup \{\bullet\}$ and $\bar{\Sigma}^{\bullet} = \mathcal{T} \cup \{\ell^{\bullet} \mid \ell \in \bar{\Sigma} \setminus \{\odot, \neq\}\}$ with $arity(\ell^{\bullet}) = arity(\ell) + 2$; $\mathcal{N}^{\bullet} = \{\ell^{\bullet} \mid \ell \in \mathcal{N}\}$ denotes the set of *threaded nonterminals*.

Let $G \in \mathcal{G}_{\Sigma^{\bullet}}$. A node $v \in \dot{G}$ is a *thread node* if $lab_G(v) = \bullet$, and a *kernel node* otherwise. $\lceil \dot{G} \rfloor$ and $\dot{G}^{\bullet}$ denote the sets of all kernel nodes and thread nodes of G, respectively. An edge is *threaded* if its label is of the form $\ell^{\bullet}$, and *unthreaded* otherwise.

A graph $G \in \mathcal{G}_{\Sigma^{\bullet}}$ is *threaded* if all of its edges except the $\odot$- and $\neq$-edges are threaded and the following additional conditions hold:

1. For every threaded edge $e \in \bar{G}$ with $att_G(e) = u_1 \ldots u_k u_{k+1} u_{k+2}$, the nodes $u_1, \ldots, u_k$ are kernel nodes of G and u_{k+1}, u_{k+2} are thread nodes of G.
2. $\dot{G}^{\bullet}$ can be ordered as $\dot{G}^{\bullet} = \{v_0, \ldots, v_n\}$ for some $n \in \mathbb{N}$ such that, for every $i \in [n]$, $\bar{G}$ contains exactly one threaded edge e so that $att_G(e)$ ends in $v_{i-1} v_i$, and there are no further threaded edges than these.

We call v_0 the *first* and v_n the *last* thread node of G.

The *kernel graph* of G is the graph $\lceil G \rfloor \in \mathcal{G}_{\Sigma}$ obtained by replacing every edge label $\ell^{\bullet}$ by ℓ, and removing the thread nodes and their attached tentacles. In this case, G is a *threading* of $\lceil G \rfloor$. Note that a threading of a graph in $\mathcal{G}_{\Sigma}$ is uniquely determined by an ordering of its non-$\{\odot, \neq\}$-edges and vice versa.

We use a set $\Sigma_{\mathrm{aux}} = \{\otimes, \mathsf{bind}\}$ of *auxiliary edge labels*, disjoint with Σ, where $\otimes$ is unary and labels edges that mark *unmatched nodes* in a configuration, whereas bind is binary, and represents the *binding* of a stack node to a node in the input.

Definition 13 (Configuration). A graph C over $\Sigma^{\bullet} \cup \mathcal{T} \cup \Sigma_{\mathrm{aux}}$ is a *configuration* if

- only $\Sigma^{\bullet}$-edges are attached to thread nodes and
- the subgraph $stack(C)$ of C induced by its $\Sigma^{\bullet}$-edges and thread nodes is a threaded graph.

The edge attached to the first thread node is said to be *topmost* on the stack. The *input* of C is the subgraph *input*(C) induced by the (unthreaded) $\mathcal{T}$-edges. C is called

- *initial* if *stack*(C) is a handle of $Z^\bullet$, $\otimes$-edges are attached to all other nodes of C, and there are no bind-edges in C;
- *accepting* if *stack*(C) is an isolated node, all further nodes are attached to bind-edges, and all other edges in C are labeled with $\odot$ or $\neq$.

Definition 14 (Top-down parser). Let $\mathcal{R}$ be a set of conditional rules. A derivation $C \Rightarrow^*_\mathcal{R} C'$ is a *parse* if C is an initial configuration. A parse $C \Rightarrow^*_\mathcal{R} C'$ is *successful* if C' is an accepting configuration. $\mathcal{R}$ is a *top-down parser* for Γ if, for each initial configuration C, *input*$(C) \in \mathcal{L}(\Gamma)$ if and only if there is a successful parse $C \Rightarrow^*_\mathcal{R} C'$.

We define two kinds of top-down parsing rules operating on the topmost edge e on the stack:

- If e is nonterminal, expand pops e from the stack, and pushes the right-hand side of a production for e onto the stack;
- If e is terminal, e is matched with a corresponding unthreaded edge e' in the input; then e is popped from the stack, and e' is removed.

For a match rule to match e to e', we must have $lab(e) = lab(e')^\bullet$, and each pair u, v of corresponding attached nodes must either already be bound to each other by a bind-edge, or there must be $\otimes$-edges attached to both, which indicates that they are still unprocessed, or u must have both a $\otimes$-edge and a $\odot$-edge attached to it. The latter covers the case where u is a still unprocessed borrowed node which can thus be bound to a node already bound earlier. To make sure that the borrowed node is not bound to a node of the same right-hand side, an application condition checking for the absence of a $\neq$-edge is needed. Also, the condition forbids that borrowed nodes are treated as ordinary ones.

Definition 15 (Expand and match rules). For every borrowing production $(p: P \rightarrowtail R) \in \hat{\mathcal{P}}$, the *expand rule* $t_p: P' \rightarrowtail R'$ is given as follows:

- P' and R' are threadings of P and R, respectively, where every node introduced by p has a $\otimes$-edge attached in R' (and no others have).
- The interface I_{t_p} is I_p plus the last thread node of P'.

A conditional rule $t: c[\,] P \rightarrowtail R$ with configurations P and R is a *match rule* for a terminal symbol $a \in \mathcal{T} \setminus \{\odot, \neq\}$ if the following holds:

- *stack*(P) is a handle of $a^\bullet$, say with attached nodes $u_1 \cdots u_k u_{k+1} u_{k+2}$.
- *input*(P) is a handle of a, say with attached nodes $v_1 \cdots v_k$.
- For every pair (u_i, v_i), $i \in [k]$, precisely one of the following conditions holds:
  - P contains a bind-edge with attachment $u_i v_i$,
  - a $\otimes$-edge is attached to both u_i and v_i, or
  - both a $\otimes$-edge and a $\odot$-edge is attached to u_i.

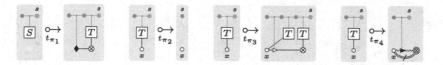

Fig. 11. Expand rules of the top-down parser for linked trees

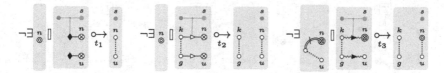

Fig. 12. Match rules for the top-down parser for linked trees

- I_t is P without the a-edge, the $a^\bullet$-edge, and the first thread node.
- R consists of the nodes $u_1, \ldots, u_{k+2}, v_1, \ldots, v_k$, the last thread node of $stack(P)$, and a bind-edge from u_i to v_i for every $i \in [k]$.
- For every $i \in [k]$, the application condition c requires the following: Let $m \colon P \to G$ be the occurrence. If $u_i \in \dot{P}_\odot$, then there is no $z \in \dot{G}$ with $(m(u_i), z) \in \neq_G$ such that z has a bind-edge to $m(v_i)$. If $u_i \notin \dot{P}_\odot$, then c requires that $m(u_i) \notin \dot{G}_\odot$.

We let $\mathcal{R}_{\Gamma^\bullet}$ denote the set of all expand and match rules of $\Gamma^\bullet$.

Example 5 (Top-down parser for linked trees). For the expand rules of the top-down parser for linked trees with detached links (Fig. 11) we have threaded the right-hand sides so that terminal edges come first, and nonterminals attached to the source node of a terminal edge next. Nodes attached to a $\otimes$-edge or a $\odot$-edge are drawn as $\otimes$ and $\odot$ respectively, rather than drawing the edge separately. A node attached to both a $\otimes$-edge and a $\odot$-edge, is drawn as $\circledast$. We draw bind-edges as dotted lines (in downward direction), and $\neq$-edges as double lines with a vertical bar.

In general, the terminal symbols of Δ lead to four match rules for $\blacklozenge$, and nine rules for $\triangleright$ and $\blacktriangleright$ each. However, inspection of Δ reveals that just three match rules are needed by the parser, which are shown in Fig. 12: A $\blacklozenge$-edge is matched when its attached node is unbound, and $\triangleright$-edges and $\blacktriangleright$-edges are matched when their first attached node is already bound. The application conditions for the match rules t_1 and t_2 are actually not needed; analysis of Δ reveals that the unbound nodes of these edges will never be attached to $\odot$-edges.

Figure 13 shows a successful parse of the linked tree T derived in Fig. 2 with these rules. We have left shades of the matched edges in the configurations to illustrate how the parse constructs the derivation with the borrowing HR grammar $\hat{\Gamma}$ in Fig. 5, which corresponds to the derivation with Γ in Fig. 2.

Note that match rules consume the thread and the matched edges. Expand rules do not modify the input, but just replace the first nonterminal on the thread by the replacement graph of a threaded production for this nonterminal.

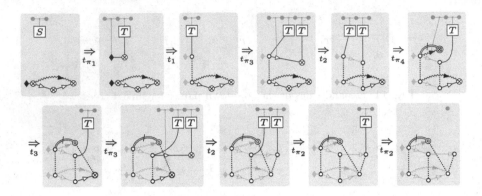

Fig. 13. A parse of the linked tree T generated in Fig. 2

In the following, we prove formally that $\mathcal{R}_{\Gamma^\bullet}$ is indeed a top-down parser for Γ, provided that Γ is acyclic.

Fact 1 (Invariants of configurations)

1. The bind-edges define an irreflexive partial function

$$\mapsto = \{(x, y) \mid e \in \bar{C},\ lab_C(e) = \mathsf{bind},\ att_C(e) = xy\},$$

between non-thread nodes, called *binding*, such that $x \mapsto y$ implies that (1) x is not in $input(C)$, and (2) y is not in $stack(C)$.
2. No node is attached to several $\otimes$-edges.
3. A kernel node of a threaded edge in $stack(C)$ is attached to a $\otimes$-edge if and only if it is not the source of a bind-edge.
4. Every node of $input(C)$ that is not the target of a bind-edge is attached to a $\otimes$-edge. (The converse is not true because a $\otimes$-edge may be attached to the target of a bind-edge if the source of that bind-edge is in $\dot{C}_\odot$; see Fig. 13.)

We now consider the first direction of the correctness of the parser.

Lemma 2. *Every graph $G \in \mathcal{L}(\Gamma)$ has a successful parse $C \Rightarrow^*_{\mathcal{R}_{\Gamma^\bullet}} C'$, where C is the initial configuration with $input(C) = G$.*

The proof relies on the following construction to obtain a successful parse.

Construction 1. Let $Z \Rightarrow_{p_1} G_1 \Rightarrow_{p_2} \cdots \Rightarrow_{p_n} G_n$ with $p_i \in \mathcal{P}$ for $i \in [n]$ be a derivation for $G = G_n$. Consider a borrowing version $Z \Rightarrow_{\hat{p}_1} H_1 \Rightarrow_{\hat{p}_2} \cdots \Rightarrow_{\hat{p}_n} H_n$ of the derivation, where G is the μ-contraction of $G' = H_n$. Let $t \in \mathbb{T}_{\hat{\Gamma}}$ be the derivation tree of $Z \Rightarrow_{\hat{p}_1} H_1 \Rightarrow_{\hat{p}_2} \cdots \Rightarrow_{\hat{p}_n} H_n$, where the ordering of subtrees corresponds to the edge ordering in rules of Γ.[1] Let $t_1 t_2 \cdots t_n$ be the sequence

---

[1] Note that, while t is a derivation tree over the borrowing HR grammar $\hat{\Gamma}$, the ordering of right-hand sides of productions in $\hat{\Gamma}$ ignores edges that are labeled with $\neq$ and $\odot$, so that it is the same as in Γ and provides the required edge ordering.

of subtrees obtained from the pre-order traversal $pre(t)$ of t by keeping only the trees with terminal and nonterminal root handles.

For $k \in \{0, \ldots, n\}$, construct the sequence $T_k \in \mathcal{H}_\Sigma^*$ from $t_1 t_2 \cdots t_k$ by keeping only the trees that are terminal handles. Let N_k be the set of nodes occurring in T_k. By definition, $N_k \subseteq \dot{G}'$. As the edges of terminal handles in T_k are exactly those in G', and $\bar{G} = \bar{G}'$, each handle in T_k identifies a unique edge in G.

Let $S_k \in \mathcal{H}_\Sigma^*$ be the sequence obtained from $t_{k+1} t_{k+2} \cdots t_n$ by removing each tree t_i that is a subtree of a tree t_j where $k < j < i$, and replacing each remaining tree t_i by its root handle $root(t_i)$. Moreover, let L_k be the subset of those handles in $\{root(t') \mid t' \prec t_i \text{ for some } i < k\}$ whose edges are labeled with $\neq$ or $\odot$.

The configuration C_k is then obtained by the following steps:

1. Define C_k^T as the threaded graph whose thread contains the (threaded versions) of edges in S_k in this order. Additionally, all nodes that occur in T_k but not in S_k, and all edges in L_k are in C_k^T.
2. Replace each kernel node u of C_k^T by a fresh copy $copy(u)$. Add a $\otimes$-edge in C_k^T to $copy(u)$ if $u \notin N_k$.
3. Obtain C_k^U from G by removing all edges occurring in handles in T_k. Add a $\otimes$-edge in C_k^U to u if $u \notin N_k$.
4. Let $C_k = C_k^\mathrm{U} \cup C_k^\mathrm{T}$.
5. For each node $u \in N_k$, add a bind-edge from $copy(u)$ to $\dot{\mu}(u)$ in C_k.   □

The following example illustrates this construction:

Example 6. Consider Example 2 again, and the illustration of the derivation tree $t_{\hat{T}}$ in Fig. 5, which results in the graph $\hat{T}$ of Fig. 4. The pre-order traversal of $t_{\hat{T}}$ is $pre(t_{\hat{T}}) = \tau_1 \cdots \tau_{12}$ where τ_i is the subtree of $t_{\hat{T}}$ whose root handle carries i as a small number in Fig. 5. Note that τ_7 and τ_8 are handles whose edges are labeled with $\neq$ or $\odot$. Construction 1 thus ignores them and considers the following sequence $t_1 \cdots t_{10}$ of the remaining ten subtrees (again, only the root handles are depicted):

Construction 1 creates the configurations in the parse $C_0 \Rightarrow_{\mathcal{R}_{\Gamma^\bullet}}^* C_{10}$ shown in Fig. 13 from this sequence. Figure 14 displays the sequences T_k and S_k and the set L_k of handles used for creating each C_k for $k \in \{0, \ldots, 10\}$.

We are now ready to sketch a proof of Lemma 2:

Proof Sketch. We build graphs $C_0, C_1, \ldots, C_n$ following Construction 1. C_0 and C_n are clearly an initial and an accepting configuration, respectively, with $input(C_0) = G$. To see that $C_0 \Rightarrow_{\mathcal{R}_{\Gamma^\bullet}} C_1 \Rightarrow_{\mathcal{R}_{\Gamma^\bullet}} \cdots \Rightarrow_{\mathcal{R}_{\Gamma^\bullet}} C_n$, consider C_{k-1} and C_k for some $k \in [n]$. We can distinguish two cases for $H = root(t_k)$:

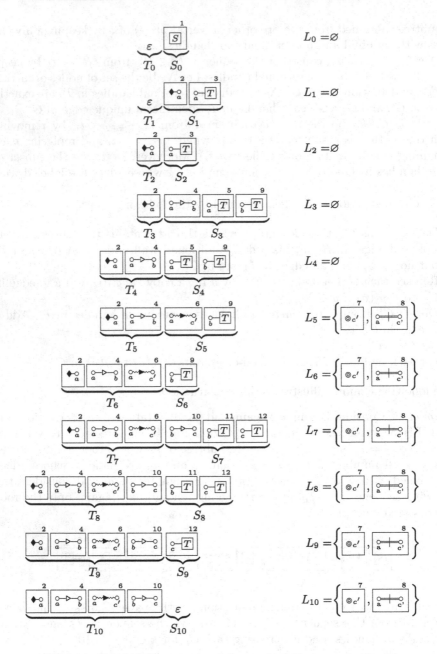

Fig. 14. Steps of Construction 1 for creating the parse in Fig. 13

Case 1: H is terminal.

This implies $T_k = T_{k-1}H$, $S_{k-1} = HS_k$, and $N_k = N_{k-1} \cup \dot{H}$. The reader can confirm by following the steps in Construction 1 that $C_{k-1} \Rightarrow_{\mathcal{R}_{\Gamma^\bullet}} C_k$ using the match rule for the edge label in H.

Case 2: H is nonterminal.
Then $t_{k-1} = \langle H, p, t'_1 t'_2 \cdots t'_l \rangle$ for derivation trees $t'_1 t'_2 \cdots t'_l$, and we have $T_k = T_{k-1}$, $N_k = N_{k-1}$, $S_{k-1} = t_{k-1}R$, and $S_k = root(t'_1)\, root(t'_2) \cdots root(t'_l)\, R$ with $R \in \mathbb{T}^*_\Gamma$. By Definition 6, $H \Rightarrow_p H^\circ \cup \bigcup_{i=1}^l root(t'_i)$. Construction 1 makes sure that $C_{k-1} \Rightarrow_{\mathcal{R}_{\Gamma^\bullet}} C_k$ using the expand rule for p.   □

The next lemma covers the other direction of the correctness of the parser.

Lemma 3. *If Γ is acyclic, then the existence of a successful parse $C \Rightarrow^*_{\mathcal{R}_{\Gamma^\bullet}} C'$ implies $input(C) \in \mathcal{L}(\Gamma)$.*

Proof. Let $C_0 \Rightarrow_{t_1} C_1 \Rightarrow_{t_2} \cdots \Rightarrow_{t_n} C_n$ be any successful parse with $t_1, \ldots, t_n \in \mathcal{R}_{\Gamma^\bullet}$. Let $G = input(C_0)$. For $i = 0, \ldots, n$, consider the following subgraphs Top_i and Bot_i of C_i: Top_i is the subgraph induced by $\dot{C}_i \setminus \dot{G}$, and Bot_i the subgraph obtained from C_i by deleting all edges in Top_i and all thread nodes. Note that $C_i = Top_i \cup Bot_i$ and, since C_0 is initial, $Top_0 = Z^\bullet$ and $Bot_0 = G$.

Bot_i may contain bind-edges, which define the binding relation $\mapsto$ of Fact 1. For every graph H containing Bot_i as a subgraph (for any i), we obtain $merge(H)$ as the homomorphic image of H without its bind-edges by mapping node x to node y for each $(x, y) \in \mapsto$.

Let us now consider a parse step $C_{i-1} \Rightarrow_{t_i} C_i$ for $i \in [n]$. There are two cases:

Case 1: t_i is a match rule.
Then C_i is obtained by deleting a terminal edge as well as its threaded version from C_{i-1}, and by adding some bind-edges. As a consequence, there is a handle Del_i of a terminal edge with $\dot{Del}_i \subseteq X(\lceil Top_{i-1}\rfloor)$ so that

$$\lceil Top_{i-1}\rfloor = \lceil Top_i\rfloor \cup Del_i \quad \text{and} \quad merge(Bot_{i-1}) = merge(Bot_i \cup Del_i).$$

Case 2: t_i is an expand rule for a rule $p_i \in \hat{P}$.
By the definition of the expand rule we have

$$\lceil Top_{i-1}\rfloor \underset{p_i}{\Rightarrow} \lceil Top_i\rfloor \quad \text{and} \quad Bot_{i-1} = Bot_i.$$

Let $Del = \bigcup_{j-1}^n Del_j$ where Del_j is the empty graph if t_j is an expand rule. Making use of the fact that $G = merge(Bot_0)$ and $\lceil Top_0\rfloor = Z$, a straightforward induction yields

$$(1)\; Z \overset{*}{\underset{\hat{P}}{\Rightarrow}} \lceil Top_n\rfloor \cup Del \quad \text{and} \quad (2)\; G = merge(Bot_n \cup Del).$$

Let $F = \lceil Top_n\rfloor \cup Del$. Note that for every node $x \in \dot{F}_\circledcirc$ (see Definition 8), there is a unique node $y \in \dot{F} \setminus \dot{F}_\circledcirc$ so that G has a node u with $x \mapsto u$ and $y \mapsto u$ in Bot_n. Now consider the morphism $\mu \colon F \to F'$ where $\bar{\mu}$ and the restriction of $\dot{\mu}$ to $\dot{F} \setminus \dot{F}_\circledcirc$ are inclusions, and $\dot{\mu}$ maps each node $x \in \dot{F}_\circledcirc$ to the corresponding node $y \in \dot{F} \setminus \dot{F}_\circledcirc$ as described above. The application conditions of match rules make sure that $(v, \dot{\mu}(v)) \notin \neq_F$ for every $v \in \dot{F}_\circledcirc$. Hence, μ is a joining morphism, and $core(F') \in \mathcal{L}(\Gamma)$ because of (1) and the assumption that Γ is acyclic. However, $core(F')$ is isomorphic to $merge(Bot_n \cup Del)$, and (2) implies $G \in \mathcal{L}(\Gamma)$.   □

Corollary 1. *If Γ is acyclic, then $\mathcal{R}_{\Gamma^\bullet}$ is a top-down parser for Γ.*

5   Conclusions

In this paper, we have shown that our rule-based definition of top-down parsing for CHR grammars is correct if the dependencies arising from the use of context nodes in these grammars are acyclic. This extends our correctness proof for HR grammars in [8, Theorem 2]. Our result can be specialized for predictive top-down parsing by equipping expand rules with application conditions that allow to predict the only promising production for a nonterminal; cf. Theorem 4 of that paper.

The language of all graphs over Σ, unrestricted flowcharts of imperative programs, statecharts [5, Ex. 1, 2 & Sect. 3], and graphs representing object-oriented programs [4] cannot be generated with HR grammars; however, they can be generated with CHR grammars. This indicates that the extension is practically relevant. Moreover, the mentioned CHR grammars are acyclic and PTD-parsable. This suggests that these restrictions will not be too strong in practice.[2] The cyclic grammar Λ for dags in Example 4 is not PTD-parsable. (We conjecture that there is no acyclic CHR grammar for dags at all.)

Much of the related work on parsing for HR grammars follows the well-known Cocke-Younger-Kasami algorithm. An implementation for unrestricted HR grammars (plus edge-embedding rules) in DiaGen [18] works for practical input with hundreds of nodes and edges, although their worst-case complexity is exponential. D. Chiang et al. [1] have implemented a polynomial algorithm for a subclass of HR grammars (based on the work of C. Lautemann [17]). S. Gilroy, A. Lopez, and S. Maneth [12] have proposed a linear parsing algorithm for Courcelle's "regular" graph grammars [2]. Both algorithms apply to graphs as they occur in computational linguistics.

To our knowledge, early approaches to parsing for context-free node replacement grammars [10] like [15,19] are no longer pursued.

Like many scientific efforts, this paper raises more questions than it answers: (i) Is there a decidable sufficient and necessary condition for acyclicity? (ii) Can parsing for CHR grammars be extended to cyclic CHR grammars? (iii) Can PSR parsing [7] be defined by graph transformation in a similar way? (iv) Is PSR parsing more powerful than PTD parsing? All this remains for future work.

Acknowledgments. We thank Annegret Habel, Verone Stillger, and the anonymous reviewers for their advice.

---

[2] The mentioned CHR grammars can be downloaded at www.unibw.de/inf2/grappa. The *graph parser distiller* GRAPPA developed by Mark Minas, generates PTD parsers that run in quadratic time, and often even in linear time [6]. The website also contains specifications of the PTD parser for linked trees with the AGG system [11] along the lines of Example 5.

References

1. Chiang, D., Andreas, J., Bauer, D., Hermann, K.M., Jones, B., Knight, K.: Parsing graphs with hyperedge replacement grammars. In: Proceedings of 51st Annual Meeting of the Association for Computational Linguistics (vol. 1: Long Papers), Sofia, Bulgaria, pp. 924–932. Association for Computational Linguistics, August 2013
2. Courcelle, B.: The monadic second-order logic of graphs V: on closing the gap between definability and recognizability. Theoret. Comput. Sci. **80**, 153–202 (1991)
3. Drewes, F., Habel, A., Kreowski, H.J.: Hyperedge replacement graph grammars. In: Rozenberg [20], chapter. 2, pp. 95–162
4. Drewes, F., Hoffmann, B.: Contextual hyperedge replacement. Acta Informatica **52**(6), 497–524 (2015). https://doi.org/10.1007/s00236-015-0223-4
5. Drewes, F., Hoffmann, B., Minas, M.: Contextual hyperedge replacement. In: Schürr, A., Varró, D., Varró, G. (eds.) AGTIVE 2011. LNCS, vol. 7233, pp. 182–197. Springer, Heidelberg (2012). https://doi.org/10.1007/978-3-642-34176-2_16
6. Drewes, F., Hoffmann, B., Minas, M.: Predictive top-down parsing for hyperedge replacement grammars. In: Parisi-Presicce, F., Westfechtel, B. (eds.) ICGT 2015. LNCS, vol. 9151, pp. 19–34. Springer, Cham (2015). https://doi.org/10.1007/978-3-319-21145-9_2
7. Drewes, F., Hoffmann, B., Minas, M.: Formalization and correctness of predictive shift-reduce parsers for graph grammars based on hyperedge replacement. J. Log. Algebraic Methods Program. (JLAMP) **104**, 303–341 (2019). https://doi.org/10.1016/j.jlamp.2018.12.006
8. Drewes, F., Hoffmann, B., Minas, M.: Graph parsing as graph transformation. In: Gadducci, F., Kehrer, T. (eds.) ICGT 2020. LNCS, vol. 12150, pp. 221–238. Springer, Cham (2020). https://doi.org/10.1007/978-3-030-51372-6_13
9. Ehrig, H., Ehrig, K., Prange, U., Taentzer, G.: Fundamentals of Algebraic Graph Transformation. MTCSAES, Springer, Heidelberg (2006). https://doi.org/10.1007/3-540-31188-2
10. Engelfriet, J., Rozenberg, G.: Node replacement graph grammars. In: Rozenberg [20], chapter. 1, pp. 1–94
11. Ermel, C., Rudolf, M., Gabriele, T.: The AGG approach: language and environment. In: Engels, G., Ehrig, H., Kreowski, H.J., Rozenberg, G. (eds.) Handbook of Graph Grammars and Computing by Graph Transformation, Vol. II: Applications, Languages, and Tools, pp. 551–603. World Scientific, Singapore (1999)
12. Gilroy, S., Lopez, A., Maneth, S.: Parsing graphs with regular graph grammars. In: Proceedings of the 6th Joint Conference on Lexical and Computational Semantics (*SEM 2017), Vancouver, Canada, pp. 199–208. Association for Computational Linguistics, August 2017. https://doi.org/10.18653/v1/S17-1024
13. Habel, A.: Hyperedge Replacement: Grammars and Languages. LNCS, vol. 643. Springer, Heidelberg (1992). https://doi.org/10.1007/BFb0013875
14. Habel, A., Pennemann, K.H.: Correctness of high-level transformation systems relative to nested conditions. Math. Struct. Comput. Sci. **19**(2), 245–296 (2009)
15. Kaul, M.: Practical applications of precedence graph grammars. In: Ehrig, H., Nagl, M., Rozenberg, G., Rosenfeld, A. (eds.) Graph Grammars 1986. LNCS, vol. 291, pp. 326–342. Springer, Heidelberg (1987). https://doi.org/10.1007/3-540-18771-5_62
16. Lange, K.J., Welzl, E.: String grammars with disconnecting or a basic root of the difficulty in graph grammar parsing. Discret. Appl. Math. **16**, 17–30 (1987)

17. Lautemann, C.: The complexity of graph languages generated by hyperedge replacement. Acta Informatica **27**, 399–421 (1990)
18. Minas, M.: Diagram editing with hypergraph parser support. In: Proceedings of 1997 IEEE Symposium on Visual Languages (VL 1997), Capri, Italy, pp. 226–233. IEEE Computer Society Press (1997)
19. Pavlidis, T.: Linear and context-free graph grammars. J. ACM **19**(1), 11–22 (1972). https://doi.org/10.1145/321679.321682
20. Rozenberg, G. (ed.): Handbook of Graph Grammars and Computing by Graph Transformation, Vol. I: Foundations. World Scientific, Singapore (1997)

Nets with Mana: A Framework for Chemical Reaction Modelling

Fabrizio Genovese[1] , Fosco Loregian[2] , and Daniele Palombi[3] (✉)

[1] University of Pisa, Pisa, Italy
[2] Tallinn University of Technology, Tallin, Estonia
[3] Sapienza University of Rome, Rome, Italy

Abstract. We use categorical methods to define a new flavor of Petri nets where transitions can only fire a limited number of times, specified by a quantity that we call mana. We do so with chemistry in mind, looking at ways of modelling the behavior of chemical reactions that depend on enzymes to work. We prove that such nets can be either obtained as a result of a comonadic construction, or by enriching them with extra information encoded into a functor. We then use a well-established categorical result to prove that the two constructions are equivalent, and generalize them to the case where the firing of some transitions can "regenerate" the mana of others. This allows us to represent the action of catalysts and also of biochemical processes where the byproducts of some chemical reaction are exactly the enzymes that another reaction needs to work.

1 Introduction

Albeit they have found great use outside their original domain, Petri nets were invented to describe chemical reactions [21]. The interpretation is as simple as it can get: places of the net represent types of compounds (be it atoms or molecules); tokens represent the amount of each combination we have available; transitions represent reactions transforming compounds.

Still, things are not so easy in real-world chemistry: reactions often need "context" to happen, be it a given temperature, energy, presence of enzymes and catalysts. This is particularly true in biochemical processes, where enzymes of all sorts mediate rather complicated reactions. Importantly, these enzymes tend to degrade over time, resulting in reactions that do not keep happening forever [14]. This is one of the (many) reasons why organisms wither and die, but it is not captured by the picture above, where the transition can fire every time it is enabled.

F. Gadducci and T. Kehrer (Eds.): ICGT 2021, LNCS 12741, pp. 185–202, 2021.
https://doi.org/10.1007/978-3-030-78946-6_10

Borrowing the terminology from the popular Turing machine *Magic: The gathering* [7, 24] we propose a possible solution to this problem by endowing transitions in a net with *mana* [23], representing the "viability" of reactions: once a reaction is out of mana, it cannot fire anymore.

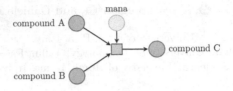

Now, we could just represent mana by adding another place for each transition in a net. Indeed, this is the idea we will start with. Still, being accustomed to the *yoga* of type-theoretic reasoning, we are also aware that throwing everything in the same bucket is rarely a good idea: albeit mana can be a chemical compound, it is more realistic to consider it as conceptually separated from the reactions it catalyzes.

Resorting to categorical methods, we show how we can axiomatize the idea of mana in a better way. We do so by relaxing the definitions in the categorical approach to coloured nets already developed in [13], defining a functorial semantics representing the equipment of a net with mana. Then, we will prove how categorical techniques allow us to internalize such a semantics, exactly obtaining what we represented in the picture above.

Finally, we will show how the categorical semantics naturally leads to a further generalization, where transitions not only need mana to function but also provide byproducts that can be used as mana for other transitions. This allows us to represent *catalysts*[1] (i.e. cards 'adding ∞ to the mana pool', or more precisely mana that does not deteriorate over time) and in general nets apt to describe *two-layered* chemical processes, the first layer being the usual one represented by Petri nets and the second layer being the one of enzymes and catalysts being consumed and exchanged by different reactions.

2   Nets and Their Executions

Before presenting the construction itself, it is worth recapping the main points about categorical semantics for Petri nets. The definition of net commonly used in the categorical line of work is the following:

Notation 1. *Let S be a set; denote with $S^{\oplus}$ the set of* finite *multisets over S. Multiset sum will be denoted with $\oplus$, multiplication with $\odot$ and difference (only partially defined) with $\ominus$. $S^{\oplus}$ with $\oplus$ and the empty multiset is isomorphic to the free commutative monoid on S.*

---

[1] An unrelated categorical approach to nets with catalysts can be found in [2].

Definition 1 (Petri net). We define a *Petri net* as a couple functions $T \xrightarrow{s,t} S^\oplus$ for some sets T and S, called the set of places and transitions of the net, respectively.

A *morphism of nets* is a couple of functions $f : T \to T'$ and $g : S \to S'$ such that the following square commutes, with $g^\oplus : S^\oplus \to S'^\oplus$ the obvious lifting of g to multisets:

$$
\begin{array}{ccccc}
S^\oplus & \xleftarrow{\ s\ } & T & \xrightarrow{\ t\ } & S^\oplus \\
{\scriptstyle g^\oplus}\downarrow & & \downarrow{\scriptstyle f} & & \downarrow{\scriptstyle g^\oplus} \\
S'^\oplus & \xleftarrow{\ s'\ } & T' & \xrightarrow{\ t'\ } & S'^\oplus
\end{array}
$$

Petri nets and their morphisms form a category, denoted **Petri**. The reader can find additional details in [18].

Definition 2 (Markings and firings). A *marking* for a net $T \xrightarrow{s,t} S^\oplus$ is an element of $S^\oplus$, representing a distribution of tokens in the net places. A transition u is *enabled* in a marking M if $M \ominus s(u)$ is defined. An enabled transition can *fire*, moving tokens in the net. Firing is considered an atomic event, and the marking resulting from firing u in M is $M \ominus s(u) \oplus t(u)$.

Category theory provides a slick definition to represent all the possible executions of a net – all the ways one can fire transitions starting from a given marking – as morphisms in a category. There are various ways to do this [3,11,12,17,18,22], depending if we want to consider tokens as indistinguishable (common-token philosophy) or not (individual-token philosophy). In this work, we focus on chemical reactions. Since we consider atoms and molecules of the same kind to be physically indistinguishable, we will adopt the common-token perspective. In this case, the category of executions of a net is a *commutative monoidal category* – a monoidal category whose monoid of objects is commutative.

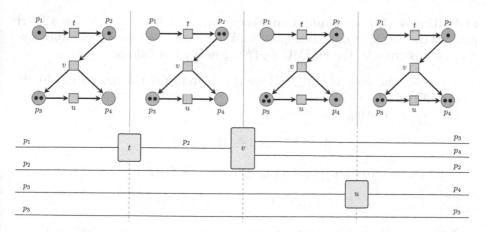

Definition 3 (Category of executions – common-token philosophy). Let $N : T \xrightarrow{s,t} S^\oplus$ be a Petri net. We can generate a *free commutative strict monoidal category (FCSMC)*, $\mathfrak{C}(N)$, as follows:

- The monoid of objects is $S^\oplus$. Monoidal product of objects A, B, denoted with $A \oplus B$, is given by the multiset sum;
- Morphisms are generated by T: each $u \in T$ corresponds to a morphism generator (u, su, tu), pictorially represented as an arrow $su \xrightarrow{u} tu$; morphisms are obtained by considering all the formal (monoidal) compositions of generators and identities.

The readers can find a detailed description of this construction in [17].

As shown in the picture above, objects in $\mathfrak{C}(N)$ represent markings of a net: $A \oplus A \oplus B$ means "two tokens in A and one token in B". Morphisms represent executions of a net, mapping markings to markings. A marking is reachable from another one if and only if there is a morphism between them.

The correspondence between Petri nets and their executions is categorically well-behaved, defining an adjunction between the category **Petri** and the category **CSMC** of commutative strict monoidal categories, with Definition 3 building the left-adjoint **Petri** → **CSMC**. The readers can find additional details in [17].

3   The Internal Mana Construction

The idea presented in the introduction can naïvely be formalised by just attaching an extra input place to any transition in a net, representing the mana a given transition can consume. We call the following construction *internal* because it builds a category directly, in contrast with an external equivalent construction given in Definition 7.

Definition 4 (Internal mana construction). Let $N : T \xrightarrow{s,t} S^\oplus$ be a Petri net, and consider $\mathfrak{C}(N)$, its corresponding FCSMC. The *internal mana construction of N* is given by the FCSMC $\mathfrak{C}_M(N)$ generated as follows:

- The generating objects of $\mathfrak{C}_M(N)$ are the coproduct of the generating objects of $\mathfrak{C}(N)$ and T;
- For each generating morphism

$$A_1 \oplus \cdots \oplus A_n \xrightarrow{u} B_1 \oplus \cdots \oplus B_m$$

in $\mathfrak{C}(N)$, we introduce a morphism generator in $\mathfrak{C}_M(N)$:

$$A_1 \oplus \cdots \oplus A_n \oplus u \xrightarrow{u} B_1 \oplus \cdots \oplus B_m$$

Notice that the writing above makes sense because u is an element of T.

Because of the adjunction between **Petri** and **CSMC**, we can think every FCSMC as being presented by a Petri net. The category of Definition 4 is presented precisely by the net obtained from N as we did in Sect. 1: the additional generating objects of $\mathfrak{C}_M(N)$ represent the places containing the mana associated with each transition.

Example 1. Performing the construction in Definition 4 on the category of executions of the net on the left gives the category of executions of the net on the right, as we expect:

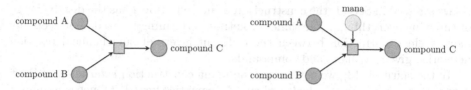

Proposition 1. The assignment $\mathfrak{C}(N) \mapsto \mathfrak{C}_M(N)$ defines a comonad[2] in the category of FCMSCs and strict monoidal functors between them, **FCSMC**.

Proof. First of all, we have to prove that the procedure is functorial. For any strict monoidal functor $F : \mathfrak{C}(N) \to \mathfrak{C}(M)$ we define the action on morphisms $\mathfrak{C}_M(F) : \mathfrak{C}_M(N) \to \mathfrak{C}_M(M)$ as the following monoidal functor:

- $\mathfrak{C}_M(F)$ agrees with F on generating objects coming from $\mathfrak{C}(N)$. If u is a generating morphism of $\mathfrak{C}(N)$ and it is $Fu = f$, then $\mathfrak{C}_M(F)\,u = f^{\mathbb{N}}$, with $f^{\mathbb{N}}$ being the multiset[3] counting how many times each generating morphism of $\mathfrak{C}(M)$ is used in f.
- $\mathfrak{C}_M(F)$ agrees with F on generating morphisms.

Identities and compositions are clearly respected, making $\mathfrak{C}_M(\_)$ an endofunctor in **FCSMC**. As a counit, on each component N we define the strict monoidal functor $\epsilon_N : \mathfrak{C}_M(N) \to \mathfrak{C}(N)$ sending:

- Generating objects coming from $\mathfrak{C}(N)$ to themselves, and every other generating object to the monoidal unit.
- Generating morphisms to themselves.

The procedure is natural in the choice of N, making ϵ into a natural transformation $\mathfrak{C}_M(\_) \to \mathrm{id}_{\textbf{FCSMC}}$.

As for the comultiplication, on each component N we define the strict monoidal functor $\delta_N : \mathfrak{C}_M(N) \to \mathfrak{C}_M(\mathfrak{C}_M(N))$ sending:

---

[2] Given a category $\mathcal{C}$, a *comonad* on $\mathcal{C}$ is an endofunctor S endowed with two natural transformations $\delta : S \Rightarrow S \circ S$ and $\epsilon : S \Rightarrow 1_{\mathcal{C}}$ such that δ is coassociative and has ϵ as a counit. More succinctly, δ is a comonoid in the monoidal category $[\mathcal{C}, \mathcal{C}]$ of endofunctors of $\mathcal{C}$. We give a precise definition in our Appendix, see Definition 9. See also [20, §5.3] for the definition and a variety of examples.

[3] This makes sense since $\mathfrak{C}(M)$ is free, hence decomposition of morphisms in terms of (monoidal) compositions of generators and identities is unique modulo the axioms of monoidal categories, which do not introduce nor remove generating objects.

- Generating objects coming from $\mathfrak{C}(N)$ to themselves, every other generating object u is sent to $u \oplus u$.
- Generating morphisms are again sent to themselves.

The naturality of δ and the comonadicity conditions are a straightforward check.

□

4   The External Mana Construction

As we stressed in Sect. 1, the construction as in Definition 4 has the disadvantage of throwing everything in the same bucket: in performing it, we do not keep any more a clear distinction between the different layers of our chemical reaction networks, given by mana and compounds.

In the spirit of [13], we now recast the mana construction *externally*, as Petri nets with a *semantics* attached to them. A semantics for a Petri net is a functor from its category of executions to some other monoidal category $\mathcal{S}$.

A huge conceptual difference is that in [13] this functor was required to be strict monoidal. This point of view backed up the interpretation that a semantics "attaches extra information to tokens", to be used by the transitions somehow. In here, we require this functor to be *lax-monoidal*:[4] lax-monoidality amounts to saying that we can attach *non-local* information to tokens: tokens may "know" something about the overall state of the net and the laxator represents the process of "tokens joining knowledge".

In terms of mana construction, we want to endow each token with a local "knowledge" of how much mana each transition has available. Laxating amounts to consider ensembles of tokens together – as entangled, if you wish – where their knowledge is merged.

Example 2.

If token a knows that transition u has 3 mana left, and token b knows that transitions u and v have 1 and 8 mana left, respectively, then tokens a and b, considered together, know that transitions u and v have $3 + 1 = 4$ and $0 + 8 = 8$ mana left, respectively.

Definition 5 (Non-local semantics – common-token philosophy). Let N be a Petri net and let $\mathcal{S}$ be a monoidal category. A *Petri net with a non-local*

---

[4] A *lax monoidal* functor between two monoidal categories $(\mathfrak{C}, \boxtimes J)$, $(\mathcal{D}, \otimes, I)$ is a functor $F : \mathfrak{C} \to \mathcal{D}$ endowed with maps $m : FA \otimes FB \to F(A \boxtimes B)$ and $u : I \to FJ$ satisfying suitable coherence conditions; see [16, Def. 3.1]. If m, u are isomorphisms in $\mathcal{D}$, F is called *strong monoidal*. If just u is an isomorphism, F is called *normal monoidal*.

commutative semantics is a couple $(N, N^\sharp)$, with $N^\sharp$ a lax-monoidal functor $\mathfrak{C}(N) \to \mathcal{S}$. A morphism $(N, N^\sharp) \to (M, M^\sharp)$ of Petri nets with commutative semantics is a strict monoidal functor $\mathfrak{C}(N) \xrightarrow{F} \mathfrak{C}(M)$.

We denote the category of Petri nets with non-local commutative semantics with **Petri**$^\mathcal{S}$.

We now provide an external version of the mana construction.

Notation 2. *We denote with* **Span** *the bicategory of sets, spans and span morphisms between them.*[5] *Recall that a morphism $A \to B$ in* **Span** *consists of a set S and a pair of functions $A \leftarrow S \to B$. When we need to notationally extract this information from f, we write $A \xleftarrow{f_1} S_f \xrightarrow{f_2} B$. We sometimes consider a span as a morphism $f : S_f \to A \times B$, thus we may write $f(s) = (a, b)$ for $s \in S_f$ with $f_1(s) = a$ and $f_2(s) = b$. Recall moreover that a 2-cell in* **Span** *$f \Rightarrow g$ is a function $\theta : S_f \to S_g$ such that $f = g \circ \theta$.*

Observe that there is nothing in the previous definition of **Span** that requires the objects to be mere sets; in particular, we will later employ the following variation on Notation 2:

Definition 6 (Spans of pointed sets). Define a bicategory **Span.** *of spans of pointed sets* objects the pointed sets, (A, a) where $a \in A$ is a distinguished element; composition of spans is as expected

Remark 1. This is in turn just a particular case of a more general construction: let $\mathcal{C}$ be a category with pullbacks; then, there is a bicategory **Span** $\mathcal{C}$ having 1-cells the spans $A \leftarrow X \to B$ of morphisms of $\mathcal{C}$, and where a pullback of their adjacent legs defines the composition of 1-cells. Evidently, **Span** = **Span(Set)** and **Span.** = **Span(Set.)**, where **Set.** is the category of pointed sets (A, a) and maps that preserve the distinguished elements of the domain and codomain. See [9, §2] and [8] for a way more general perspective on bicategories of the form **Span** $\mathcal{C}$ and the universal property of the **Span** construction.

Definition 7 (External mana construction). Given a Petri net $N : T \xrightarrow{s,t} S^\oplus$, define the following functor $N^\sharp : \mathfrak{C}(N) \to$ **Span**:

- Each object A of $\mathfrak{C}(N)$ is mapped to the set $T^\oplus$, the set of multisets over the transitions of N;
- Each morphism $A \xrightarrow{f} B$ is sent to the span $N^\sharp f$ defined as:

$$T^\oplus \xleftarrow{-\oplus f^N} T^\oplus = T^\oplus$$

With f^N being the multiset counting how many times each generating morphism of $\mathfrak{C}(M)$ is used in f.

---

[5] See [6, Def. 1.1] for the definition of bicategory; intuitively, in a bicategory, one has objects (0-cells), 1-cells and 2-cells, and composition of 1-cells is associative and unital up to some specified invertible 2-cells $F(GH) \cong (FG)H$ and $F1 \cong F \cong 1F$.

Proposition 2. The functor of Definition 7 is lax monoidal. Functors as in Definition 7 form a subcategory of $\mathbf{Petri}^{\mathbf{Span}}$, which we call $\mathbf{Petri}^{\mathcal{M}}$.

Proof. Functor laws are obvious: $\mathrm{id}_A^{\mathbb{N}}$ is the empty multiset for each object A, hence $N^{\sharp}\mathrm{id}_A = \mathrm{id}_{T^{\oplus}}$. This correspondence preserves composition since

$$
\begin{array}{ccccc}
T^{\oplus} & = & T^{\oplus} & = & T^{\oplus} \\
{\scriptstyle -\oplus g^{\mathbb{N}}}\downarrow & & \downarrow{\scriptstyle -\oplus g^{\mathbb{N}}} & & \\
T^{\oplus} & = & T^{\oplus} & & \\
{\scriptstyle -\oplus f^{\mathbb{N}}}\downarrow & & & & \\
T^{\oplus} & & & &
\end{array}
\qquad = \quad -\oplus f^{\mathbb{N}}\oplus g^{\mathbb{N}}
\qquad
\begin{array}{ccc}
T^{\oplus} & = & T^{\oplus} \\
& & \downarrow \\
& & T^{\oplus}
\end{array}
$$

The laxator is the morphism $S^{\oplus} \times S^{\oplus} \xrightarrow{\oplus} S^{\oplus}$ that evaluates two multisets to their sum, embedded in a span. The naturality condition for the laxator reads:

$$
\begin{array}{ccc}
T^{\oplus} \times T^{\oplus} & \xrightarrow{N^{\sharp}f \times N^{\sharp}g} & T^{\oplus} \times T^{\oplus} \\
{\scriptstyle \oplus}\downarrow & & \downarrow{\scriptstyle \oplus} \\
T^{\oplus} & \xrightarrow[N^{\sharp}(f\oplus g)]{} & T^{\oplus}
\end{array}
$$

And the two morphisms from $T^{\oplus} \times T^{\oplus} \to T^{\oplus}$ are:

$$
\begin{array}{ccccc}
T^{\oplus} \times T^{\oplus} & = & T^{\oplus} \times T^{\oplus} & \xrightarrow{\oplus} & T^{\oplus} \\
\| & & \| & & \\
T^{\oplus} \times T^{\oplus} & = & T^{\oplus} \times T^{\oplus} & & \\
{\scriptstyle (-\oplus f^{\mathbb{N}})\times(-\oplus g^{\mathbb{N}})}\downarrow & & & & \\
T^{\oplus} \times T^{\oplus} & & & &
\end{array}
\qquad = \quad
\begin{array}{ccccc}
T^{\oplus} \times T^{\oplus} & \xrightarrow{\oplus} & T^{\oplus} & = & T^{\oplus} \\
{\scriptstyle (-\oplus f^{\mathbb{N}})\times(-\oplus g^{\mathbb{N}})}\downarrow & & & & \downarrow{\scriptstyle -\oplus f^{\mathbb{N}}\oplus g^{\mathbb{N}}} \\
T^{\oplus} \times T^{\oplus} & \xrightarrow{\oplus} & T^{\oplus} & & \\
\| & & & & \\
T^{\oplus} \times T^{\oplus} & & & &
\end{array}
$$

which evidently coincide. Interaction with the associators, unitors and symmetries of the monoidal structure is guaranteed by the fact that they are all identities in $\mathfrak{C}(N)$. $\qquad\square$

The external mana construction has the advantage of keeping the reaction layer and the mana layer separated completely. In this setting, we say that a marking of the net is a couple (X, u), with X an object of $\mathfrak{C}(N)$ and $u \in T^{\oplus}$ representing the initial distribution of mana for our transitions. A transition $X \xrightarrow{f} Y$ is again a generating morphism of $\mathfrak{C}(N)$, and we say that it is enabled if $N^{\sharp}f_1$ hits u, or, more explicitly, if $u \ominus f^{\mathbb{N}}$ is defined. Since $f^{\mathbb{N}}$ for f a morphism generator is defined to be 0 everywhere and 1 on f, this amounts to say that f is enabled when $u(f) - 1 \geq 0$. In that case, the resulting marking after the firing is $(Y, u(f) - 1)$: Each firing just decreases the mana of the firing transition by 1.

Example 3. Consider the net

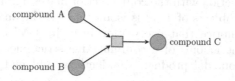

compound A

compound C

compound B

In the marking $(A \oplus B, 2)$, the transition is enabled. The resulting marking will be $(C, 1)$. The transition is *not* enabled in the marking $(A \oplus B, 0)$ or $(A, 4)$.

4.1 Internalization

Having given two different definitions of endowing a net with mana, it seems fitting to say how the two are connected. As we already stressed, we abide by the praxis already established in [13] and prove that the external and internal mana constructions describe the same thing from different points of view:

Theorem 1. *Let $(N, N^\sharp)$ be an object of* **Petri**$^{\mathcal{M}}$. *The category $\mathfrak{C}_M(N)$ of Definition 4 is isomorphic to the category of elements $\int N^\sharp$.*[6] *Explicitly:*

- *Objects of $\int N^\sharp$ are couples (X, x) where X is a object of $\mathfrak{C}_M(N)$ and $x \in N^\sharp X$.*
- *Morphisms $(X, x) \to (Y, y)$ of $\int N^\sharp$ are morphisms (f, s) with $f : X \to Y$ of $\mathfrak{C}_M(N)$ and s such that $N^\sharp f s = (x, y)$.*

Proof. First of all, we need to define a commutative strict monoidal structure on $\int N^\sharp$. Given the particular shape of $N^\sharp$, the objects of its category of elements are pairs where the first component is a multiset on the places of N and the second one is a multiset on its transition. Hence we can define:

$$(C, x) \boxtimes (D, y) := (C \oplus D, x \oplus y)$$

(Note that in order to obtain an element in $N^\sharp(C \oplus D)$, we have implicitly applied the laxator $\oplus : N^\sharp C \times N^\sharp D \to N^\sharp(C \oplus D)$ to the elements in the second coordinate.) Commutativity of $\boxtimes$ follows from the commutativity of $\oplus$. On morphisms, if we have $(A_1, x_1) \xrightarrow{(f_1, s_1)} (B_1, y_1)$ and $(A_2, x_2) \xrightarrow{(f_2, s_2)} (B_2, y_2)$ then it is $N^\sharp f_1 s_1 = (x_1, y_1)$ and $N^\sharp f_2 s_2 = (x_2, y_2)$, and hence by naturality of the laxator $N^\sharp(f_1 \oplus f_2)(s_1, s_2) = ((x_1 \oplus x_2), (y_1 \oplus y_2))$, allowing us to set $f_1 \boxtimes f_2 = f_1 \oplus f_2$. Associators and unitors are defined as in $\mathfrak{C}(N)$.

Now we prove freeness: by definition, objects are a free monoid generated by couples (p, I) and (I, u) with p a generating object of $\mathfrak{C}(N)$ (a place of N), u a

---

[6] The category of elements of a functor $F : \mathcal{C} \to$ **Set** is defined having objects the pairs (C, x), where $x \in FC$, and morphisms $(C, x) \to (C', x')$ the morphisms $u : C \to C'$ such that Fu sends x into x'. See [4, §12.2], where this is called the *Grothendieck construction* performed on F. Here we need to tweak this construction in order for it to make sense for *lax* functors valued in **Span**, using essentially the same technique in [19].

generating morphism of $\mathfrak{C}_M(N)$ (a transition of N), and I the tensor unit. These generators are in bijection with the coproduct of places and transitions of N. As such, the monoid of objects of $\int N^\sharp$ is isomorphic to the one of $\mathfrak{C}_B(N)$.

On morphisms, notice that every morphism in $\int N^\sharp$ can be written univocally – modulo the axioms of a commutative strict monoidal category – as a composition of monoidal products of identities and morphisms of the form $(A, u) \xrightarrow{(f,u)} (B, u')$, with f a morphism generator in $\mathfrak{C}(N)$ and $u = u' \oplus f^{\mathbb{N}}$.

The isomorphism between $\int N^\sharp$ and $\mathfrak{C}_B(N)$ follows by observing that the following mappings between objects and morphism generators are bijections:

$$(A, u) \mapsto A \oplus u$$
$$(A, u) \xrightarrow{(f,u)} (B, u') \mapsto A \oplus u \xrightarrow{f} B \oplus u'$$

$\square$

Example 4. The internalization of the net in Example 3 gives exactly the net of Example 1.

5   Extending the Mana Construction

Focusing more on the external mana construction of Definition 7, we realize that it is somehow restrictive: it makes sense to map each generating object of an FCSMC $\mathfrak{C}_M(N)$ to the set of multisets over the transitions of N. This construction captures the idea of endowing each transition with an extra place representing its mana. On the other hand, the only requirement we would expect on morphisms is that, to fire, a transition must consume mana only from its mana pool. In Definition 7 we do much more than this, hardcoding that "one firing = one mana" in the structure of the functor.

The act of replacing the mapping on morphisms in Definition 7 with the following span provides a reasonable generalization of the previous perspective:

$$T^\oplus \xleftarrow{-\oplus(\alpha \odot f^{\mathbb{N}})} T^\oplus \xrightarrow{-\oplus(\beta_f)} T^\oplus$$

with α and β_f arbitrary multisets. In doing so, the only thing we are disallowing in our new definition is for transitions to consume mana of other transitions: each transition may use only the mana in its pool. Still, it is now possible for transitions to:

- Fire without consuming mana;
- Consume more than 1 unit of mana to fire;
- Produce mana – also for other transitions – upon firing.

These are all good conditions in practical applications. The first models chemical reactions that do not need any additional compound to work; the second aims to model reactions that need more than one molecule of a given compound to work; the third models both catalysts – which completely regenerate their mana at the end of the reactions they aid – and reactions that produce, as byproducts, enzymes needed by other reactions.

Example 5. It is worth giving an explicit description of how the internalized version of a net, as in our attempted generalized definition, looks like. In the picture below, each transition has its mana, but now this mana does not have to be necessarily used, as for transition u_1, or can be used more than once, as for transition u_2. Furthermore, transitions such as u_3 regenerate their mana after firing (catalysts), while transitions such as u_2 and u_4 produce mana for each other in a closed loop. u_4 also produces more than one kind of mana as a byproduct of its firing. It is worth noticing that this formalism allows to model nets that never run out of mana, and that we think of as "self-sustaining" [14].

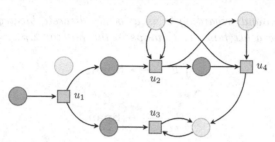

When looking at technicalities, unfortunately, things are not so easy. Defining α and the family β_f so that functorial laws are respected is tricky. Luckily enough, we do not need to do so explicitly. Indeed, we can generalize the internal mana-net construction of Definition 4 to the following one, that subsumes nets as in Example 5:

Definition 8 (Generalized internal mana construction). Let $N : T \xrightarrow{s,t} S^{\oplus}$ be a Petri net, and consider $\mathfrak{C}(N)$, its corresponding FCSMC. A *generalized internal mana construction for N* is any FCSMC $\mathfrak{C}_M(N)$ such that:

- The generating objects of $\mathfrak{C}_M(N)$ are the coproduct of the generating objects of $\mathfrak{C}(N)$ and T;
- Generating morphisms

$$A_1 \oplus \cdots \oplus A_n \xrightarrow{u} B_1 \oplus \cdots \oplus B_m$$

in $\mathfrak{C}(N)$ are in bijection with generating morphisms in $\mathfrak{C}_M(N)$:

$$A_1 \oplus \cdots \oplus A_n \oplus U_1 \xrightarrow{u} U_2 \oplus B_1 \oplus \cdots \oplus B_m$$

With U_1 a multiset over T being 0 on any $u' \neq u$, and U_2 being an arbitrary multiset over T.

Notice moreover that, for each generalized mana-net $\mathfrak{C}_M(N)$, we obtain a strict monoidal functor $F : \mathfrak{C}_M(N) \to \mathfrak{C}(N)$ as in Proposition 1: we send generating objects of $\mathfrak{C}(N)$ to themselves, all the other generating objects to the monoidal unit and generating morphisms to themselves. We keep calling F the *counit* of $\mathfrak{C}_M(N)$, even if it won't be in general true that we still get a comonad.

Counits can be turned into functors $\mathfrak{C}(N) \to \mathbf{Span}$ using a piece of categorical artillery called the *Grothendieck construction* (or the *category of elements* construction).

Theorem 2 (Grothendieck construction, [19]). *Let $\mathcal{C}$ be a category. Then, there is an equivalence $\mathbf{Cat}/\mathcal{C} \simeq \mathbf{Cat}_l[\mathcal{C}, \mathbf{Span}]$, with $\mathbf{Cat}_l[\mathcal{C}, \mathbf{Span}]$ being the category of lax functors $\mathcal{C} \to \mathbf{Span}$. A functor $F : \mathcal{D} \to \mathcal{C}$ defines a functor $\Gamma F : \mathcal{C} \to \mathbf{Span}$ as follows:*

- *On objects, C is mapped to the set $\{D \in \mathcal{D} \mid FD = C\}$;*
- *On morphisms, $C \xrightarrow{f} C'$ is mapped to the span*

$$\{D \in \mathcal{C} \mid FD = C\} \xleftarrow{s} \{g \in \mathcal{D} \mid Ff = g\} \xrightarrow{t} \{D \in \mathcal{C} \mid FD = C'\}$$

The other way around, regarding $\mathcal{C}$ as a locally discrete bicategory and letting $F : \mathcal{C} \to \mathbf{Span}$ be a lax functor, F maps to the functor Σ_F, from the pullback (in $\mathbf{Cat}$) below:

$$
\begin{array}{ccc}
\int F & \longrightarrow & \mathbf{Span_\bullet} \\
{\scriptstyle \Sigma_F} \downarrow & \lrcorner & \downarrow {\scriptstyle U} \\
\mathcal{C} & \xrightarrow{\ F\ } & \mathbf{Span}
\end{array}
$$

where $\mathbf{Span_\bullet}$ is the bicategory of spans between pointed sets, and U is the forgetful functor.

More concretely, $\int F$ is defined as the category (all 2-cells are identities, due to the 2-discreteness of $\mathcal{C}$) having

- *0-cells of $\int F$ are couples (X, x) where X is a 0-cell of $\mathcal{C}$ and $x \in FX$;*
- *1-cells $(X, x) \to (Y, y)$ of $\int F$ are couples (f, s) where $f : X \to Y$ is a 1-cell of $\mathcal{C}$ and $s \in S_{Ff}$ with $Ff(s) = (x, y)$. Representing a span as a function $(S, s) \to (X \times Y, (x, y))$ between (pointed) sets, a morphism is a pair (f, s) such that $Ff : s \mapsto (x, y)$.*

Finally, the categories $\int F$ and $\mathcal{D}$ are isomorphic.

This result is a particular case of a more general correspondence between slice categories and lax normal functors to the category of profunctors [15], which is well-known in category theory and dates back to Bénabou [5,6]. It gives an entirely abstract way to switch from/to and define internal/external semantics for mana-nets. Indeed, with a proof partly similar to the one carried out in our Theorem 1, we can show that:

Proposition 3. Monoidality of $\mathfrak{C}_M(N) \xrightarrow{F} \mathfrak{C}(N)$ implies ΓF is lax-monoidal.

We can thus define the external semantics of a generalized mana-net by applying Γ to F. A generalization of Theorem 1 then holds by definition.

Summing up, we showed that the mana-net construction can be generalized to more practical applications and the correspondence between a "naïve" internal semantics and a "type-aware" external one is still preserved. The evident price we have to pay for our generalization is that our external semantics is now not just lax-monoidal but lax-monoidal-lax.

6   Conclusion and Future Work

In this work, we introduced a new notion of Petri net where transitions come endowed with "mana", a quality representing how many times a transition will be able to fire before losing its effectiveness. We believe this may be especially useful in modelling chemical processes mediated by enzymes that degrade over time.

Importantly, we showed how a categorical point of view on the matter allows to give two different definitions: A naïve, "hands-on" one, that we called *internal*, and a type-aware, functorial one, that we called *external*, which we proved to be two sides of the same coin.

Indeed, the equivalence between internal and external semantics is the consequence of a much more profound result in category theory, connecting slice categories and categories of lax monoidal functors. We were able to rely on this result to generalize our mana-nets further, while keeping the equivalence between the internal and external points of view.

We believe that further generalizations of the external semantics presented here may prove valuable to produce categorical semantics for nets with *inhibitor arcs* [1]. An inhibitor arc is an input arc to a transition that is enabled only when there are no tokens in their place. This concept is powerful enough to turn Petri nets into a Turing-complete model of computation [25, 26].

Indeed, we notice that by relaxing Definition 7 to allow *any* span $T^{\oplus} \rightarrow T^{\oplus}$, we can model situations where a transition can fire just if it has no mana (e.g., we can map transition f to a span that is only defined when its source multiset has value 0 on f). The similarities in behaviour with inhibitor arcs are evident and constitute a direction of future work that we will surely pursue. The various technicalities involved are nevertheless tricky and necessitate a careful investigation.

Another direction of future work is about implementing the ideas at this moment presented using already available category theory libraries, such as [10].

Acknowledgements. The first author was supported by the project MIUR PRIN 2017FTXR7S "IT-MaTTerS" and by the Independent Ethvestigator Program.

The second author was supported by the ESF funded Estonian IT Academy research measure (project 2014-2020.4.05.19-0001).

A video presentation of this paper can be found on Youtube at 9sxVBJs1okE.

A   Category Theory

Here we collect the category-theoretic definitions on which this work relies.

Definition 9 (Monad, comonad). Let $\mathcal{C}$ be a category; a *monad* on $\mathcal{C}$ consists of an endofunctor $T : \mathcal{C} \rightarrow \mathcal{C}$ endowed with two natural transformations

- $\mu : T \circ T \Rightarrow T$, the *multiplication* of the monad, and
- $\eta : \mathrm{id}_{\mathcal{C}} \Rightarrow T$, the *unit* of the monad,

such that the following axioms are satisfied:

– the multiplication is associative, i.e. the diagram

$$T \circ T \circ T \xrightarrow{T*\mu} T \circ T$$

is commutative, i.e. the equality of natural transformations $\mu \circ (\mu * T) = \mu \circ (T * \mu)$ holds;
– the multiplication has the transformation η as unit, i.e. the diagram

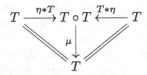

is commutative, i.e. the equality of natural transformations $\mu \circ (\eta * T) = \mu \circ (T * \eta) = \mathrm{id}_T$ holds.

Dually, let $\mathcal{C}$ be a category; a *comonad* on $\mathcal{C}$ consists of an endofunctor $T : \mathcal{C} \to \mathcal{C}$ endowed with two natural transformations

– $\sigma : T \Rightarrow T \circ T$, the *comultiplication* of the comonad, and
– $\epsilon : T \Rightarrow \mathrm{id}_{\mathcal{C}}$, the *counit* of the comonad,

such that the following axioms are satisfied:

– the comultiplication is coassociative, i.e. the diagram

$$T \circ T \circ T \xleftarrow{T*\sigma} T \circ T$$

is commutative.
– the comultiplication has the transformation ϵ as counit, i.e. the diagram

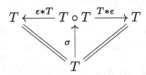

is commutative.

Definition 10 (Bicategory). A *(locally small) bicategory* $\mathcal{B}$ consists of the following data.

1. A class $\mathcal{B}_o$ of *objects*, denoted with Latin letters like $A, B, \ldots$.
2. A collection of (small) categories $\mathcal{B}(A, B)$, one for each $A, B \in \mathcal{B}_o$, whose objects are called *1-cells* or *arrows* with *domain* A and *codomain* B, and whose morphisms $\alpha : f \Rightarrow g$ are called *2-cells* or *transformations* with domain f and codomain g; the composition law $\circ$ in $\mathcal{B}(A, B)$ is called *vertical composition* of 2-cells.
3. A *horizontal composition* of 2-cells

$$\boxminus_{\mathcal{B}, ABC} : \mathcal{B}(B, C) \times \mathcal{B}(A, B) \to \mathcal{B}(A, C) : (g, f) \mapsto g \boxminus f$$

defined for any triple of objects A, B, C. This is a family of functors between hom-categories.
4. For every object $A \in \mathcal{B}_o$ there is an arrow $\mathrm{id}_A \in \mathcal{B}(A, A)$.

To this basic structure we add

1. a family of invertible maps $\alpha_{fgh} : (f \boxminus g) \boxminus h \cong f \boxminus (g \boxminus h)$ natural in all its arguments f, g, h, which taken together form the *associator* isomorphisms;
2. a family of invertible maps $\lambda_f : \mathrm{id}_B \boxminus f \cong f$ and $\varrho_f : f \boxminus \mathrm{id}_A \cong f$ natural in its component $f : A \to B$, which taken together form the *left unitor* and *right unitor* isomorphisms.

Finally, these data are subject to the following axioms.

1. For every quadruple of 1-cells f, g, h, k we have that the diagram

commutes.
2. For every pair of composable 1-cells f, g,

commutes.

Definition 11 (Pseudofunctor, (co)lax functor). Let $\mathcal{B}, \mathcal{C}$ be two bicategories; a *pseudofunctor* consists of

1. a function $F_o : \mathcal{B}_o \to \mathcal{C}_o$,

2. a family of functors $F_{AB} : \mathcal{B}(A, B) \to \mathcal{C}(FA, FB)$,

3. an invertible 2-cell $\mu_{fg} : Ff \circ Fg \Rightarrow F(fg)$ for each $A \xrightarrow{g} B \xrightarrow{f} C$, natural in f (with respect to vertical composition) and an invertible 2-cell $\eta : \eta_f : \mathrm{id}_{FA} \Rightarrow F(\mathrm{id}_A)$, also natural in f.

These data are subject to the following commutativity conditions for every 1-cell $A \to B$:

$$
\begin{array}{ccc}
Ff \circ \mathrm{id}_A \xrightarrow{\varrho_{Ff}} Ff & \qquad & \mathrm{id}_B \circ Ff \xrightarrow{\lambda_{Ff}} Ff \\
\downarrow^{Ff*\eta} \qquad \downarrow^{F(\varrho_f)} & & \downarrow^{\eta*Ff} \qquad \downarrow^{F(\lambda_f)} \\
Ff \circ F(\mathrm{id}_A) \xrightarrow{\mu_{f,\mathrm{id}_A}} F(f \circ \mathrm{id}_A) & & F(\mathrm{id}_B) \circ Ff \xrightarrow{\mu_{\mathrm{id}_B,f}} F(\mathrm{id}_B \circ f)
\end{array}
$$

$$
\begin{array}{ccc}
(Ff \circ Fg) \circ Fh & \xrightarrow{\alpha_{Ff,Fg,Fh}} & Ff \circ (Fg \circ Fh) \\
\downarrow^{\mu_{fg}*Fh} & & \downarrow^{Ff*\mu_{gh}} \\
F(fg) \circ Fh & & Ff \circ F(gh) \\
\downarrow^{\mu_{fg}*Fh} & & \downarrow^{\mu_{f,gh}} \\
F((fg)h) & \xrightarrow{F\alpha_{fgh}} & F(f(gh))
\end{array}
$$

(we denote invariably α, λ, ϱ the associator and unitor of $\mathcal{B}, \mathcal{C}$).

A *lax* functor is defined by the same data, but both the 2-cells $\mu : Ff \circ Fg \Rightarrow F(fg)$ and $\eta : \mathrm{id}_{FA} \Rightarrow F(\mathrm{id}_A)$ can be non-invertible; the same coherence diagrams in Definition 11 hold. A *colax* functor reverses the direction of the cells μ, η, and the commutativity of the diagrams in Definition 11 changes accordingly.

Example 6. Here we collect a few examples of bicategories and 2-categories;

1. Let $\mathcal{J}$ be the terminal category, having only one object and identity morphism, and let $\mathcal{C}$ be a 2-category; a lax functor $T : \mathcal{J} \xrightarrow{L} \mathcal{C}$ is a correspondence that sends
    - the unique object $*$ into a 0-cell X of $\mathcal{C}$;
    - the unique 1-cell id_* into a endo-1-cell $X \xrightarrow{T} X$;
   the laxity cells of T are two 2-cells $\mu : T \circ T \Rightarrow T$ and $\eta : \mathrm{id}_X \Rightarrow T$, and the axioms of a lax functor in Definition 11 correspond exactly to the monad axioms in Definition 9.

2. The class of (small) categories, functors and natural transformations is a *strict* 2-category: the 2-cells $\alpha_{fgh}, \lambda_f, \rho_f$ are all identities. The same is true for the class of (small) $\mathcal{V}$-categories, $\mathcal{V}$-functors, and $\mathcal{V}$-natural transformations, enriched over a monoidal category $\mathcal{V}$.

References

1. Agerwala, T.: Complete model for representing the coordination of asynchronous processes (1974)
2. Baez, J.C., Foley, J., Moeller, J.: Network models from petri nets with catalysts. Compositionality **1**, 4 (2019)
3. Baez, J.C., Genovese, F., Master, J., Shulman, M.: Categories of Nets. arXiv: 210 1.04238 [cs, math] (2021)
4. Barr, M., Wells, C.: Category Theory for Computing Science, vol. 49. Prentice Hall, New York (1990)
5. Bénabou, J., Streicher, T.: Distributors at work. Lecture notes written by Thomas Streicher (2000)
6. Bénabou, J.: Introduction to bicategories. In: Bénabou, J., et al. (eds.) Reports of the Midwest Category Seminar, vol. 47, pp. 1–77. Springer, Heidelberg (1967)
7. Churchill, A., Biderman, S., Herrick, A.: Magic: The gathering is Turing complete. arXiv preprint arXiv:1904.09828 (2019)
8. Dawson, R.M., Paré, R., Pronk, D.A.: Universal properties of span. Theory Appl. Categ. **13**(4), 61–85 (2004)
9. Dawson, R., Paré, R., Pronk, D.: The span construction. Theory Appl. Categ. **24**(13), 302–377 (2010)
10. Genovese, F., et al.: Idris-Ct: A Library to Do Category Theory in Idris. arXiv: 1912.06191 [cs, math] (2019)
11. Genovese, F., Gryzlov, A., Herold, J., Perone, M., Post, E., Videla, A.: Computational Petri Nets: Adjunctions Considered Harmful. arXiv: 1904.12974 [cs, math] (2019)
12. Genovese, F., Herold, J.: Executions in (semi-)integer petri nets are compact closed categories. In: Electronic Proceedings in Theoretical Computer Science, vol. 287, pp. 127–144 (2019)
13. Genovese, F., Spivak, D.I.: A categorical semantics for guarded petri nets. In: Gadducci, F., Kehrer, T. (eds.) ICGT 2020. LNCS, vol. 12150, pp. 57–74. Springer, Cham (2020). https://doi.org/10.1007/978-3-030-51372-6_4
14. Letelier, J., Soto-Andrade, J., Guinez, F., Cornish-Bowden, A., Cárdenas, M.: Organizational invariance and metabolic closure: analysis in terms of M; R systems. J. Theor. Biol. **238**, 949–961 (2006)
15. Loregian, F.: Coend Calculus. London Mathematical Society Lecture Note Series, vol. 468 (2021). ISBN 9781108746120
16. Marcelo Aguiar, S.M.: Monoidal Functors, Species and Hopf Algebras. Centre de Recherches Mathématiques Monograph Series, vol. 29. American Mathematical Society (2010)
17. Master, J.: Petri nets based on Lawvere theories. Math. Struct. Comput. Sci. **30**(7), 833–864 (2020). arXiv: 1904.09091
18. Meseguer, J., Montanari, U.: Petri nets are monoids. Inf. Comput. **88**(2), 105–155 (1990)
19. Pavlović, D., Abramsky, S.: Specifying interaction categories. In: Moggi, E., Rosolini, G. (eds.) CTCS 1997. LNCS, vol. 1290, pp. 147–158. Springer, Heidelberg (1997). https://doi.org/10.1007/BFb0026986
20. Perrone, P.: Notes on Category Theory with examples from basic mathematics. arXiv:1912.10642 (2019)
21. Petri, C., Reisig, W.: Petri Net. Scholarpedia (2008). http://www.scholarpedia.org/article/Petri_net

22. Sassone, V.: On the category of Petri net computations. In: Mosses, P.D., Nielsen, M., Schwartzbach, M.I. (eds.) CAAP 1995. LNCS, vol. 915, pp. 334–348. Springer, Heidelberg (1995). https://doi.org/10.1007/3-540-59293-8_205
23. Wikipedia. Magic (Game Terminology) (2020). https://en.wikipedia.org/wiki/Magic_(game_terminology)
24. Wikipedia. Magic: The Gathering (2020). https://en.wikipedia.org/wiki/Magic:_The_Gathering
25. Zaitsev, D.A.: Universal petri net. Cybern. Syst. Anal. **48**(1), 498–511 (2012)
26. Zaitsev, D.A.: Toward the minimal universal petri net. IEEE Trans. Syst. Man Cybern. Syst. **44**(1), 47–58 (2014)

A Case Study
on the Graph-Transformational
Modeling and Analysis of Puzzles

Hans-Jörg Kreowski and Aaron Lye[✉]

Department of Computer Science, University of Bremen,
P.O. Box 33 04 40, 28334 Bremen, Germany
{kreo,lye}@uni-bremen.de

Abstract. In this paper, we start a case study on the use and usefulness of graph transformation in modeling and analyzing games and puzzles beginning with logic puzzles. More explicitly, we consider Sudoku,
Hashiwokakero, Arukone, and Maze aka Labyrinth. In these cases, the
underlying data structures can be represented by graphs and the puzzles have start configurations and goals besides the solving rules. Sometimes it is meaningful to regulate the rule application by some control conditions. These are the ingredients of graph transformation units
which are therefore applied as modeling framework. Based on the graph-
transformational models, one can show that Labyrinth can be solved in
polynomial time and solvability of the other three is NP-complete.

1 Introduction

Playing games and solving puzzles are typical activities of human beings. One
encounters a wealth of games and puzzles all over the world. As many games
and all puzzles challenge human intelligence, they are obvious candidates to be
investigated in the area of Artificial Intelligence. Indeed, the development of programs for Chess, Go, Poker, and many other games have been very successful in
the last decades. Several games are played on boards and many puzzles have certain diagrammatic or geometric patterns as underlying structures. Such boards
and patterns can be nicely represented by graphs so that the rules of a game
and the instructions to solve a puzzle become graph transformation rules. In the
literature of graph transformation, one encounters some singular examples in
this direction concerning the modeling of Pacman, Ludo and a few other games
(see [1–6]). In this paper, we start a more systematic case study on the use and
usefulness of graph transformation in modeling and analyzing games and puzzles
beginning with logic puzzles as they are made popular by the Japanese company
and publisher Nicoli since 1980 and as they appear in puzzle corners of various
tabloids and magazines since then. More explicitly, we consider Sudoku, Hashiwokakero, Arukone, and Maze aka Labyrinth in Sects. 3, 4, 5 and 6, respectively.
In these cases, each puzzle has a start configuration and a goal besides the solving rules. Sometimes it is meaningful to regulate the rule application by some

© Springer Nature Switzerland AG 2021
F. Gadducci and T. Kehrer (Eds.): ICGT 2021, LNCS 12741, pp. 203–220, 2021.
https://doi.org/10.1007/978-3-030-78946-6_11

control conditions. Provided that the underlying structures are graphs, these are exactly the ingredients of graph transformation units as they are introduced in [7] (see [8] for a comprehensive overview). We recall the notion of graph transformation units in Sect. 2. Based on the graph-transformational models, one can show that Labyrinth can be solved in polynomial time (cf. Sect. 7) and solvability of the other three is NP-complete.

From the point of view of puzzle solving, random solving steps are not recommendable. Much more interesting and expedient is to invent heuristics that improve the chances to find a solution fast although the search space is exponential. In Sect. 8, this topic is considered with respect to Sudoku. Section 9 concludes the paper by discussing the lessons learned.

2   Graph Transformation Units

In this section, the basic notions and notations of graph transformation units are recalled. The concept is approach-independent meaning that one can choose a class of graphs and class of rules, a fixed type of rule application, the types of graph class expressions to specify particular initial and terminal graphs, and types of control conditions to regulate the derivation process. For the purposes of this paper, we use edge-labeled directed graphs and spans of inclusions as rules, an explicit construction of rule application, which is needed for correctness proofs, and a variety of graph class expressions and control conditions.

2.1   Directed Edge-Labeled Graphs

Let Σ be a set of labels with $* \in \Sigma$. A (directed edge-labeled) *graph* over Σ is a system $G = (V, E, s, t, l)$ where V is a finite set of *vertices*, E is a finite set of *edges*, $s, t \colon E \to V$ are mappings assigning a *source* $s(e)$ and a *target* $t(e)$ to every edge $e \in E$, and $l \colon E \to \Sigma$ is a mapping assigning a *label* to every edge $e \in E$.

An edge e with $s(e) = t(e)$ is called a *loop*. An edge with label $*$ is also called an *unlabeled edge*. Undirected edges are pairs of edges between the same vertices in opposite directions. In drawings of graphs, the label $*$ is omitted. The components V, E, s, t, and l of G are also denoted by V_G, E_G, s_G, t_G, and l_G, respectively. The empty graph is denoted by $\emptyset$. The class of all directed edge-labeled graphs is denoted by $\mathcal{G}_\Sigma$.

For graphs $G, H \in \mathcal{G}_\Sigma$, a *graph morphism* $g \colon G \to H$ is a pair of mappings $g_V \colon V_G \to V_H$ and $g_E \colon E_G \to E_H$ that are structure-preserving, i.e., $g_V(s_G(e)) = s_H(g_E(e))$, $g_V(t_G(e)) = t_H(g_E(e))$, and $l_G(e) = l_H(g_E(e))$ for all $e \in E_G$.

If the mappings g_V and g_E are bijective, then G and H are *isomorphic*, denoted by $G \cong H$. If they are inclusions, then G is called a *subgraph* of H, denoted by $G \subseteq H$. For a graph morphism $g \colon G \to H$, the image of G in H is called a *match* of G in H, i.e., the match of G with respect to the morphism g is the subgraph $g(G) \subseteq H$. If the mappings g_V and g_E are injective, the match $g(G)$ is also called *injective*. In this case, G and $g(G)$ are isomorphic.

Let $G = (V, E, s, t, l) \in \mathcal{G}_\Sigma$. A sequence $p = e_1 \ldots e_k$ of edges for some $k \in \mathbb{N}$ is a *path* of length k from v to v' for $v, v' \in V$ if $v = s(e_1), t(e_i) = s(e_{i+1})$ for $i = 1, \ldots, k - 1$, and $v' = t(e_k)$. It is a *cycle* if $v = v'$ and $k > 0$. The set of visited vertices v and $t(e_i)$ for $i = 1, \ldots, k$ is denoted by $VISIT_G(p)$. Given $v_0 \in V$, the set of vertices that can be reached by paths from v_0 is denoted by $REACH_G(v_0)$. Let $\overline{E} = \{\overline{e} \mid e \in E\}$ be a disjoint copy of E. Then the *symmetric closure* of G is the graph $S(G) = (V, E + \overline{E}, \overline{s}, \overline{t}, \overline{l})$ with $G \subseteq S(G)$ and $\overline{s}(\overline{e}) = t(e)$, $\overline{t}(\overline{e}) = s(e)$, and $\overline{l}(\overline{e}) = l(e)$ for $\overline{e} \in \overline{E}$. G is *connected* if each two vertices v and v' are connected by a path in $S(G)$.

2.2 Rules and Rule Application

A *rule* $r = (L \supseteq K \subseteq R)$ consists of three graphs $L, K, R \in \mathcal{G}_\Sigma$ such that K is a subgraph of L and R. The components L, K, and R are called *left-hand side*, *gluing graph*, and *right-hand side*, respectively.

The application of $r = (L \supseteq K \subseteq R)$ to a graph $G = (V, E, s, t, l)$ consists of the following three steps.

1. Choose a match $g(L)$ of L in G.
2. Remove the vertices of $g_V(V_L) - g_V(V_K)$ and the edges of $g_E(E_L) - g_E(E_K)$ yielding Z, i.e., $Z = G - (g(L) - g(K))$.
3. Add the right-hand side R to Z by gluing Z with R in $g(K)$ yielding H.

The construction is subject to the dangling condition that Z becomes a subgraph of G so that H becomes a graph automatically. Moreover, we allow that g is non-injective, but only inside K (identification condition).

The application of r to G w.r.t. the graph morphism g is denoted by $G \underset{r}{\Longrightarrow} H$. It is called a *direct derivation* from G to H. A *derivation* from G to H is a sequence of direct derivations $G_0 \underset{r_1}{\Longrightarrow} G_1 \underset{r_2}{\Longrightarrow} \cdots \underset{r_n}{\Longrightarrow} G_n$ with $G_0 = G$, $G_n \cong H$ and $n \geq 0$. If $r_1, \ldots, r_n \in P$, then the derivation is also denoted by $G \overset{n}{\underset{P}{\Longrightarrow}} H$. If n does not matter, we write $G \overset{*}{\underset{P}{\Longrightarrow}} H$ and if the underlying rule set is clear from the context, the subscript P may be omitted.

Using $\#S$ for the cardinality of a finite set S, it is worth noting that the application of a given (fixed) rule to a graph G with $\#V_G$ vertices and $\#E_G$ edges can be performed in polynomial time provided that the equality of labels can be checked in polynomial time. This is due to the fact that any left-hand side of size k has at most $(\#V_G + \#E_G)^k$ matches in G. Moreover, the further steps of the rule application can be done in linear time.

It may also be noted that the chosen notion of rule applications fits into the DPO framework as introduced in [9] (see, e.g., [10] for a comprehensive survey).

2.3 Graph Class Expressions and Control Conditions

Sometimes it is desirable to restrict the class $\mathcal{G}_\Sigma$ of graphs to some subclass. For example, one may want to start derivations only from specific initial graphs or filter out a subclass of all derived graphs as output. To this aim *graph class expressions* restrict the class $\mathcal{G}_\Sigma$ to subclasses, i.e., each graph class expression e specifies a set $SEM(e) \subseteq \mathcal{G}_\Sigma$. The class of all graph class expressions is denoted by $\mathcal{E}$.

Control conditions can reduce the nondeterminism given by several matches of a rule and the possible applicability of several rules to the current graph. The class of all control conditions is denoted by $\mathcal{C}$. Every control condition $C \in \mathcal{C}$ specifies a binary relation $SEM(C) \subseteq \mathcal{G}_\Sigma \times \mathcal{G}_\Sigma$.

The particular graph class expressions and control conditions used in this paper are introduced and explained where needed.

2.4   Graph Transformation Units

Graph transformation units were introduced in [7] (see [8] for a comprehensive overview).

By definition, a rule r provides a binary relation on graphs $\underset{r}{\Longrightarrow} \subseteq \mathcal{G}_\Sigma \times \mathcal{G}_\Sigma$ so that a set of rules P induces two further relations $\underset{P}{\Longrightarrow}$ and $\underset{P}{\overset{*}{\Longrightarrow}}$ where the first one is the union of the relations of the single rules in P and the second one is the reflexive and transitive closure of $\underset{P}{\Longrightarrow}$, called *derivation relation*. A graph transformation unit provides rules, a control condition to restrict the derivation relation, and two graph class expressions to specify initial and terminal graphs.

A *graph transformation unit* is a system $gtu = (I, P, C, T)$ where $I \in \mathcal{E}$ is the *initial graph class expression*, P is a finite set of *rules*, $C \in \mathcal{C}$ is a control condition over P and $T \in \mathcal{E}$ is the *terminal graph class expression*. The *semantics* of gtu is the binary relation $SEM(gtu) = (SEM(I) \times SEM(T)) \cap \underset{P}{\overset{*}{\Longrightarrow}} \cap SEM(C)$.

In examples, a graph transformation unit is presented schematically where the components I, P, C, and T are listed after respective keywords *initial*, *rules*, *cond*, and *terminal*. We omit the control condition if it does not impose any restriction on the order of rule applications. We omit the specification of terminal graphs if all graphs are accepted as terminal.

It is worth noting that each graph transformation unit gtu may serve as a graph class expression with *generated language* $L(gtu) = pr_2(SEM(gtu))$ as semantics where pr_2 denotes the projection to the second component.

Moreover, each graph transformation unit gtu specifies a decidability problem $SOLV(gtu) : \mathcal{G}_\Sigma \rightarrow \{TRUE, FALSE\}$ defined by $SOLV(gtu)(G) = TRUE$ if $(G; H) \in SEM(gtu)$ for some H and $FALSE$ otherwise.

3   Sudoku

Sudoku is one of the most popular of the Nicoli puzzles. A start pattern is a 9×9 grid which is subdivided in 9 3×3 subgrids. Some of the 81 fields are prescribed by digits between 1 and 9. The goal is to enter digits into all the empty fields in such a way that each two fields in a horizontal line and in a vertical line as well as in a subgrid have different entries. Figure 1 shows a typical example.

5	3			7				
6			1	9	5			
	9	8					6	
8				6				3
4			8		3			1
7				2				6
	6					2	8	
			4	1	9			5
				8			7	9

5	3	4	6	7	8	9	1	2
6	7	2	1	9	5	3	4	8
1	9	8	3	4	2	5	6	7
8	5	9	7	6	1	4	2	3
4	2	6	8	5	3	7	9	1
7	1	3	9	2	4	8	5	6
9	6	1	5	3	7	2	8	4
2	8	7	4	1	9	6	3	5
3	4	5	2	8	6	1	7	9

Fig. 1. A sample Sudoku puzzle (left) and its solution (right)

In our graph-transformational generalization, the grid is represented by an unlabeled undirected graph where the vertices represent the fields and the edges the horizontal, vertical and subgrid relation. An entry of a field becomes a loop labeled with the entry where we allow numbers between 1 and k for some $k \in \mathbb{N}$. Initially, some vertices have such loops. For technical reasons, all other vertices get 0-loops. Then the goal is to replace the 0-loops by loops with labels from $[k] = \{1, \ldots, k\}$ such that no two direct neighbors have loops with the same label.

γ-*sudoku(k)-configs*
  initial: $\emptyset$

  rules: $\emptyset \supseteq \emptyset \subseteq \overset{i}{\circlearrowleft}$
  for $i \in [k]$ or $i = 0$

  $\circ \ \circ \supseteq \circ \ \circ \subseteq \circ\!\!-\!\!\circ$

γ-*sudoku(k)*
  initial: γ-*sudoku(k)-configs*

  rules: $\overset{0}{\circlearrowleft} \supseteq \circ \subseteq \overset{i}{\circlearrowleft}$ for $i \in [k]$

  terminal: *forbidden*$(\ \overset{0}{\circlearrowleft}, \overset{i}{\circlearrowleft}\!\overset{i}{\circlearrowleft} \mid i \in [k])$

The units use the terminal graph class expression *all* on one hand and forbidden structures on the other hand. The constant *all* does not impose any restriction, i.e., $SEM(all) = \mathcal{G}_\Sigma$. This graph class condition is considered as default (and hence omitted in the left-hand side unit). Given some graph F, *forbidden*(F) specifies the class of all graphs that do not have any subgraph isomorphic to F. Given a set $\mathcal{F}$ of graphs, *forbidden*$(\mathcal{F})$ specifies the class of all graphs that do not have any subgraph isomorphic to some $F \in \mathcal{F}$. The elements of $\mathcal{F}$ are enumerated in special cases as above.

To formulate and prove correctness and complexity of γ-*sudoku(k)*, an unlabeled and undirected graph with an extra i-loop for some $i \in \{0, \ldots, k\}$ at each vertex is called *Sudoku*-graph. Given two such graphs G and H, $G \leq H$ denotes $U(G) = U(H)$ and $G^- \subseteq H$, where U removes all labeled loops from the argument graph and G^- is G without 0-loops. If $G \leq H$ and H is a terminal graph of γ-*sudoku(k)*, then H is called *solution* of G.

It can be shown that γ-*sudoku(k)-configs* generates the class of *Sudoku*-graphs and that γ-*sudoku(k)* computes all solutions of *Sudoku*-graphs. Moreover, the solvability problem of γ-*sudoku(k)* turns out to be NP-complete. To prove this, k-colorability is reduced to γ-*sudoku(k)*.

Theorem 1 (Correctness and complexity)

1. $L(\gamma$-*sudoku(k)-configs*) *is the class of all Sudoku-graphs.*
2. $(G, H) \in SEM(\gamma$-*sudoku(k))* *if and only if* H *is a solution of* G.
3. $SOLV(\gamma$-*sudoku(k))* *is NP-complete.*

Proof. 1. Let $\emptyset \stackrel{n}{\Longrightarrow} G$ be a derivation in γ-*sudoku(k)-configs*. By a simple induction on n, one can show that G is a *Sudoku*-graph as rule applications preserves this property.

Conversely, a *Sudoku*-graph G is either $\emptyset$ or one can apply one of the inverse rules. Therefore, by induction on the size of G, one can show that G is derivable from $\emptyset$.

2. Let G be initial and $G \stackrel{n}{\Longrightarrow} H$ be a derivation in γ-*sudoku(k)*. By a simple induction on n, one can show that H is a *Sudoku*-graph and $G \leq H$. If additionally, $(G, H) \in SEM(\gamma$-*sudoku(k))*, then H is terminal and, therefore, a solution of G.

Conversely, let H be a solution of G and $m = \#(G(0)$ be the number of 0-loops of G. Then H is terminal by definition and G is initial as a *Sudoku*-graph. Moreover, one can prove by induction on m that there is a derivation $G \stackrel{m}{\Longrightarrow} H$ so that $(G, H) \in SEM(\gamma$-*sudoku(k))*.

Induction base: $\#G(0) = 0$ implies $G = H$ and $H \stackrel{0}{\Longrightarrow} H$.

Induction step for $\#G(0) = m + 1$: Let $v_0 \in V_G$ that has an 0-loop in G and an i-loop in H which exists as $G \leq H$. Then the rule for i can be applied to G yielding $G \Longrightarrow G'$. Obviously, one gets $G' \leq H$ and $\#G'(0) = m$ by induction hypothesis, and a composition derivation $G \Longrightarrow G' \stackrel{*}{\Longrightarrow} H$ as desired.

3. Let $G \stackrel{n}{\Longrightarrow} H$ be a derivation in γ-*sudoku(k)*. Then $n \leq \#G(0)$ as each rule application consumes a 0-loop. Together with Point 2, this proves $SOLV(\gamma$-*sudoku(k))* \in NP. Consider, particularly, G without i-loops for $i \in [k]$, i.e., $\#G(0) = \#V_G$. G is k-*colorable* if there is a mapping $col: V_G \to [k]$ with $col(v) \neq col(v')$ for each two neighbors in G. If one replaces the 0-loops by a $col(v)$-loop for each $v \in V_G$, then the resulting graph H is a solution of G such that $(G, H) \in SEM(\gamma$-*sudoku(k))* and $SOLV(\gamma$-*sudoku(k))(G) = TRUE$.

Conversely, $SOLV(\gamma$-*sudoku(k))(G) = TRUE$, by definition, yields $(G, H) \in SEM(\gamma$-*sudoku(k))* for some H. As $G \leq H$, one can assume $V_G = V_H$. Consider the mapping $col: V_G \to [k]$ with $col(v) = i$ if v has an i-loop in H. Because H is terminal, there are no two neighbors with i-loops so that col is a k-coloring of G. Summarizing, this proves that k-colorability can be seen as a special case of solvability of γ-*sudoku(k)* so that $SOLV(\gamma$-*sudoku(k))* turns out to be NP-complete.

Related Work. As a computational problem, the general problem of solving Sudoku puzzles on $n^2 \times n^2$ grids of $n \times n$ blocks is known to be NP-complete (cf. [11]). As the graph representations of such grids are *Sudoku*-graphs, this implies the NP-completeness of $SOLV(\gamma\text{-}sudoku(k))$. Our reduction of k-colorability emphasises the intuition that Sudoku is a coloring problem. It does not apply to the special case of square grids.

4   Hashiwokakero

The underlying patterns of Hashiwokakero (which has also been published in English under the name Bridges, Chopsticks and Hashi) are rectangular grids where some fields are circles with inscribed numbers and all other fields are empty such that there is at least one empty field between each two horizontal and vertical circles. The left part of Fig. 2 displays a typical instance. The task is to add one or two straight lines between horizontal and vertical neighbor circles such that the number of lines incident to each circle equals the inscribed number, crossing lines are forbidden, and each two circles are connected by sequences of lines. The right part of Fig. 2 shows a solution of the puzzle given by the left pattern.

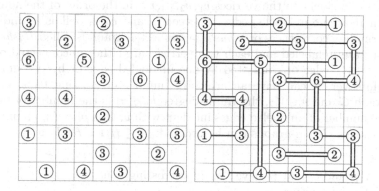

Fig. 2. A sample Hashi puzzle (left) and its solution (right)

To model Hashi by means of graph transformation, the circles are represented by vertices, the inscribed numbers by the same number of unlabeled loops, and the neighborhood of circles by ?-edges. The choice of lines is defined by three rules the application of which replaces a ?-edge by one unlabeled edge or two unlabeled edges or removes a ?-edge, respectively. The geometric property of forbidden crossings is expressed by adding extra vertices incident with four edges which are labeled with $1, 2, 3, 4$, respectively. These structures, called *crossing bans*, allow to forbid unlabeled edges between 1 and 2 as well as 3 and 4. To guarantee the connectedness of the results (if there are some), the initial patterns are (by construction) connected, and a ?-edge is only removed if the removal does not break the connectedness.

γ-hashi-configs
  initial: ○

  rules: ○ ⊇ ○   ⊆ ○⁻⁰

    ○ ○ ⊇ ○ ○ ⊆ ○⁻⁰

    ○ ⊇ ○   ⊆ ♀

    ○⁻⁰ ⊇ ○⁻⁰ ⊆ ⟨graph with vertices 1,2,3,4⟩

  cond: *forbidden*(○○○)

γ-hashi
  initial: *γ-hashi-configs*

  rules: (*zero*) ○⁻⁰ ⊇ ○ ○ ⊆ ○ ○

    (*one*) ⟨graph⟩ ⊇ ○ ○ ⊆ ○—○

    (*two*) ⟨graph⟩ ⊇ ○ ○ ⊆ ∞

  cond: *zero* is only applied if the matched
       ?-edge lies on a cycle

  terminal: *forbidden*(♀, ⟨graph 1,2,3,4⟩)

The control condition of *γ-hashi-configs* forbids parallel ?-edges. The control condition of *γ-hashi* guarantees that a ?-edge is only removed if this does not break connectedness. A graph is terminal if it has no loops and if, for each crossing structure incident to the vertices v_1, v_2, v_3, v_4 (in the order of the numbered edges), there are no unlabeled edges between v_1 and v_2 as well as v_3 and v_4.

To formulate correctness of *γ-hashi*, a connected graph is called *Hashi-graph* if all loops are unlabeled and each two vertices are either unconnected or connected by one ?-edge or connected by one unlabeled edge or by two unlabeled edges. Moreover, it can have an arbitrary number of crossing bans. Given a *Hashi*-graph G and $v \in V_G$, the *Hashi*-degree $hd_G(v)$ is the number of loops at v and the number of unlabeled edges incident with v. G is *initial* if all unlabeled edges are loops. *Hashi*-graphs can be ordered: $G \leq H$ for *Hashi*-graphs G and H if

1. $V_G = V_H$, and G and H have the same crossing bans,
2. $hd_G(v) = hd_H(v)$ for all $v \in V_G$,
3. if there is a ?-edge connecting v and v' in H, then this edge is in G, too,
4. if there is one unlabeled edge connecting v and v' in H, then there is this one edge in G, too, or there is a ?-edge instead, and
5. the same for two unlabeled edges.

H is a *solution* of an initial G if $G \leq H$ and H is terminal.

Using these notions, one can show that *γ-hashi-configs* generates initial *Hashi*-graphs and that *γ-hashi* computes all solutions of initial *Hashi*-graphs and that the solvability of *γ-hashi* is NP-complete.

Theorem 2 (Correctness and complexity)

1. $L(γ\text{-}hashi\text{-}configs)$ *is the set of initial Hashi-graphs.*
2. $(G, H) \in SEM(γ\text{-}hashi)$ *if and only if H is a solution of G.*
3. $SOLV(γ\text{-}hashi)$ *is NP-complete.*

Proof (sketch). Points 1 and 2 can be proved analogously to the corresponding proofs of Theorem 1. Concerning Point 3, $SOLV(\gamma\text{-}hashi) \in$ NP because each derivation is linearly bounded by the number of ?-edges in the initial graph. Moreover, let G be a connected loop-free unlabeled undirected graph. This is turned into an initial *Hashi*-graph $\overline{G}$ by labeling all edges with ? and adding two unlabeled loops at each vertex. Then, obviously, $\overline{G}$ has a solution if and only if G has a Hamiltonian cycle, i.e., a cycle of length $\#V_G$ that visits all vertices.

Related Work. The solvability of Hashiwokakero puzzles is shown to be NP-complete in [12].

5   Arukone

A typical sample of Arukone is given by a 9×9 grid where 14 fields are inscribed by the numbers from 1 to 7 such that each number occurs twice. The other fields are initially empty. The task is to connect equal numbers by lines composed of horizontal and vertical sections through empty fields such that all fields are visited but none twice. Figure 3 shows an Arukone start pattern (left) and the solution (right).

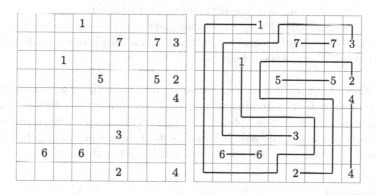

Fig. 3. An Arukone start pattern (left) and the solution (right)

In the graph-transformational model of Arukone, the underlying patterns are chosen as simple undirected graphs where, for some $k \geq 1, 2k$ vertices get an e-loop, an r-loop and an i-loop for $i \in [k]$ each such that every $i \in [k]$ occurs twice. The label i correspond to the number of a field. The role of the labels e and r is explained later. The other vertices get an unlabeled loop each. The ordinary edges are unlabeled, too. These graphs are generated by the unit $\gamma\text{-}arukone(k)\text{-}configs$. Based on the Arukone configurations, one can model the puzzle by the unit $\gamma\text{-}arukone(k)$.

$\gamma\text{-}arukone(k)\text{-}configs$
   initial: $\emptyset$

   rules: $\emptyset \quad \supseteq \emptyset \quad \subseteq \;\; $ (graph rule diagrams)

   $ \;\; \supseteq \;\; \subseteq \;\; $ (graph rule diagrams)
   for $i \in [k]$
cond: second rule with injective
   match only & third rule is
   applied once for each $i \in [k]$
terminal: $forbidden(\; \infty \;)$

$\gamma\text{-}arukone(k)$
   initial: $\gamma\text{-}arukone(k)\text{-}configs$

   rules: (graph rule diagrams) $\supseteq \;\; \circ \subseteq$ (graph rule diagrams)

   (graph rule diagrams) $\supseteq \;\; \subseteq$ (graph rule diagrams)

terminal: $forbidden(\; , \;)$

In $\gamma\text{-}arukone(k)\text{-}configs$, the first rule allows creating arbitrarily many isolated vertices with an unlabeled loop. The second rule allows connecting two vertices by an unlabeled edge. The first part of the control condition prevents that further loops are generated. And the third rule replaces the unlabeled loops of two vertices by an i-loop, an r-loop and an e-loop each, where $i \in [k]$, and r, e stands for *run* and *end*, respectively. The second part of the control condition guarantees that there are exactly two vertices with i-loops for each $i \in [k]$. The terminal graphs are simple as parallel edges are forbidden.

Correctness and complexity of $\gamma\text{-}arukone(k)$ can be shown analogously to Sudoku and Hashi. A solution of $\gamma\text{-}arukone(1)$ is a Hamiltonian path with fixed ends so that NP-completeness follows.

Related Work. Arukone is a stricter variation of Numberlink, which allows fields to remain unvisited. As a computational problem, finding a solution to a given Numberlink puzzle is NP-complete (cf. [13]). NP-completeness is maintained even if longer paths are allowed. Informally, this means paths may have unnecessary bends in them (see [14] for a more technical explanation).

6   Labyrinths

Labyrinths are popular puzzles with a long history. A famous Greek myth features the Labyrinth of Crete. It was inhabited by the men-eating Minotaur. The monster was slaughtered by Theseus who could safely leave the labyrinth by means of the Ariadne thread. The usual geometric structure of labyrinths with corridors and walls can be represented by undirected graphs where exits are marked by ε-loops. The walking in a labyrinth is properly modeled by traveling along edges. The current position of the walker is marked by a θ-loop (θ for Theseus). The unit $\gamma\text{-}labyrinth\text{-}configs$ generates these graphs. The walking is modeled by the unit $\gamma\text{-}labyrinth$ with the goal that the walker reaches an exit. This is specified by using the graph class expression $requested(U)$ for some graph U the semantics of which is the class of all graphs with a subgraph isomorphic to U.

γ-*labyrinth-configs*
    initial: $\circ\!\!=\!\!\circ\,\theta$
    rules: $\circ \quad \supseteq \circ \quad \subseteq \circ\!\!=\!\!\circ$
         $\circ\ \circ \supseteq \circ\ \circ \subseteq \circ\!\!=\!\!\circ$
         $\circ \quad \supseteq \circ \quad \subseteq \circ\!\!=\!\!\circ\,\varepsilon$

γ-*labyrinth*
    initial: γ-*labyrinth-configs*
    rules: $\theta\,\circ\!\!=\!\!\circ \supseteq \circ\ \circ \subseteq \circ\!\!=\!\!\circ\,\theta$
    terminal: *requested*$(\ \theta\,\circ\!\!=\!\!\circ\,\varepsilon\ )$

Obviously, the unit γ-*labyrinth-configs* generates all non-empty unlabeled undirected graphs extended by a single θ-loop and arbitrarily many ε-loops. It is not difficult to show that the unit γ-*labyrinth* can find an exit if an exit is reachable.

Theorem 3 (Correctness). *Let $G \in L(\gamma$-labyrinth-configs$)$ and v_0 be the vertex with the θ-loop. Then there is a terminal graph G' with $G \stackrel{*}{\Longrightarrow} G'$ if and only if there is a path in G from v_0 to a vertex with an ε-loop.*

Proof. Let $G \stackrel{n}{\Longrightarrow} G'$ be a derivation in γ-*labyrinth*. Then a simple induction on n reveals that the sequence of matched edges applying the rule one after the other forms a path starting from v_0 so that a vertex with ε-loop is reached if G' is terminal. Conversely, let $e_1 \cdots e_m$ be a path in G from v_0 to a vertex with an ε-loop. Then a simple induction on m shows that the rule can be applied to $e_1, \ldots, e_m$ successively yielding a derivation from G to a terminal graph.

Note that the unit describes random walks so that there are derivations of infinite length provided that there is at least one edge. Therefore, the labyrinth puzzle is not safely solved by γ-*labyrinth*. But one can do much better as the next section proves.

7   Ariadne Thread

Considering γ-*labyrinth*, the search for an exit means to find some path which is equivalent to find the shortest path. In other words, the polynomial shortest-path algorithms by Dijkstra, Floyd and Warshall and others are applicable. See, e.g., [15] for graph-transformational models of shortest-path algorithms. But to adapt these algorithms to a single walker would mean some extra work because they employ often best-first and breadth-first strategies. It happens that a next step starts from a position which is not the current one. In contrast to those, a backtracking or depth-first strategy is well-suited for a single walker in a labyrinth as the following unit shows that mimics the use of an Ariadne thread.

γ-*labyrinth-with-Ariadne-thread*

    initial: γ-*labyrinth-configs* $+\ \substack{1 \\ \circ}$

    rules: (*forward*) $\theta\,\circ\!\!=\!\!\circ\ \substack{i \\ \circ} \supseteq \circ\ \circ\ \circ \subseteq \circ\!\!\stackrel{i}{=}\!\!\circ\,\theta\ \substack{i+1 \\ \circ}$ for $i \in \mathbb{N}$
        (*backward*) $\circ\!\!\stackrel{i}{=}\!\!\circ\,\theta \supseteq \circ\ \circ \subseteq \theta\,\circ\!\!=\!\!\circ\ \circ$ for $i \in \mathbb{N}$
    cond: *forward** $|$ (*forward* !; *backward*(*max*); *forward**)*
    terminal: *requested*$(\ \theta\,\circ\!\!=\!\!\circ\,\varepsilon\ )$

The initial graphs are disjoint unions of the initial graphs of γ-*labyrinth* and an extra vertex with an 1-loop. The latter is an initial counter. In each application of a forward rule, the matched edge is replaced by a backward edge labeled with the counter value and this is increased by 1. A backward rule application matches such a backward edge and moves the θ-loop along the edge removing it at the same time. Although the set of rules is infinite formally, for each initial graph, the number of applied rules is never larger than twice the number of its edges. The control condition is a regular expression extended by the as-long-as-possible operator !. It accepts a derivation of arbitrarily many applications of forward rules or of applications of forward rules as long as possible follows by an application of a backward rule followed by arbitrarily many applications of forward rules. In this context, *backward(max)* means that the backward rule out of all applicable backward rules with the maximal label is applied. A graph is terminal if there is a vertex with a θ-loop and an ε-loop.

The unit γ-*labyrinth-with-Ariadne-thread* specifies the following algorithm.

exitsearch
  input: $G \in L(\gamma\text{-}labyrinth\text{-}configs)$,
  procedure: *add the initial counter disjointly; iterate rule applications according to the control condition as long as possible,*
  output: *all vertices with ε-loops that get a θ-loop intermediately.*

It turns out that *exitsearch* finds all reachable exits and runs in quadratic time with respect to the number of edges of the input.

Theorem 4 (Correctness and complexity). *Let $G \in L(\gamma\text{-}labyrinth\text{-}configs)$ be an input of exitsearch and exitsearch(G) a corresponding output. Let $v_0 \in V_G$ be the vertex with the θ-loop and $m = \#E_G$. Then the following holds.*

1. $exitsearch(G) = REACH_G(v_0) \cap SEM(requested(\ \theta \hspace{-0.2em}\circ\hspace{-0.3em}\circ\hspace{-0.3em}\circ\ \varepsilon\ ))$.
2. $exitsearch \in O(m^2)$.

Proof. 1. Let $G \overset{*}{\Longrightarrow} H$ be a derivation generated by *exitsearch* and *seq* be the sequence of edges matched one after the other. If one replaces each backward edge by its original, then one get a path p in G so that $exitsearch(G) = VISIT_G(p) \cap SEM(requested(\ \theta \hspace{-0.2em}\circ\hspace{-0.3em}\circ\hspace{-0.3em}\circ\ \varepsilon\ ))$. According to the following lemma, $VISIT_G(p) = REACH_G(v_0)$ proving the statement.

2. The length of p is $\leq 2 \cdot \#E_G$ as each forward rule application replaces an edge by a backward edge and each backward rule application consumes a backward edge. In particular, there are never more than $\#E_G$ edges present during the derivation so that $\#E_G$ is a bound for the time needed to find a match. Both together prove $exitsearch \in O(m^2)$.

It may be noted that *exitsearch* is in $O(m)$ if the input graphs are restricted to those with a bounded degree and if it is possible to find a match by checking the incident edges of the vertex with the θ-loop only.

Lemma 1. *Let $G \in L(\gamma\text{-}labyrinth\text{-}configs)$ and $G + \ell^1_\theta \overset{*}{\Longrightarrow} H$ be the longest derivation in γ-labyrinth-with-Ariadne-thread. Let p be the corresponding path in G consisting of the matched edges in the order of the derivation where each backward edge is replaced by its original and let $v_0 \in V_G$ be the vertex with the θ-loop. Then $VISIT_G(p) = REACH_G(v_0)$.*

Proof. The lemma is proved by induction on $m = \#E_G$.

Induction base for $m = 0$: Then p has length 0 and is the only path starting in v_0 meaning that $VISIT_G(p) = \{v_0\} = REACH_G(v_0)$.

Induction step for $m + 1$: In the simple case, there is no unlabeled edge incident with v_0. Then the same arguments apply as in the induction base. Otherwise, the derivation decomposes into $G \overset{*}{\Longrightarrow} \overline{G}$ and $\overline{G} \overset{*}{\Longrightarrow} H$ where the last step of the former is a forward step $G' \Longrightarrow \overline{G}$ and the first step of the latter is a backward step $\overline{G} \Longrightarrow H'$ and all remaining steps are backward rule applications. Let $p = e_1 \cdots e_n$ and $e_1 \cdots e_k$ the initial section corresponding to $G \overset{*}{\Longrightarrow} \overline{G}$. Then $e_k = e_{k+1}$ according to the control condition. Consider G^- obtained by removing e_k from G and p^- obtained by removing $e_k e_{k+1}$ from p. Moreover, H' and G' can be seen as equal up to e_k in G' and the counter values. The counter can be ignored as no forward step follows after G'. And as all backward edges of H' are in G', one gets a derivation $G \overset{*}{\Longrightarrow} G' \overset{*}{\Longrightarrow} H$. As none of its rule applications matches e_k, one can restrict this derivation to G^- yielding $G^- \overset{*}{\Longrightarrow} H^-$ with p^- as corresponding path (now in G^-). Obviously, this derivation is the longest possible one meeting the control condition. By construction, $\#E_{G^-} = \#E_G - 1 = m$ so that the induction hypothesis applies to G^- yielding $VISIT_{G^-}(p) = \{v_0\} = REACH_{G^-}(v_0)$. The two rule applications matching e_k and e_{k+1} respectively move the θ-loop from some vertex v_{k-1} to some vertex v_k and back. As p^- and p coincide up to $e_k e_{k+1}$ one get $VISIT_G(p) = VISIT_{G^-}(p^-) \cup \{v_k\} = REACH_{G^-}(v_0) \cup \{v_k\} \subseteq REACH_G(v_0)$. The latter inclusion holds because $G^- \subseteq G$ and $e_1 \cdots e_k$ reaches v_k. It remains to prove the converse inclusion. Let $v \in REACH_G(v_0)$ such that there is a path $\overline{p} = \overline{e}_1 \cdots \overline{e}_l$ from v_0 to v. Then (1) $v = v_k$ or (2) $\overline{p}$ in G^- or (3) $\overline{p}$ passes e_k to enter or to leave v_k. Without loss of generality, one can assume that $\overline{p}$ is simple, i.e., $\overline{p}$ does not contain cycles. In the latter leaving case, $\overline{p}$ decomposes into $\overline{e}_1 \cdots \overline{e}_i$ from v_0 to v_k and $\overline{e}_{i+1} = e_k$ from v_k to v_{k-1} and $\overline{e}_{i+2} \cdots \overline{e}_l$ from v_{k+1} to v. As e_k appears in $\overline{p}$ only once, $\overline{e}_{i+2} \cdots \overline{e}_l$ is a path in G^-. Moreover, $v_{k-1} \in REACH_{G^-}(p^-)$ so that there is a path p' in G^- from v_0 to v_{k-1}. The composition $p' \overline{e}_{i+2} \cdots \overline{e}_l$ provides a path in G^- from v_0 to v. In the entering case, $\overline{p}$ decomposes into $\overline{e}_1 \cdots \overline{e}_{i-1}$ from v_0 to v_{k-1} and $\overline{e}_i = e_k$ from v_{k-1} to v_k and $\overline{e}_{i+1} \cdots \overline{e}_l$ from v_k to v. The latter lies also in G^-. In p, e_k is the last entrance into v_k. If there is an earlier one, the p has an initial section $\hat{p}$ from v_0 to v_k which is also in G^- as it avoids e_k. Consequently, the composition $\hat{p} \overline{e}_{i+1} \cdots \overline{e}_l$ is a path in G^- from v_0 to v. If there is no earlier entrance into v_k, then e_k is the last and the first entering of v_k so that it is the only edge incident with v_k according to the control condition. This means that $\overline{e}_{i+1}$ does not exits and $\overline{p}$ is a path from v_0 to $v = v_k$. As one gets in all cases, $v = v_k$ or $v \in REACH_{G^-}(v_0)$, the lemma is proved.

8   Sudoku with a Bit of Cleverness

The unit $\gamma\text{-}sudoku(k)$ models the puzzle with the goal to complete a given partial k-coloring by rules that allow to choose a color for an uncolored vertex in each step. Whether the goal is reachable or not, is only checked at the very end. Such random way of solving the puzzle is not at all recommendable. The challenge (and fun) is to find best or at least good solving steps among all possible ones in each situation. Two possibilities are pointed out.

8.1   Restricted Choice

Obviously, it is totally meaningless to choose a color for a vertex that appears already in the neighborhood. Therefore, the second forbidden structure in terminal graphs should be checked before at each step.

More interesting restrictions are to find patterns that force a certain choice of a color. For instance, if an uncolored vertex has $k-1$ neighbors with $k-1$ colors, then only one color is left. Another situation is that an uncolored vertex belongs to a k-clique (i.e., each two of k vertices are neighbors) where some neighbors have already colors and the other neighbors have neighbors all colored with a further color. The further color is the only possible one left for the considered vertex. Both restrictions can be expressed by positive context conditions given in Fig. 4.

Fig. 4. Two positive context conditions for $\gamma\text{-}sudoku(k)$

In both cases, the central vertex must be colored by i because otherwise a forbidden structure appears immediately or somewhen later. It may be noted that the Sudoku in Fig. 1 can be solved by applying the restricted rules only. It is advisable to apply the restricted rules whenever possible because they provide unique solving steps that are always right and do not interfere with other steps. Clearly, none of the rules is applicable in many situations. Hence, an interesting question is under which conditions the two restricted rules (and may be some further rules like them) are enough to solve a Sudoku instance. In puzzle corners, Sudoku instances are often characterized as "easy", "middle", "difficult", and "very difficult". We guess that the two restricted rules cover most easy Sudoku instances.

8.2   Bookkeeping

The choice of colors can be supported if one stores for each uncolored vertex the colors that are available without resulting in a forbidden structure. This idea is modeled by the following unit. The uncolored vertices get additional loops labeled from $1?, \ldots, k?$ initially. The coloring process is modeled by three types of rules: *clean1* removes $i?$ for $i \in [k]$ if a neighbor has color i, *clean2* removes $j?$ from a vertex with color i for $i, j \in [k], i \neq j$, and *choose* colors a vertex with i if it has a loop labeled with $i?$.

γ-*sudoku2(k)-configs*
initial: $\emptyset$

rules: $\emptyset \quad \supseteq \emptyset \quad \subseteq$
for $i \in [k]$ or $i = 0$

$\emptyset \quad \supseteq \emptyset \quad \subseteq$

$\circ \circ \supseteq \circ \circ \subseteq \circ\!\!-\!\!\circ$

terminal: *forbidden*( )

γ-*sudoku2(k)*
initial: γ-*sudoku2(k)-configs*

rules: (*choose*) $\supseteq \circ \quad \subseteq$
for $i \in [k]$

(*clean1*) $\supseteq$ $\subseteq$
for $i \in [k]$

(*clean2*) $_{j?} \supseteq$ $\subseteq$
for $i, j \in [k], i \neq j$

control: $((clean1 \mid clean2)!; choose)^*$

terminal: *forbidden*(, $\mid i \in [k]$)

The control condition requires that the cleaning rules are applied as long as possible before a color for a vertex is chosen; repeating this arbitrary often.

Correctness and complexity of γ-*sudoku2(k)* can be shown in quite the similar way as of γ-*sudoku(k)*.

9   Lessons Learned

In this paper, we have started a systematic study of graph-transformational modeling and analysis of games and puzzles considering four logic puzzles exemplarily. Labyrinths can be seen as shortest-path problems so that always a polynomial solution strategy works. In contrast to that, the solvability of the graph-transformational generalizations of Sudoku, Hashi and Arukone have turned out to be NP-complete. In all cases, the puzzle instances are generated by graph transformation units, and the puzzles with their allowed solving steps and goals are modeled by graph transformation units supporting the proofs of correctness and complexity. Which lessons are learned? As this is our very first attempt on the matter, the answers must be preliminary. But a few points can be addressed.

1. The considered logic puzzles (and many like them) are defined on some plane patterns often in form of grids, their solving requires to make some entries, to select from some alternatives or to move in some way sometimes under certain conditions, and the goals are to reach certain configurations. All this fits very well to graph transformation and graph transformation units, in particular.

2. In our examples, the puzzle instances to be solved are given by generative graph transformation units. They may be seen as syntax-directed editors to provide particular instances. The classes of initial graphs could also be specified by monadic 2nd order formulas – but less constructive.

3. We have tried to design the graph-transformational puzzle-solving rules as close as possible to the original ones while we have been quite general with respect to the initial graph configurations. On one hand, this allows to solve unexpected but nevertheless challenging instances of puzzles. On the other hand, one may argue that we are too generous allowing meaningless inputs like Sudoku instances where two neighbors carry the same color from the very beginning. If one wants to work with more meaningful initial graphs, then one can replace the units that generate initial configurations by more restrictive ones by refining rules, restricting the terminal graphs or adding control conditions.

4. In many graph-transformational generalizations of logic puzzles, the lengths of derivations are linearly bounded so that the corresponding solvability problems belong to the class NP.

5. Looking at our examples and at the examples discussed in the literature (see, e.g., [11–14,16]), it seems to be typical that the solvability problems are even NP-complete. Our complexity results are not new, but we demonstrate that they can be proved systematically within the framework of graph transformation. In contrast to that, the proofs in the literature are rather based on ad hoc formalizations. Moreover, we back the trustworthiness of our puzzle models up by correctness proofs while the formalizations in the literature are mainly based on plausibility considerations.

6. Our correctness proofs follow a basic schema. By an induction on the lengths of derivations, an invariant property is shown such that derived terminal graphs turn out to be solutions of the initial graphs. Conversely, by induction on the size of solutions of initial graphs, it is shown that they are derivable. We wonder how much of such proofs can be automatically deduced or at least supported by proof tools.

7. All the graph class expressions and control conditions used in this paper can be expressed in monadic 2nd order logic. This observation may help to come up with tool support for the analysis of graph transformation systems as considered.

8. From the point of view of puzzle solving, random solving steps are not recommendable (and boring). Much more interesting is to invent heuristics that improve the chances to find a solution fast although the search space is exponential as we sketch in Sect. 8 with respect to Sudoku. An interesting question in this context is whether the graph-transformational modeling supports the inference ("learning") of such heuristics.

9. As the Points 2 to 8 indicate, the graph-transformational modeling of logical puzzles seems to follow certain principles. But it may be too early to come up with guidelines based on these observations.

10. In [6], Zambon and Rensink have modeled the *N-queens* problem by specifying initial and terminal graphs, a set of rules and some control conditions.

This can be seen as an example of a graph transformation unit because the concept is approach-independent. Like the labyrinth model, the *N-queens* model is polynomial. One very important aspect of this model is that it is analyzed and evaluated using the GROOVE tool. It is a desirable future task to employ GROOVE or other graph transformation tools for experiments on logic puzzles like the ones run on the *N-queens* model.

11. If one encounters logic puzzles in puzzle corners of magazines or newspapers, then they are solvable in general while the initial configurations in all our examples include unsolvable instances, too. This is not an essential difference as one may replace the units generating initial instances by units that generate solvable instances only. In the case of *γ-labyrinth-configs*, one may start from a simple path leading from a θ-looped vertex to an ϵ-looped vertex. Or in the case of *γ-sudoku*, one may first generate k-colored graphs where each vertex carries a loop labeled with the respective color and then recolor some loops by 0.

12. In many applications of rule-based modeling, determinism, confluence and functionality are desirable properties so that quite a lot of research is done in this respect. In contrast to that, games and puzzles are of particular interest as they are essentially nondeterministic and rule applications may be not only in conflict with each other, but also good or bad. These features may deserve more attention.

We plan to continue the graph-transformational case study on games and puzzles by shedding more light on the fascinating issues of correctness, complexity and heuristics and by extending considerations to multi-player games. Interested readers are welcome to join us in this adventure.

Acknowledgment. We are very grateful to the anonymous reviewers for their helpful comments that led to various improvements.

References

1. Heckel, R.: Graph transformation in a nutshell. Electron. Notes Theor. Comput. Sci. **148**(1), 187–198 (2006)
2. Hölscher, K., Kreowski, H.-J., Kuske, S.: Autonomous units to model interacting sequential and parallel processes. Fundamenta Informaticae **92**, 233–257 (2009)
3. Kreowski, H.-J., Kuske, S., Tönnies, H.: Autonomous units to model games. In: Fischer, S., Maehle, E., Reischuk, R. (eds.) Informatik 2009, Im Focus das Leben, volume 154 of Lecture Notes in Informatics, pp. 3465–3472 (2009)
4. Priemer, D., George, T., Hahn, M., Raesch, L., Zündorf, A.: Using graph transformation for puzzle game level generation and validation. In: Echahed, R., Minas, M. (eds.) ICGT 2016. LNCS, vol. 9761, pp. 223–235. Springer, Cham (2016). https://doi.org/10.1007/978-3-319-40530-8_14
5. Rensink, A., et al.: Ludo: a case study for graph transformation tools. In: Schürr, A., Nagl, M., Zündorf, A. (eds.) AGTIVE 2007. LNCS, vol. 5088, pp. 493–513. Springer, Heidelberg (2008). https://doi.org/10.1007/978-3-540-89020-1_34

6. Zambon, E., Rensink, A.: Solving the N-Queens problem with GROOVE - towards a compendium of best practices. Electron. Commun. Eur. Assoc. Softw. Sci. Technol. **67**, 13 p. (2014)
7. Kreowski, H.-J., Kuske, S.: On the interleaving semantics of transformation units— a step into GRACE. In: Cuny, J., Ehrig, H., Engels, G., Rozenberg, G. (eds.) Graph Grammars 1994. LNCS, vol. 1073, pp. 89–106. Springer, Heidelberg (1996). https://doi.org/10.1007/3-540-61228-9_81
8. Kreowski, H.-J., Kuske, S., Rozenberg, G.: Graph transformation units – an overview. In: Degano, P., De Nicola, R., Meseguer, J. (eds.) Concurrency, Graphs and Models. LNCS, vol. 5065, pp. 57–75. Springer, Heidelberg (2008). https://doi.org/10.1007/978-3-540-68679-8_5
9. Ehrig, H., Pfender, M., Schneider, H.-J.: Graph grammars: an algebraic approach. In: IEEE Conference on Automata and Switching Theory, Iowa City, pp. 167–180 (1973)
10. Corradini, A., Ehrig, H., Heckel, R., Löwe, M., Montanari, U., Rossi, F.: Algebraic approaches to graph transformation part I: basic concepts and double pushout approach. In: Rozenberg, G. (ed.) Handbook of Graph Grammars and Computing by Graph Transformation, vol. 1: Foundations, pp. 163–245. World Scientific, Singapore (1997)
11. Yato, T., Seta, T.: Complexity and completeness of finding another solution and its application to puzzles. IEICE Trans. Fundam. Electron. Commun. Comput. Sci. **86**, 1052–1060 (2003)
12. Andersson, D.: Hashiwokakero is NP-complete. Inf. Process. Lett. **109**(19), 1145–1146 (2009)
13. Kotsuma, K., Takenaga, Y.: NP-completeness and enumeration of Number Link puzzle. IEICE Tech. Rep. **109**(465), 1–7 (2010)
14. Adcock, A., et al.: Zig-zag numberlink is NP-complete. J. Inf. Process. **23**(3), 239–245 (2015)
15. Kreowski, H.-J., Kuske, S.: Modeling and analyzing graph algorithms by means of graph transformation units. J. Object Technol. **19**(3), 3:1–14 (2020). https://doi.org/10.5381/jot.2020.19.3.a9
16. Kendall, G., Parkes, A.J., Spoerer, K.: A survey of NP-complete puzzles. J. Int. Comput. Games Assoc. **31**(1), 13–34 (2008)

Interval Probabilistic Timed Graph Transformation Systems

Maria Maximova(✉) ⓘ, Sven Schneider ⓘ, and Holger Giese ⓘ

University of Potsdam, Hasso Plattner Institute, Potsdam, Germany
{maria.maximova,sven.schneider,holger.giese}@hpi.de

Abstract. For complex distributed embedded probabilistic real-time systems, ensuring correctness of their software components is of great importance. The rule-based formalism of Probabilistic Timed Graph Transformation Systems (PTGTSs) allows for modeling and analysis of such systems where states can be represented by graphs and where timed and probabilistic behavior is important. In PTGTSs, probabilistic behavior is specified by assigning precise probabilities to rules. However, for embedded systems, only lower and upper probability bounds may be estimated because unknown physical effects may influence the probabilities possibly changing them over time.

In this paper, we (a) introduce the formalism of Interval Probabilistic Timed Graph Transformation Systems (IPTGTSs) in which rules are equipped with probability intervals rather than precise probabilities and (b) extend the preexisting model checking approach for PTGTSs to IPTGTSs w.r.t. worst-case/best-case probabilistic timed reachability properties using an encoding of probability intervals. Moreover, we ensure that this adapted model checking approach is applicable to IPT-GTSs for which the finiteness of the state space may only be a consequence of the timing constraints. Finally, in our evaluation, we apply an implementation of our model checking approach in AUTOGRAPH to a running example.

Keywords: Cyber-physical systems · Graph transformation systems · Interval probabilistic timed systems · Qualitative analysis · Quantitative analysis · Model checking

1 Introduction

Software correctness plays an important role in the ever growing area of complex distributed embedded probabilistic real-time systems. In this context, modeling formalisms allowing for formal analysis while capturing relevant system aspects are required for designing, understanding, and improving the behavior of such systems.

Funded by the Deutsche Forschungsgemeinschaft (DFG, German Research Foundation) - 241885098, 148420506.

F. Gadducci and T. Kehrer (Eds.): ICGT 2021, LNCS 12741, pp. 221–239, 2021.
https://doi.org/10.1007/978-3-030-78946-6_12

Many probabilistic real-time systems with complex coordination behavior or spatial movements of components can be modeled using the formalism of PTGTSs [8] when the states of the system can be represented by graphs. The formalism of PTGTSs allows for structure dynamics by means of rule-based graph transformation (cf. [3]), timed behavior (employing clocks as in Timed Automata (TA) [1]), and probabilistic choices (as in Probabilistic Automata (PA) [13]) among alternative outcomes of graph transformation steps. As usual, nondeterministic models such as PTGTSs, where the passage of time competes with possibly multiple rule applications, implicitly cover real-time systems by the resolution of their nondeterminism. However, assigning precise probabilities to the alternative outcomes of graph transformation steps in PTGTSs is insufficient when unknown physical effects may affect the actual probabilities (possibly even over time). Hence, PTGTSs may only be employed as a suitable modeling formalism when (at least) pseudo-random variables are used to decide probabilistic choices.

The subformalisms of Timed Graph Transformation Systems (TGTSs) [2,10] and Probabilistic Graph Transformation Systems (PGTSs) [5,6] including tool support for their formal analysis have been introduced before. Essentially, tool-based analysis support is obtained by translating a PTGTS, TGTS, or PGTS into the corresponding Probabilistic Timed Automaton (PTA) [7], TA, or PA, respectively, and by then reusing the existing model checking support for the resulting automata. In particular, the tools PRISM [6] and UPPAAL [14] support the model checking of PTA and TA, respectively, with varying feature sets.

As for PTGTSs, the precise probabilities required in PTA may not be appropriate for the system at hand. To relax this precision, Interval Probabilistic Timed Automata (IPTA) [4,15] have been defined as an extension of PTA where probability intervals are used instead of precise probabilities. These probability intervals are then resolved to precise probabilities nondeterministically at use-time to derive steps.

In this paper, we introduce the formalism of IPTGTSs as an extension of PTGTSs by integrating the handling of probability intervals from IPTA to allow for the modeling of systems where only lower and upper probability bounds can be estimated for some probabilistic steps. Following our work on PTGTSs and IPTA [4], we present a formal translation of IPTGTSs into PTA via IPTA (for the case of finite state spaces) to support the modeling of *structure dynamics*, *timed behavior*, and *interval probabilistic behavior* using IPTGTSs and their analysis w.r.t. best-case/worst-case probabilistic reachability properties using PRISM. Hereby, we improve upon our earlier work in [8] by constructing the state space of the TGTS underlying the given IPTGTS while using UPPAAL to ensure that all obtained states are reachable for the given IPTGTS. Hence, we enable the analysis of an IPTGTS with a finite state space even when the state space of its underlying Graph Transformation System (GTS) is infinite.

As a running example, we consider a gossiping protocol where all agents in a directed wireless network have a local Boolean value. The Boolean value true represents the information that must be propagated to all agents. At run-time, agents with Boolean value true attempt to send this value along the directed physical channels given by edges. Each sending operation is subject to probabilistic choice where the message is transported successfully through the channel with a probability between 0.7 and 0.8 possibly being affected by e.g. the available energy of the sender or the spatial distance between agents. Moreover, due to limited energy and imperfect local clocks, each agent may send a message at most every 2 to 5 time units. Lastly, we evaluate the described system in terms of e.g. the best-case/worst-case probability that each agent adopts the Boolean value true within a given time bound but also in terms of the number of sending errors processed by an observer that counts and deletes them.

This paper is structured as follows. In Sect. 2, we introduce preliminaries for our approach including graph transformation and IPTA. In Sect. 3, we introduce the novel formalism of IPTGTSs as an extension of PTGTSs widening probabilistic choices from precise probabilities to probability intervals. In Sect. 4, we present the steps of our translation-based model checking approach for IPTGTSs. In Sect. 5, we evaluate our approach by applying its implementation in the tool AUTOGRAPH to our running example. Finally, in Sect. 6, the paper is closed with a conclusion and an outlook on future work.

2   Preliminaries

In this section, we introduce graphs, graph transformation, IPTA, and probabilistic timed reachability problems to be solved for IPTA as preliminaries for the subsequent presentation of IPTGTSs and our model checking approach.

Employing the variation of symbolic graphs [11] from [12], we consider typed attributed graphs (such as the graph G_0 in Fig. 1d), which are typed over a type graph TG (such as the one in Fig. 1a). The values of the variables connected to attributes are specified using Attribute Conditions (ACs) over a many sorted first-order attribute logic. The AC $\bot$ (false) in TG means thereby that the type graph does not restrict attribute values. Graph Transformation (GT) is then executed by applying a GT rule $\rho = (\ell : K \hookrightarrow L, r : K \hookrightarrow R, \gamma)$ for a match $m : L \hookrightarrow G$ on the graph to be transformed (see [12] for technical details). A GT rule specifies (a) that the graph elements in $L - \ell(K)$ are to be deleted and the graph elements in $R - r(K)$ are to be added using the monomorphisms ℓ and r, respectively, according to a Double Pushout (DPO) diagram and (b) that the values of variables in the resulting graph are derived from those of G using the AC γ (e.g. $x' = x + 2$) in which the variables from L and R are used in unprimed and primed form, respectively. Nested application conditions are straightforwardly supported by our approach but, to improve readability, they are not used in the running example and omitted subsequently.

We now recall IPTA [4,15], which subsume TA [1] where clocks are used to capture real-time phenomena and PTA [7] where probabilistic choices are used additionally to approximate/describe the likelihood of outcomes of certain steps. First, we provide required notions on clocks and (intervals of) probabilities.

For a set of clock variables X, clock constraints $\psi \in \mathsf{CC}(X)$ are finite conjunctions of clock comparisons of the form $c_1 \sim n$ and $c_1 - c_2 \sim n$ where $c_1, c_2 \in X$, $\sim \in \{<, >, \leq, \geq\}$, and $n \in \mathbf{N} \cup \{\infty\}$. A clock valuation $v \in \mathsf{CV}(X)$ of type

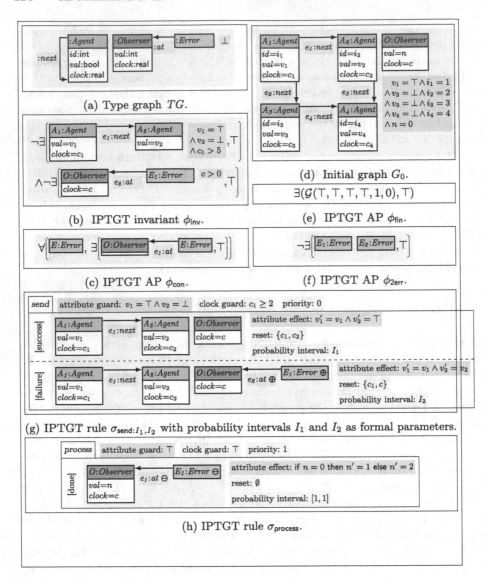

Fig. 1. Elements of the IPTGTS for the running example.

$v : X \longrightarrow \mathbf{R}_0^+$ satisfies a clock constraint ψ, written $v \models \psi$, as expected. The initial clock valuation $\mathsf{ICV}(X)$ maps all clocks to 0. For a clock valuation v and a set of clocks X', $v[X' := 0]$ is the clock valuation mapping the clocks from X' to 0 and all other clocks according to v. For a clock valuation v and a duration $\delta \in \mathbf{R}_0^+$, $v + \delta$ is the clock valuation mapping each clock x to $v(x) + \delta$.

For a countable set A, a Discrete Interval Probability Distribution (DIPD) characterizes a non-empty set of (discrete) Probability Mass Functions (PMFs)

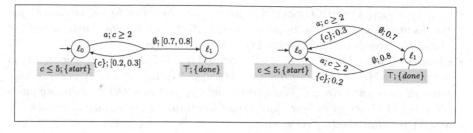

Fig. 2. IPTA A_1 (left) and PTA A_2 induced by it (right).

$\mu : A \to [0,1]$ by assigning to each $a \in A$ an interval $[x_a, y_a]$ such that it is possible to choose one element from each interval to obtain a sum of 1. Formally, a pair $(\mu_1 : A \to [0,1], \mu_2 : A \to [0,1])$ is a DIPD, written $(\mu_1, \mu_2) \in \mathsf{DIPD}(A)$, if the following two properties hold: (i) $\mu_1(a) \leq \mu_2(a)$ for each $a \in A$ and (ii) $\sum \{\mu_1(a) \mid a \in A\} \leq 1 \leq \sum \{\mu_2(a) \mid a \in A\}$ using summation over multisets. These two properties ensure non-emptiness of intervals and non-emptiness of characterized PMFs, respectively. $\mu : A \to [0,1]$ is then such a characterized PMF in the semantics of (μ_1, μ_2), written $\mu \in \langle (\mu_1, \mu_2) \rangle$, if $\mu_1(a) \leq \mu(a) \leq \mu_2(a)$ for each $a \in A$. Note that (μ, μ) is also a DIPD over A. Lastly, the *support* of (μ_1, μ_2), written $\mathsf{supp}((\mu_1, \mu_2))$, contains all $a \in A$ for which the right interval border $\mu_2(a)$ is non-zero. For example, the semantics of the DIPD $(\mu_1 = \{(\ell_0, 0.2), (\ell_1, 0.7)\}, \mu_2 = \{(\ell_0, 0.3), (\ell_1, 0.8)\})$ contains, among others, the two PMFs $\mu = \{(\ell_0, 0.3), (\ell_1, 0.7)\}$ and $\mu' = \{(\ell_0, 0.2), (\ell_1, 0.8)\}$.

An IPTA (such as A_1 from Fig. 2) consists of (a) a set of locations with a distinguished initial location (such as ℓ_0), (b) a set of clocks (such as c) which are initially set to 0, (c) an assignment of a set of Atomic Propositions (APs) (such as $\{done\}$) to each location (for subsequent analysis of e.g. reachability properties), (d) an assignment of constraints over clocks to each location as invariants (such as $c \leq 5$), and (e) a set of interval probabilistic timed edges. Each interval probabilistic timed edge consists thereby of (i) a single source location, (ii) at least one target location, (iii) an action (such as a), (iv) a clock constraint (such as $c \geq 2$) specifying as a guard when the edge is enabled based on the current values of the clocks, and (v) a DIPD assigning an interval probability (such as $[0.2, 0.3]$) to each pair consisting of a set of clocks to be reset (such as $\{c\}$) and a target location to be reached.

Definition 1 (Interval Probabilistic Timed Automaton (IPTA)). *An interval probabilistic timed automaton (IPTA) A is a tuple with the following components.*

- $\mathsf{locs}(A)$ *is a finite set of locations,*
- $\mathsf{iloc}(A)$ *is the unique initial location from* $\mathsf{locs}(A)$,
- $\mathsf{acts}(A)$ *is a finite set of actions disjoint from* $\mathbf{R}_0^+$,
- $\mathsf{clocks}(A)$ *is a finite set of clocks,*

- $\mathsf{invs}(A)$: $\mathsf{locs}(A) \rightarrow \mathsf{CC}(\mathsf{clocks}(A))$ *maps each location to an invariant for that location such that the initial clock valuation satisfies the invariant of the initial location (i.e., $\mathsf{ICV}(\mathsf{clocks}(A)) \models \mathsf{invs}(A)(\mathsf{iloc}(A)))$,*
- $\mathsf{edges}(A) \subseteq \mathsf{locs}(A) \times \mathsf{acts}(A) \times \mathsf{CC}(\mathsf{clocks}(A)) \times \mathsf{DIPD}(2^{\mathsf{clocks}(A)} \times \mathsf{locs}(A))$ *is a finite set of IPTA edges of the form $(\ell_1, a, \psi, (\mu_1, \mu_2))$ where ℓ_1 is the source location, a is an action, ψ is a guard, and (μ_1, μ_2) is a DIPD mapping pairs (Res, ℓ_2) of clocks to be reset and target locations to probability intervals,*
- $\mathsf{aps}(A)$ *is a finite set of APs, and*
- $\mathsf{lab}(A)$: $\mathsf{locs}(A) \rightarrow 2^{\mathsf{aps}(A)}$ *maps each location to a set of APs.*

Moreover, we define PTA and TA as IPTA restrictions as follows.

- *A is a PTA if for all $(\ell, a, \psi, (\mu_1, \mu_2)) \in \mathsf{edges}(A)$: $\langle (\mu_1, \mu_2) \rangle$ is a singleton.*
- *A is a TA if for all $(\ell, a, \psi, (\mu_1, \mu_2)) \in \mathsf{edges}(A)$: $\mathsf{supp}((\mu_1, \mu_2))$ is a singleton.*

The semantics of an IPTA is given in terms of the induced Probabilistic Timed System (PTS), which defines timed probabilistic paths as expected. The states of the induced PTS are pairs of locations and clock valuations. The steps between such states then either (a) nondeterministically advance time or (b) nondeterministically select an enabled IPTA edge, nondeterministically determine a PMF from the given DIPD, and probabilistically determine a reset set and target location based on that PMF.

Definition 2 (PTS Induced by IPTA). *Every IPTA A induces a unique probabilistic timed system (PTS) $\mathsf{IPTAtoPTS}(A) = P$ consisting of the following components.*

- $\mathsf{states}(P) = \{(\ell, v) \in \mathsf{locs}(A) \times \mathsf{CV}(\mathsf{clocks}(A)) \mid v \models \mathsf{invs}(A)(\ell)\}$ *contains as PTS states pairs of locations and clock valuations satisfying the location's invariant,*
- $\mathsf{istate}(P) = (\mathsf{iloc}(A), \mathsf{ICV}(\mathsf{clocks}(A)))$ *is the unique initial state from $\mathsf{states}(P)$,*
- $\mathsf{acts}(P) = \mathsf{acts}(A)$ *is the same set of actions,*
- $\mathsf{steps}(P) \subseteq \mathsf{states}(P) \times (\mathsf{acts}(P) \cup \mathbf{R}_0^+) \times \mathsf{DIPD}(\mathsf{states}(P))$ *is the set of PTS steps where $((\ell, v), a, (\mu, \mu)) \in \mathsf{steps}(P)$, if one of the two following cases applies.*
  - ○ TIMED STEP:
    - ▷ $a \in \mathbf{R}_0^+$ *is a duration,*
    - ▷ $(\ell, v + t') \in \mathsf{states}(P)$ *for all $t' \in [0, a]$, i.e., the invariant of ℓ is also satisfied for all intermediate time points t', and*
    - ▷ $\mu(\ell, v + a) = 1$ *identifies the unique PTS target state $(\ell, v + a)$.*
  - ○ DISCRETE STEP:
    - ▷ $a \in \mathsf{acts}(A)$ *is an action of the IPTA,*
    - ▷ $(\ell, a, \psi, (\mu_1', \mu_2')) \in \mathsf{edges}(A)$ *is an edge of the IPTA,*
    - ▷ $v \models \psi$, *i.e., the guard ψ of the edge is satisfied by the valuation v,*
    - ▷ $\mu_i(\ell', v') = \sum \{\mu_i'(X, \ell') \mid X \subseteq \mathsf{clocks}(A), v' = v[X := 0]\}$ *for $i \in \{1, 2\}$ is the DIPD on the PTS target states (ℓ', v'), and*
    - ▷ $\mu \in \langle (\mu_1, \mu_2) \rangle$ *is a PMF from the semantics of the DIPD (μ_1, μ_2).*
- $\mathsf{aps}(P) = \mathsf{aps}(A)$ *is the same set of APs, and*

- $\mathsf{lab}(P)(\ell, v) = \mathsf{lab}(A)(\ell)$ *labels states in P according to the location labeling of the IPTA.*

For the special case of PTA, we will use in Sect. 5 the PRISM model checker [6] to solve the following analysis problems defined for the induced PTSs. Intuitively, these analysis problems ask for the probability with which states labeled with a given AP can be reached (possibly within a given time bound). However, the probability to be computed depends on how the nondeterminism in the PTS is resolved. Technically, this nondeterminism is resolved using an adversary, which selects for the finite path constructed so far the next timed/discrete step where the target state of a discrete step is subject to an additional probabilistic choice. The probability values obtained for all such adversaries result in a unique lower and a unique upper bound. These unique lower and upper bounds intuitively correspond to the worst-case or the best-case probabilities depending on whether reaching a state labeled with the given AP is desirable. The worst-case and the best-case probabilities can then be computed using PRISM.

Definition 3 (Min/Max Probabilistic Timed Reachability Problems).
Evaluate $\mathcal{P}_{op=?}(\mathsf{F}_{\sim c}\ ap)$ for a PTS P with $op \in \{\min, \max\}$, $\sim\ \in \{\leq, <\}$, $c \in \mathbb{N} \cup \{\infty\}$, and $ap \in \mathsf{aps}(P)$ to obtain the infimal/supremal probability (depending on op) over all adversaries to reach some state in P labeled with ap within $t \sim c$ time units.

For example, for the PTS $P = \mathsf{IPTAtoPTS}(A_1)$ induced by the IPTA A_1 from Fig. 2, (a) $\mathcal{P}_{\max=?}(\mathsf{F}_{\leq 5}\ done)$ is evaluated to probability $0.96 = 0.8 + 0.2 \times 0.8$ since the probability maximizing adversary would enable two discrete steps e.g. at time points 2 and 4 with the maximal probability of 0.8 to reach ℓ_1 in each case and (b) $\mathcal{P}_{\min=?}(\mathsf{F}_{\leq 5}\ done)$ is evaluated to probability 0.7 since the probability minimizing adversary would enable only one discrete step (e.g. at time point 5) where the minimal probability to reach ℓ_1 would be 0.7.

3   Interval Probabilistic Timed GTSs

In this section, we introduce the new formalism of IPTGTSs, which allows for the modeling and analysis of systems exhibiting structure dynamics, timed behavior, and interval probabilistic behavior. IPTGTSs extend PTGTSs, which are a combination of PGTSs and TGTSs, by allowing interval probabilities as in IPTA instead of precise probabilities as in PTA.

As usual, we assume that all graphs in IPTGTSs are typed over some fixed type graph TG. Moreover, we denote the set of all variables of sort real that represent clocks of a given graph G by $\mathsf{C}(G)$ (for the graph G_0 from Fig. 1d, $\mathsf{C}(G_0) = \{c_1, c_2, c_3, c_4, c\}$). Note that, in this paper, we employ variables of the symbolic graphs to represent clocks rather than clock nodes as in [8] to simplify the technical presentation.

For our running example, introduced in Sect. 1, the type graph TG is given in Fig. 1a and the initial graph G_0 is given in Fig. 1d. In the following, we use the abbreviation of the form $\mathcal{G}(v_1, v_2, v_3, v_4, n, e)$ for all reachable graphs where v_1, v_2, v_3, v_4, and n correspond to the values of variables and e is the number of *Error* nodes connected to the *Observer* node. Using this abbreviation, we denote the initial graph G_0 by $\mathcal{G}(\top, \bot, \bot, \bot, 0, 0)$.

An IPTGT rule σ contains a set of GT rules $\mathsf{rules}(\sigma)$ with a common left-hand side graph $\mathsf{lhs}(\sigma)$, which is matched into the graph under transformation. A clock constraint over the clocks from $\mathsf{lhs}(\sigma)$ is used as a guard and is evaluated w.r.t. a considered match. A DIPD assigns a non-empty probability interval to each GT rule ρ. Each GT rule ρ is equipped with a set of clocks to be reset ranging over the clocks from the right-hand side graph $\mathsf{rhs}(\rho)$ of ρ.

Definition 4 (IPTGT Rule). *An* interval probabilistic timed graph transformation rule (IPTGT rule) σ *is a tuple with the following components.*

- $\mathsf{lhs}(\sigma)$ *is a common left-hand side graph,*
- $\mathsf{rules}(\sigma)$ *is a finite set of GT rules ρ with $\mathsf{lhs}(\rho) = \mathsf{lhs}(\sigma)$ where $\mathsf{lhs}(\rho)$ is the left-hand side graph of the GT rule ρ,*
- $\mathsf{guard}(\sigma) \in \mathsf{CC}(\mathsf{C}(\mathsf{lhs}(\sigma)))$ *is a guard defined as a clock constraint over the clocks from the left-hand side graph $\mathsf{lhs}(\sigma)$,*
- $\mathsf{dipd}(\sigma) \in \mathsf{DIPD}(\mathsf{rules}(\sigma))$ *is a DIPD on $\mathsf{rules}(\sigma)$ with $\mathsf{supp}(\mathsf{dipd}(\sigma)) = \mathsf{rules}(\sigma)$,*
- $\mathsf{reset}(\sigma)(\rho) \subseteq \mathsf{C}(\mathsf{rhs}(\rho))$ *identifies the clocks to be reset for each $\rho \in \mathsf{rules}(\sigma)$, and*
- $\mathsf{prio}(\sigma) \in \mathbf{N}$ *is the priority assigned to σ.*

For our running example, consider the two IPTGT rules $\sigma_{\mathsf{send}:I_1,I_2}$[1] and $\sigma_{\mathsf{process}}$ in Fig. 1g and Fig. 1h, respectively. The two GT rules $\rho_{\mathsf{send:success}}$ and $\rho_{\mathsf{send:failure}}$ of $\sigma_{\mathsf{send}:I_1,I_2}$ and the single GT rule $\rho_{\mathsf{process:done}}$ of $\sigma_{\mathsf{process}}$ are given in integrated notation where graph elements to be added/removed are marked with $\oplus/\ominus$. To limit rule application, each IPTGT rule has an *attribute guard* on the current attribute values, a *clock guard* on the current clock values, and a *priority*. Also, each of the underlying GT rules has an AC describing an attribute modification (called *attribute effect*), a *reset* set of clocks to be reset after rule application, and a *probability interval*. Intuitively, the IPTGT rule $\sigma_{\mathsf{send}:I_1,I_2}$ is used to attempt the sending of the Boolean value true ($\top$) from one agent to another agent with Boolean value $\bot$ when the clock value of the sending agent is at least 2. If sending succeeds, the receiving agent adopts the Boolean value $\top$ and may then send that value as well. If sending fails, an error is created and connected to the observer. The IPTGT rule $\sigma_{\mathsf{process}}$ (which has a higher priority than $\sigma_{\mathsf{send}:I_1,I_2}$) is used to allow the observer to process (and delete) pending errors counting processed errors up to a maximal number of 2.

An IPTGT invariant ϕ is a nested graph condition over the empty graph $\emptyset$. IPTGT invariants are used to rule out invalid potential IPTGT configurations. Potential IPTGT configurations (G, v) are given by a finite graph G and

---

[1] In our evaluation in Sect. 5, we consider different instantiations of the two probability intervals I_1 and I_2.

a clock valuation $v \in \mathsf{CV}(\mathsf{C}(G))$ on its clocks. For our running example, the IPTGT invariant ϕ_{inv} in Fig. 1b states that (a) a sending agent must send its Boolean value $\top$ after waiting not longer than 5 time units and (b) an observer with a pending error must process that error urgently. Note that the clocks of the agents and the observer are reset to 0 whenever they become eligible to send their Boolean value or to process an error in the two IPTGT rules. An IPTGT invariant ϕ is satisfied by a potential IPTGT configuration (G, v), written $(G, v) \models \phi$, if $v \models \exists V.\, \mathsf{ac}(G) \wedge \gamma$ where V is the set of all non-clock variables of G, $\mathsf{ac}(G)$ is the AC of G, and γ is an attribute constraint on the variables of G obtained by evaluating ϕ for G. For our running example, the potential IPTGT configuration $(\mathcal{G}(\top, \bot, \bot, \bot, 0, 0), v)$ where $v = \mathsf{ICV}(\{c_1, c_2, c_3, c_4\})$ satisfies the PTGT invariant ϕ_{inv} since v satisfies the derived AC equivalent to $c_1 \leq 5$ where V contains all id and val variables, $\mathsf{ac}(G)$ is the AC given in Fig. 1d, and γ states for each of the four clocks c_1-c_4 that their value must be less equal 5 when the corresponding val attribute equals $\top$.

Similarly, an IPTGT Atomic Proposition (IPTGT AP) ϕ is a nested graph condition over the empty graph $\emptyset$ labeling a (potential) IPTGT configuration (G, v) if G satisfies ϕ, written $G \models \phi$. Note that IPTGT APs may not depend on the clock valuation v as for the labeling in IPTA. For our running example, we employ the IPTGT APs ϕ_{fin}, ϕ_{con}, and ϕ_{2err} from Fig. 1. ϕ_{fin} (given using the previously introduced abbreviation) checks whether the value $\top$ has been successfully adopted by all agents, precisely one error was processed by the observer, and no errors are pending. ϕ_{con} checks whether all errors are connected to some observers and ϕ_{2err} checks whether no two errors are present.

We now define IPTGTSs based on the notions introduced above for a fixed type graph. For our running example, the components of the considered IPTGTS are given in Fig. 1.

Definition 5 (IPTGTS). *An interval probabilistic timed graph transformation system (IPTGTS) S is a tuple with the following components.*

- $\mathsf{iG}(S)$ *is a finite initial graph,*
- $\mathsf{rules}(S)$ *is a finite set of IPTGT rules,*
- $\mathsf{invs}(S)$ *is a finite set of IPTGT invariants, which are all satisfied for the initial graph and the initial clock valuation* $(\mathsf{iG}(S), \mathsf{ICV}(\mathsf{C}(\mathsf{iG}(S))))$, *and*
- $\mathsf{aps}(S)$ *is a finite set of IPTGT APs.*

Moreover, we define the following notions.

- *A potential IPTGT configuration (G, v) given by a finite graph G and a clock valuation $v \in \mathsf{CV}(\mathsf{C}(G))$ is an IPTGT configuration of S, written $(G, v) \in \mathsf{Confs}(S)$, if (G, v) satisfies all IPTGT invariants of S, i.e., $(G, v) \models \phi$ for each $\phi \in \mathsf{invs}(S)$.*
- *Two given IPTGT configurations (G_1, v_1) and (G_2, v_2) are equivalent, written $(G_1, v_1) \equiv (G_2, v_2)$, if there is some isomorphism $m : G_1 \to G_2$ such that $v_2 \circ m = v_1$. The equivalence relation $\equiv$ also induces equivalence classes denoted by $[(G_1, v_1)]_\equiv$.*

The semantics of IPTGTSs is defined below in terms of an induced PTS, for which we first define a step relation. As for IPTA, IPTGTSs can execute, on the one hand, *timed steps* advancing all clocks while respecting the invariants and, on the other hand, *discrete steps* by applying some IPTGT rule where the IPTGT rule, the match, and the Probability Mass Function (PMF) of the DIPD are chosen nondeterministically and the GT rule to be used is chosen probabilistically. Moreover, for the discrete steps, we ensure that (a) the guard of the IPTGT rule is satisfied by the current clock valuation and match, (b) no discrete step using an IPTGT rule with higher priority can be applied, and (c) all GT rules of the IPTGT rule are applicable using the same match. Then, considering a GT rule ρ of the IPTGT rule σ, we define a discrete step based on the corresponding GT step and ensure that the clock valuation is adapted as expected also enforcing the clock resets specified in the IPTGT rule.

Definition 6 (IPTGT Step). *An IPTGTS S defines the following two kinds of steps.*

- TIMED STEP: $(G, v) \xrightarrow{\delta} (G, v + \delta)$, *if*
  - $\delta \in \mathbf{R}_0^+$ *is a duration and*
  - $(G, v + \delta') \in \mathsf{Confs}(S)$ *for each* $\delta' \in [0, \delta]$, *i.e., the IPTGT invariants are also satisfied for all intermediate time points.*
- DISCRETE STEP: $(G_1, v_1) \xrightarrow{\sigma, \rho, m} (G_2, v_2)$, *if*
  - $\sigma \in \mathsf{rules}(S)$ *is an IPTGT rule,*
  - $m : \mathsf{lhs}(\sigma) \hookrightarrow G_1$ *is a match,*
  - $v_1 \models \mathsf{guard}(\sigma)$, *i.e., the guard of the IPTGT rule σ is satisfied by the given valuation v_1,*
  - $\rho \in \mathsf{rules}(\sigma)$ *is a GT rule of σ,*
  - *no IPTGT rule σ' with higher priority is applicable, i.e., there are no G_2', v_2', σ', ρ', and m' such that $(G_1, v_1) \xrightarrow{\sigma', \rho', m'} (G_2', v_2')$ and $\mathsf{prio}(\sigma') > \mathsf{prio}(\sigma)$,*
  - *the GT rule ρ is applicable and results in the IPTGT configuration (G_2, v_2), i.e., $(G_1, v_1) \xrightarrow{\sigma, \rho, m} (G_2, v_2)$,*
  - *every GT rule of σ is applicable for the match m, i.e., for all $\rho' \in \mathsf{rules}(\sigma)$ there are G_2' and v_2' such that $(G_1, v) \xrightarrow{\sigma, \rho', m} (G_2', v_2')$,*

  where $(G_1, v_1) \xrightarrow{\sigma, \rho, m} (G_2, v_2)$, *if*
  - $(G_1, v_1), (G_2, v_2) \in \mathsf{Confs}(S)$,
  - $\rho = (\ell : K \hookrightarrow L, r : K \hookrightarrow R, \gamma)$ *is a GT rule[2],*
  - $G_1 \xRightarrow{\rho, m} G_2$ *is the DPO GT step from Fig. 3a,*
  - v_2 *is obtained from v_1 by preserving values of preserved clocks unless they are reset to 0, i.e., for each $X \in \mathsf{C}(G_1)$ and each $Y \in \mathsf{C}(D)$ satisfying $\ell'(c) = c'$ it holds that $v_2(r'(c)) = v_1[m(\mathsf{reset}(\sigma)(\rho)) := 0](c')$ (see Fig. 3b), and*
  - *the clock value 0 is assigned to all clocks created by the GT step, i.e., for each $c \in \mathsf{C}(G_2) - r'(\mathsf{C}(D))$ it holds that $v_2(c) = 0$.*

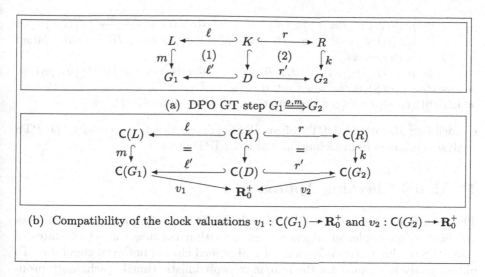

(a) DPO GT step $G_1 \xRightarrow{\rho, m} G_2$

(b) Compatibility of the clock valuations $v_1 : C(G_1) \to \mathbf{R}_0^+$ and $v_2 : C(G_2) \to \mathbf{R}_0^+$

Fig. 3. Visualizations for Definition 6

We now define the semantics of IPGTSs in terms of an induced PTS as before for IPTA. Note that, for the following definition, IPTGT steps as well as the notions of IPTGT satisfaction for APs, guards, and invariants are preserved by equivalence of IPTGT configurations. Hence, the choice of a representant from an equivalence class is not important.

Definition 7 (PTS Induced by IPTGTS). *Every IPTGTS S induces a unique PTS IPTGTStoPTS$(S) = P$ consisting of the following components.*

- states(P) *is given by the smallest set of equivalence classes $[(G, v)]_\equiv$ where $(G, v) \in$ Confs(S) such that steps(P) below is well-defined and istate(P) (see the next item) is in states(P),*
- istate$(P) = [(\text{iG}(S), \text{ICV}(C(\text{iG}(S))))]_\equiv$ *is the unique initial state from states(P) given by the equivalence class containing the initial configuration of S,*
- acts(P) *is the smallest set of tuples (σ, m) where $\sigma \in$ rules(S) and m is a match such that steps(P) below is well-defined,*
- $([(G, v)]_\equiv, a, (\mu, \mu)) \in$ steps(P), *if one of the two following cases applies.*
  - ○ TIMED STEP:
    - ▷ $a \in \mathbf{R}_0^+$ *is a duration,*
    - ▷ $(G, v) \xrightarrow{a} (G, v + a)$ *is a timed step of S, and*
    - ▷ $\mu([G, v + a]_\equiv) = 1$ *identifies the unique target state $[(G, v + a)]_\equiv$.*
  - ○ DISCRETE STEP:
    - ▷ $a = (\sigma, m) \in$ acts(P) *is a partial step label,*
    - ▷ $(G, v) \xrightarrow{\sigma, \rho, m} (G', v')$ *for some $\rho \in$ rules(σ) is a discrete step of S,*
    - ▷ dipd$(\sigma) = (\mu_1', \mu_2')$ *is the DIPD of σ,*

---

[2] We omit here the handling of attribute modifications given by γ for brevity.

▷ $\mu_i([(\bar{G}, \bar{v})]_{\equiv}) = \sum \{\!|\mu_i'(\rho') \mid \rho' \in \mathsf{rules}(\sigma), (G, v) \xrightarrow{\sigma, \rho', m} \equiv (\bar{G}, \bar{v})|\!\}$ for all $[(\bar{G}, \bar{v})]_{\equiv} \in \mathsf{states}(P)$ and $i \in \{1, 2\}$ is the DIPD on the target states, and

▷ $\mu \in \langle(\mu_1, \mu_2)\rangle$ is a PMF from the semantics of the DIPD (μ_1, μ_2).

- $\mathsf{aps}(P) = \mathsf{aps}(S)$ is the same set of APs, and
- $\mathsf{lab}(P)([(G, v)]_{\equiv}) = \{\phi \in \mathsf{aps}(P) \mid G \models \phi\}$ labels states in P.

By defining the induced PTS of an IPTGTS, we can now consider the PTS analysis problems from Definition 3 also for IPTGTSs.

4   Model Checking Approach

The definition of the induced PTS of an IPTGTS from the previous section does not lead to an implementable analysis algorithm because the set of states of that PTS is (due to the valuations of real-valued clocks) not even countable. To obtain analysis support for the min/max probabilistic timed reachability properties from Definition 3, we now follow the path taken for PTGTSs in [8] and translate a given IPTGTS into its underlying automata-based model preserving its semantics in terms of the induced PTS (see Fig. 4). As a first step, in Subsect. 4.1, we introduce the operation IPTGTStoIPTA translating an IPTGTS into an IPTA. As a second step, in Subsect. 4.2, we translate the obtained IPTA into a PTA using the operation IPTAtoPTA. This operation is defined based on the translation procedure from [4], which is shown to preserve the semantics in terms of the analysis problems from Definition 3. For the resulting PTA, the analysis problems from Definition 3 can then be analyzed using the PRISM model checker. To accommodate for IPTGTSs with infinite underlying GT state spaces but finite underlying timed GT state spaces, we present in Subsect. 4.3 an online filtering technique using the UPPAAL model checker. Lastly, we briefly discuss the analysis of non-probabilistic properties for IPTGTSs in Subsect. 4.4 before revisiting the min/max probabilistic timed reachability properties from Definition 3 in our evaluation in Sect. 5. For more details on our model checking approach see [9].

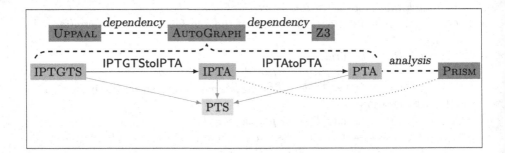

Fig. 4. Overview of the model checking approach.

4.1   Translation of IPTGTS into IPTA

We now present how IPTGTSs can be translated into IPTA using the operation IPTGTStoIPTA. This translation is an adaptation of the translation of PTGTSs into PTA presented in [8].

For a given IPTGTS S, the following four steps describe the basic idea of its translation into the corresponding IPTA. In step 1, the underlying GTS S' of the IPTGTS with the rule set $\cup\{\text{rules}(\sigma) \mid \sigma \in \text{rules}(S)\}$ and the initial graph $iG(S)$ is determined where no priorities or timing constraints of the IPTGT rules are integrated into the GT rules. In step 2, the GT state space (Q, E) of S' is constructed where Q is the set of all reachable graphs and E contains the corresponding GT steps between these graphs. In step 3, a smallest set of so-called *global clocks* Y from the GT state space (Q, E) is derived where the underlying GT spans from E are used to track such global clocks. Finally, in step 4, the resulting IPTA is obtained from the IPTGTS S and the GT state space (Q, E) where the set Y of global clocks is employed to convert and annotate GT steps from E.

Note the following for the steps 2–4 of the translation. In step 2, the GT state space (Q, E) will often contain additional steps (and also states) that are not permitted in the IPTGTS due to (a) priorities and (b) timing constraints (i.e., guards or invariants). Cases where the GT state space (Q, E) is not finite are discussed in Subsect. 4.3. Moreover, to ensure in step 3 that Y contains a finite number of clocks, the initial graph should contain all clocks to be used at some point and IPTGT rules should not add further clocks. Also, to prevent that clocks are swapped by isomorphisms during the GT state space construction, we use *id* attributes such as those in *Agent* nodes in our running example. For step 4, relying on the set of global clocks Y, (a) invariants are obtained for each graph in the GT state space and checked for satisfiability, (b) GT steps with common source location and match belonging to one IPTGT rule are grouped into one IPTA edge, and (c) the guard of an IPTA edge is obtained by ensuring that the IPTA edge is not disabled by invariants of its target locations and by also requiring the negation of the guards of IPTA edges starting in the same location but resulting from IPTGT rules with higher priorities.

Definition 8 (IPTA Induced by IPTGTS). *Every IPTGTS S induces a unique IPTA IPTGTStoIPTA$(S) = A$ consisting of the following components where (Q, E) is the GT state space of the underlying GTS of S and Y is its set of global clocks.*

- $\text{locs}(A) = Q$ *contains all graphs of the GT state space,*
- $\text{iloc}(A) = iG(S)$ *is the unique initial location,*
- $\text{acts}(A)$ *is the smallest set of tuples (σ, m) where $\sigma \in \text{rules}(S)$ and m is a match such that $\text{edges}(A)$ below is well-defined,*
- $\text{clocks}(A) = Y$ *is the set of global clocks,*
- $\text{invs}(A)(G) = \psi$ *such that $(G, v) \in \text{Confs}(S)$ iff $v \models \psi$,*
- $(G, (\sigma, m), \psi, (\mu_1, \mu_2)) \in \text{edges}(A)$, *if*

○ *the guard ψ is a clause of the disjunctive normal form[3] of the constraint $m(\mathsf{guard}(\sigma)) \wedge \psi_{inv} \wedge \neg\psi_{higher}$ where (a) ψ_{inv} is the conjunction of invariants $\mathsf{invs}(A)(G)[R := 0]$ of the target locations adapted to the source location for all $(R, G) \in \mathsf{supp}((\mu_1, \mu_2))$ and (b) ψ_{higher} is the conjunction of the guards ψ'' of all IPTA edges $(G, (\sigma'', m''), \psi'', (\mu_1'', \mu_2'')) \in \mathsf{edges}(A)$ satisfying $\mathsf{prio}(\sigma'') > \mathsf{prio}(\sigma)$[4] and*

○ $\mu_i(Y', G') = \sum \{\!|\mu_i'(\rho) \mid \rho \in \mathsf{rules}(\sigma) \wedge Y' = \mathsf{reset}(\sigma)(\rho) \wedge (G, \rho, m, G') \in E \wedge \mathsf{dipd}(\sigma) = (\mu_1', \mu_2')|\!\}$ *for $i \in \{1, 2\}$ is the DIPD on the target locations,*

● $\mathsf{aps}(A) = \mathsf{aps}(S)$ *is the same set of APs, and*

● $\mathsf{lab}(A)(G) = \{\phi \in \mathsf{aps}(A) \mid G \models \phi\}$ *labels locations in A.*

The operation $\mathsf{IPTGTStoIPTA}$ preserves the semantics in terms of the induced PTSs, i.e., the IPTGTS and the resulting IPTA induce the same PTS.

4.2 Translation of IPTA into PTA

In [4], we have implemented the IPTA model checking algorithm from [15] in PRISM. This algorithm takes an IPTA A and an analysis problem of the form $\mathcal{P}_{op=?}(\mathsf{F}_{\leq\infty} \, ap)$ from Definition 3 and computes the resulting probability value.[5] The algorithm operates on a zone-based state space where states of the form $(\ell, \psi) \in \mathsf{locs}(A) \times \mathsf{CC}(\mathsf{clocks}(A))$ represent all states (ℓ, v) of the PTS induced by A satisfying $v \models \psi$. The fixed-point computation performed by the algorithm modifies a probability vector p_i mapping states (ℓ, ψ) to probabilities. Initially, p_0 maps all *target states* (ℓ, ψ) containing locations ℓ that are labeled by the AP ap of the considered property to 1 and all other states to 0. In the fixed-point, p_i maps each state (ℓ, ψ) to the probability with which one of the target states is reached. To obtain $p_{i+1}(\ell, \psi)$ in an iteration (a) each IPTA edge used in a step from (ℓ, ψ) with DIPD (μ_1, μ_2) is considered, (b) a PMF $\mu \in \langle(\mu_1, \mu_2)\rangle$ is obtained such that the probability given by $\sum \{\!|\mu(s) \times p_i(s) \mid s \in \mathsf{locs}(A) \times \mathsf{CC}(\mathsf{clocks}(A))|\!\}$ for reaching a target state from the state (ℓ, ψ) using a path where the considered IPTA edge is taken in the first step is maximal/minimal, and (c) $p_{i+1}(\ell, \psi)$ is set to the maximal/minimal value across all IPTA edges considered for (ℓ, ψ). For the IPTA A_1 from Fig. 2 and the property $\mathcal{P}_{max=?}(\mathsf{F}_{\leq\infty} \, done)$, we obtain $p_0 = \{((\ell_0, c \leq 5), 0), ((\ell_1, \top), 1)\}$ and $p_{i+1} = \{((\ell_0, c \leq 5), 0.8 \times p_i(\ell_1, \top) + 0.2 \times p_i(\ell_0, c \leq 5)), ((\ell_1, \top), 1)\}$ for $i \in \mathbf{N}$ resulting in $p_i(\ell_0, c \leq 5) = 1$ in the limit $i \to \infty$.

Unfortunately, the described algorithm has not been integrated in the official PRISM branch. As an alternative approach to obtain model checking support for IPTA, we translate the given IPTA into a PTA (as exemplified in Fig. 2 where the IPTA A_1 is translated into the PTA A_2) and then apply PRISM to the resulting PTA. Intuitively, the PTA is obtained by replacing each IPTA edge e_1 of the

---

[3] Note that negation and disjunction are not allowed in guards.

[4] The dependency between the guards ψ and each ψ'' requires that IPTA edges of A are constructed in descending order of the priorities of the involved IPTGT rules.

[5] As usual, time bounds $\sim c$ (as in Definition 3) are encoded using an additional clock to force a step to a sink location as soon as the time bound is violated.

IPTA by a set of PTA edges. Thereby, a replacement edge e_2 is obtained from e_1 by replacing its DIPD (μ_1, μ_2) by the DIPD (μ, μ) where $\mu \in \langle\langle(\mu_1, \mu_2)\rangle\rangle$ is a PMF that may be obtained for some p_i using the described algorithm. In fact, instead of considering all such possible p_i, it suffices to consider all permutations of the target locations of e_1. The intervals from the DIPD (μ_1, μ_2) are then resolved in the order of the permutation by choosing the maximal[6] probability from the ith interval such that the sum of the first i chosen probabilities plus the sum of the minimal probabilities of the remaining intervals does not exceed 1. For the translation in Fig. 2, the permutation (ℓ_0, ℓ_1) (the permutation (ℓ_1, ℓ_0)) results in the upper (the lower) PTA edge where the interval [0.2, 0.3] (the interval [0.7, 0.8]) for the first location is resolved by choosing the maximal value 0.3 (0.8) and by then choosing 0.7 (0.2) analogously for the second location. The described translation, defined by the operation IPTAtoPTA, is correct since the same probabilities are computed for the input IPTA and the output PTA for the analysis problems from Definition 3.

Constructing a PMF for all permutations of intervals for an IPTA edge may result in an exponentially larger PTA. This means that IPTA are exponentially more concise compared to PTA w.r.t. the considered analysis problems (which correspondingly holds for IPTGTSs and PTGTSs). However, when (a) IPTGT rules contain only few GT rules implying a small set of permutations, (b) the permutations result in a small set of PMFs (since different permutations may result in the same PMF), or (c) the model checking efficiency of the PTA or IPTA at hand is dominated by the number of clocks but not by the number of PTA edges, employing the translation via the operation IPTAtoPTA can be as efficient as the IPTA model checking algorithm from [4,15].

4.3   Analysis of Timed Reachability

As a plug-in procedure, we now discuss our on-the-fly adaptation of the operation IPTGTStoIPTA presented in Subsect. 4.1. The goal of this adaptation is to allow for the generation of (finite) IPTA when the intermediate GT state space is infinite while the timed GT state space is finite. To achieve this goal, we employ UPPAAL to check whether steps constructed in the GT state space are enabled when considering the timed behavior specified by guards, invariants, and resets in the timed GT state space. For this purpose, we construct a sequence of fragments of the priority-free GT state space where all GT steps are timed reachable (i.e., each GT step of the fragment occurs in some timed path when considering all timing constraints of the IPTGTS). We start the procedure with the GT state space fragment only containing the initial graph of the IPTGTS and then, as a first step, we construct either at most n further GT steps overall or at most n further GT steps from each unfinished state. As a second step, if GT steps have been added, we construct using the operation IPTGTStoTA[7] a TA for the current

---

[6] As all permutations are considered, we can also choose the minimal probability here.

[7] The operation IPTGTStoTA is defined similarly to the operation IPTGTStoIPTA and deviates primarily by not aggregating GT steps belonging to a common IPTGT rule.

GT state space fragment. As a third step, we determine those most recently added GT steps that are not timed reachable using UPPAAL on the constructed TA removing them from the current GT state space fragment before repeating the described three steps.

Note that a GT step is not guaranteed to be timed reachable due to e.g. the guard that is created for it in the corresponding TA edge even when its source and target graphs are timed reachable. Hence, in the operation IPTGTStoTA, we split each most recently added GT step $G_1 \xrightarrow{\rho,m} G_2$ into two edges in the resulting TA. The first edge implements the GT step but has a fresh target state G' from which the second edge is taken urgently (using an additional single fresh clock employed globally in that translation) leading to G_2.

For our running example, this improvement is essential as the priority-free GT state space for the IPTGTS would be infinite since an unbounded number of errors could be created by an unbounded number of applications of the GT rule $\rho_{send:failure}$ without ever applying the GT rule $\rho_{process:done}$. The described procedure solves this problem since the priorities of the IPTGT rules are encoded in the constructed TA similarly as in operation IPTGTStoIPTA ruling out further applications of the GT rule $\rho_{send:failure}$.

4.4   Analysis of Timed and Structural Properties

The presented analysis approach is also applicable to simpler analysis problems compared to those in Definition 3, which often provide valuable insights when e.g. unexpected results are obtained for more complex properties.

On the one hand, properties not referring to probabilities but time can be analyzed based on the TA constructed in Subsect. 4.3. For our running example, UPPAAL can be used to verify the satisfaction of e.g. the timed CTL property $\mathsf{E}\,\mathsf{F}_{\leq 6}\,\phi_{fin}$ (where E and F are the *exists* and *eventually* operators, respectively), stating that the AP ϕ_{fin} can be satisfied within 6 time units.

On the other hand, properties referring to neither probabilities nor time can be analyzed based on the GT state space (Q, E) constructed in Subsect. 4.1. For our running example, the CTL property $\mathsf{A}\,\mathsf{G}\,\phi_{con}$ (where A and G are the *always* and *globally* operators, respectively), stating that all errors are always connected to an observer, can be verified based on the GT state space (Q, E) since each graph in Q would be labeled by the AP ϕ_{con}. Moreover, when using the procedure from Subsect. 4.3 where GT step generation is interleaved with the timed reachability analysis, we can also verify the CTL property $\mathsf{A}\,\mathsf{G}\,\phi_{2err}$ stating that there can never be two unprocessed errors. This property is satisfied since in the timed GT state space each graph would be labeled by the AP ϕ_{2err} due to the higher priority of the IPTGT rule $\sigma_{process}$.

5   Evaluation

To exemplify our model checking approach for IPTGTSs and to strengthen the importance of using IPTGTSs with probability intervals instead of PTGTSs

Table 1. Model checking results for the running example.

Instantiation	$\mathcal{P}_{\max=?}(\mathsf{F}_{\leq 6}\ \phi_{\mathsf{fin}})$	$\mathcal{P}_{\max=?}(\mathsf{F}_{\leq 15}\ \phi_{\mathsf{fin}})$	$\mathcal{P}_{\min=?}(\mathsf{F}_{\leq 15}\ \phi_{\mathsf{fin}})$
$S_{[0.7,0.8],[0.2,0.3]}$	0.4056	0.4056	0.2366
$S_{[0.3,0.8],[0.2,0.7]}$	0.5952	0.5952	0.0387
$S_{[0.8,0.8],[0.2,0.2]}$	0.3072	0.3072	0.3072
$S_{[0.7,0.7],[0.3,0.3]}$	0.3087	0.3087	0.3087
$S_{[0.3,0.3],[0.7,0.7]}$	0.0567	0.0567	0.0567

with precise probabilities, we now consider the IPTGTS of our running example for which the components have been discussed in Sect. 3 and visualized in Fig. 1. Note that the IPTGT rule $\sigma_{\mathsf{send}:I_1,I_2}$ contains two underlying GT rules with the assigned probability intervals I_1 and I_2. In our evaluation, we apply our implementation of the presented model checking approach in the tool AUTOGRAPH for multiple instantiations of $\sigma_{\mathsf{send}:I_1,I_2}$ by considering concrete probability intervals and different analysis problems relying on the AP ϕ_{fin}.

See Table 1 for an overview of the obtained results. The first column shows the five considered IPTGTSs instantiations where the chosen singleton probability intervals result in PTGTSs in the last three lines. The other columns show the results for computing the minimal/maximal probability to reach a graph labeled with ϕ_{fin} within 6 or 15 time units in the instantiated IPTGTS. The property $\mathcal{P}_{\min=?}(\mathsf{F}_{\leq 6}\ \phi_{\mathsf{fin}})$ is omitted in the table because its model checking results in the probability of 0 for each of the IPTGTSs instantiations as some adversary can delay sending for up to 5 time units preventing that any state labeled with ϕ_{fin} can be reached within 6 time units. Note that the probability values in the first two columns are identical since 6 time units used in the first column are already sufficient to reach a state labeled with ϕ_{fin} with maximal probability. Moreover, the results obtained for the PTGTSs differ in each case from the results obtained for the IPTGTSs, because, on every path to a graph labeled with ϕ_{fin} in the IPTGTS, the adversary chooses two different PMFs from the DIPD of the IPTGT rule $\sigma_{\mathsf{send}:I_1,I_2}$. For example, for the IPTGTS instantiation $S_{[0.7,0.8],[0.2,0.3]}$, the adversary defines three paths such that the resulting probability of $0.4056 = 0.3 \times 0.8 \times 0.8 \times 0.8 + 0.7 \times 0.3 \times 0.8 \times 0.8 + 0.7 \times 0.7 \times 0.3 \times 0.8$ is obtained in the first line of the first two columns. This resulting probability is the sum of the probabilities of the three paths each containing three successful sending steps with probabilities 0.7 or 0.8 and one unsuccessful sending step with probability 0.3 occurring as the first, second, or third step. From the different resulting probabilities, we conclude that both of the considered IPTGTSs cannot be approximated by any such PTGTSs instantiations appropriately in the sense of obtaining the same probabilities since using singleton intervals precludes adversaries that would choose different interval borders in some of their generated paths.

6   Conclusion and Future Work

We introduced the formalism of IPTGTSs as a high-level description language for the modeling and analysis of complex distributed embedded probabilistic real-time systems. IPTGTSs support, in addition to a nondeterministic passage of time (specified using clocks), a nondeterministic description of the probabilistic rule-based behavior (specified using probability intervals in rules). Moreover, we presented a model checking approach for IPTGTSs w.r.t. worst-case/best-case probabilistic timed reachability properties. This model checking approach is implemented in our tool AUTOGRAPH and is based on a translation of IPT-GTSs into PTA via IPTA. The PTA resulting from this translation can then be analyzed using the PRISM model checker.

As future work, we will extend Metric Temporal Graph Logic (MTGL) to IPTGTSs to be able to specify more complex properties on the structure dynamics, timed behavior, and probabilistic behavior of the given IPTGTSs. Such an extension is then to be included into the mapping of IPTGTSs to PTA to allow for the automated verification using PRISM.

References

1. Alur, R., Dill, D.L.: A theory of timed automata. Theor. Comput. Sci. **126**(2), 183–235 (1994). https://doi.org/10.1016/0304-3975(94)90010-8
2. Becker, B., Giese, H.: On safe service-oriented real-time coordination for autonomous vehicles. In: 11th IEEE International Symposium on Object-Oriented Real-Time Distributed Computing (ISORC 2008), Orlando, Florida, USA, 5–7 May 2008, pp. 203–210. IEEE Computer Society (2008). https://doi.org/10.1109/ISORC.2008. 13, http://ieeexplore.ieee.org/xpl/mostRecentIssue.jsp?punumber=4519543, ISBN 978-0-7695-3132-8
3. Ehrig, H., Ehrig, K., Prange, U., Taentzer, G.: Fundamentals of Algebraic Graph Transformation. MTCSAES, Springer, Heidelberg (2006). https://doi.org/ 10.1007/3-540-31188-2
4. Krause, C., Giese, H.: Model checking probabilistic real-time properties for service-oriented systems with service level agreements. In: Yu, F., Wang, C. (eds.) Proceedings 13th International Workshop on Verification of Infinite-State Systems, INFINITY 2011, 10th October 2011, Taipei, Taiwan. EPTCS, vol. 73, pp. 64–78 (2011)https://doi.org/10.4204/EPTCS.73.8
5. Krause, C., Giese, H.: Probabilistic graph transformation systems. In: Ehrig, H., Engels, G., Kreowski, H.-J., Rozenberg, G. (eds.) ICGT 2012. LNCS, vol. 7562, pp. 311–325. Springer, Heidelberg (2012). https://doi.org/10.1007/978-3-642-33654-6_21
6. Kwiatkowska, M., Norman, G., Parker, D.: PRISM 4.0: verification of probabilistic real-time systems. In: Gopalakrishnan, G., Qadeer, S. (eds.) CAV 2011. LNCS, vol. 6806, pp. 585–591. Springer, Heidelberg (2011). https://doi.org/10.1007/978-3-642-22110-1_47
7. Kwiatkowska, M.Z., Norman, G., Segala, R., Sproston, J.: Automatic verification of real-time systems with discrete probability distributions. Theor. Comput. Sci. **282**(1), 101–150 (2002). https://doi.org/10.1016/S0304-3975(01)00046-9

8. Maximova, M., Giese, H., Krause, C.: Probabilistic timed graph transformation systems. J. Log. Algebr. Meth. Program. **101**(2018), 110–131 (2018). https://doi.org/10.1016/j.jlamp.2018.09.003
9. Maximova, M., Schneider, S., Giese, H.: Interval Probabilistic Timed Graph Transformation Systems. Technical report 134. Hasso Plattner Institute at the University of Potsdam (2021)
10. Neumann, S.: Modellierung und Verifikation zeitbehafteter Graph transformations systeme mittels Groove. MA thesis. University of Paderborn (2007)
11. Orejas, F.: Symbolic graphs for attributed graph constraints. J. Symb. Comput. **46**(3), 294–315 (2011). https://doi.org/10.1016/j.jsc.2010.09.009
12. Schneider, S., Maximova, M., Sakizloglou, L., Giese, H.: Formal testing of timed graph transformation systems using metric temporal graph logic. In: Int. J. Softw. Tools Technol. Transf (2020, accepted)
13. Segala, R.: Modeling and verification of randomized distributed real-time systems. PhD thesis. Massachusetts Institute of Technology, Cambridge, MA, USA (1995). http://hdl.handle.net/1721.1/36560
14. UPPAAL. Department of Information Technology at Uppsala University, Sweden and Department of Computer Science at Aalborg University, Denmark (2021). https://uppaal.org/
15. Zhang, J., Zhao, J., Huang, Z., Cao, Z.: Model checking interval probabilistic timed automata. In: 2009 First International Conference on Information Science and Engineering, pp. 4936–4940 (2009). https://doi.org/10.1109/ICISE.2009.749

Verifying Graph Programs with Monadic Second-Order Logic

Gia S. Wulandari[1,2] and Detlef Plump[1(✉)]

[1] Department of Computer Science, University of York, York, UK
detlef.plump@york.ac.uk
[2] School of Computing, Telkom University, Bandung, Indonesia

Abstract. To verify graph programs in the language GP 2, we present a monadic second-order logic with counting and a Hoare-style proof calculus. The logic has quantifiers for GP 2's attributes and for sets of nodes or edges. This allows to specify non-local graph properties such as connectedness, k-colourability, etc. We show how to construct a strongest liberal postcondition for a given graph transformation rule and a precondition. The proof rules establish the total correctness of graph programs and are shown to be sound. They allow to verify more programs than is possible with previous approaches. In particular, many programs with nested loops are covered by the calculus.

1 Introduction

GP 2 is a programming language based on graph transformation rules which aims to facilitate formal reasoning. Graphs and rules in GP 2 can be attributed with heterogeneous lists of integers and character strings. The language has a simple formal semantics and is computationally complete [15].

The verification of graph programs with various Hoare-style calculi is studied in [16–19] based on so-called E-conditions or M-conditions as assertions. E-conditions are an extension of nested graph conditions [7,13] with attributes (list expressions). They can express first-order properties of GP 2 graphs, while M-condition can express monadic second-order properties (without counting) of non-attributed GP 2 graphs. In both cases, verification is restricted to the class of graph programs whose loop bodies and branching guards are rule-set calls.

In this paper, we introduce a monadic second-order logic (with counting) for GP 2. We define the formulas based on a standard logic for graphs enriched with GP 2 features such as list attributes, indegree and outdegree functions for nodes, etc. We prefer to use standard logic because we believe it is easier to comprehend by programmers that are not familiar with graph morphisms and commuting diagrams. Another advantage of a standard logic is the potential for using theorem proving environments such as Isabelle [10,11], Coq [12], or Z3 [1].

G. S. Wulandari—Supported by Indonesia Endowment Fund for Education (LPDP).

F. Gadducci and T. Kehrer (Eds.): ICGT 2021, LNCS 12741, pp. 240–261, 2021.
https://doi.org/10.1007/978-3-030-78946-6_13

In [21] we show how to prove programs partially correct by using closed first-order formulas as assertions. The class of graph programs that can be verified with the calculi of [21] consists of the so-called control programs. These programs may contain certain nested loops and branching commands with arbitrary loop-free programs as guards. Hence, the class of programs that can be handled is considerably larger than the class of programs verifiable with [16].

Here, we continue that work and show how to prove total correctness of control programs in the sense that programs are both partially correct and terminating. Also, to generalise the program properties that can be verified, we use closed monadic second-order formulas as assertions. This allows to prove non-local properties such as connectedness or k-colourability. Our main technical result is the construction of a strongest liberal postcondition from a given precondition and a GP 2 transformation rule. This operation serves as the axiom in the proof calculus of Sect. 5.

2   The Graph Programming Language GP 2

GP 2 programs transform input graphs into output graphs, where graphs are directed and may contain parallel edges and loops. Formally, a graph G is a system $\langle V_G, E_G, s_G, t_G, l_G, m_G, p_G \rangle$ comprising two finite sets of vertices and edges, source and target functions, a partial node labelling function, an edge labelling function, and a partial root function. Nodes v for which $l_G(v)$ or $p_G(v)$ is undefined may only exist in the interface of GP 2 rules, but not in host graphs. Nodes and edges are labelled with lists consisting of integers and character strings. This includes the special case of items labelled with the empty list which may be considered as "unlabelled".

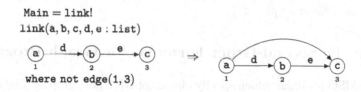

Main = link!
link(a, b, c, d, e : list)

where not edge(1, 3)

Fig. 1. Graph program `transitive-closure` [14]

The principal programming construct in GP 2 are conditional graph transformation rules labelled with expressions. For example, the rule `link` in Fig. 1 has five formal parameters of type `list`, a left-hand graph and a right-hand graph which are specified graphically, and a textual condition starting with the keyword `where`. Node identifiers are written below the nodes, and all other text in the graphs consists of labels. Parameters are typed as `list`, `atom`, `int`, `string`, or `char`, where `atom` stands for the union of integers and strings, and lists are arbitrary sequences of atoms.

Besides carrying expressions, nodes and edges can be marked red, green or blue. Also, nodes can be marked grey and edges can be dashed. An example with red and grey nodes and a dashed edge can be seen in Fig. 5 of Sect. 6.

Rules are applied to host graphs in a two-stage process. First a rule is instantiated by replacing all variables with values of the same type, evaluating all expressions in the right-hand side of the rule, and checking the application condition. This yields a standard rule in the so-called double-pushout approach with relabelling [8]. Next, the instantiated rule is applied to the host graph by constructing two suitable natural pushouts [2].

A program consists of declarations of conditional rules and (non-recursive) procedures, including a distinct procedure named **Main**. Next we briefly describe GP 2's major control constructs.

A *rule-set call* $\{r_1, \ldots, r_n\}$ non-deterministically applies one of the applicable rules to the host graph. The call *fails* if none of the rules is applicable to the host graph.

The *sequential composition* of programs P and Q is written $P; Q$.

The command **if** C **then** P **else** Q is executed on a host graph G by first executing C on a copy of G. If this results in a graph, P is executed on the original graph G; otherwise, if C fails, Q is executed on G. The command **try** C **then** P **else** Q has a similar effect, except that P is executed on the result of C's execution.

The loop command $P!$ executes the body P repeatedly until it fails. When this is the case, $P!$ terminates with the graph on which the body was entered for the last time. The **break** command inside a loop terminates that loop and transfers control to the command following the loop.

In general, the execution of a program P on a host graph G may result in different graphs, fail, or diverge. This is formally defined by the operational semantics of GP 2 which assigns to P and G the set $[\![P]\!]G$ of all possible execution outcomes. See, for example, [15].

3   Monadic Second-Order Formulas for Graph Programs

We define MSO formulas which specify classes of GP 2 host graphs. The abstract syntax of formulas is shown in Fig. 2, where type names, arithmetic operators, and special operators such as **edge**, **root**, **indeg**, **outdeg**, etc. are inherited from the GP 2 syntax. The category Char is the set of all printable ASCII characters except ''', and Digit is the set $\{0, \ldots, 9\}$. All variables are typed, with associated domains as in Table 1.

Table 1. Variable types and their domain over a graph G

Type	Node	Edge	SetNode	SetEdge	List	Atom	Int	String	Char
Domain	V_G	E_G	2^{V_G}	2^{E_G}	$(\mathbb{Z} \cup \text{Char}^*)^*$	$\mathbb{Z} \cup \text{Char}^*$	$\mathbb{Z}$	Char^*	Char

The types for labels form a subtype hierarchy, given by list ⊃ atom ⊃ int, string and string ⊃ char, where atoms are considered as lists of length one and characters are considered as strings of length one. Hence list variables may have integer, string, or character values. Such restrictions can be enforced by subtype predicates. For example, the list variable x can be constrained to hold an integer value by the predicate int(x).

Formula	::= true | false | Elem | Cond | Equal
	| Formula ('∧' | '∨') Formula | '¬'Formula | '('Formula')'
	| '∃ᵥ' NodeVar '('Formula')' | '∃ₑ' EdgeVar '('Formula')'
	| '∃ₗ' (ListVar) '('Formula')'
	| '∃ᵥ' SetNodeVar '('Formula')' | '∃ₑ' SetEdgeVar '('Formula')'
Number	::= Digit {Digit}
Elem	::= Node ('∈' | '∉') SetNodeVar | EdgeVar ('∈' | '∉') SetEdgeVar
Cond	::= (int | char | string | atom) '('Var')'
	| Lst ('=' | '≠') Lst
	| Int ('>' | '>=' | '<' | '<=') Int
	| edge '(' Node ',' Node [',' Label] [',' EMark] ')'
	| path '(' Node ',' Node [',' SetEdgeVar] ')'
	| root '(' Node ')'
Var	::= ListVar | AtomVar | IntVar | StringVar | CharVar
Lst	::= empty | Atm | Lst ':' Lst | ListVar | lᵥ '('Node')' | lₑ '('EdgeVar')'
Atm	::= Int | String | AtomVar
Int	::= ['-'] Number | '('Int')' | IntVar
	| Int ('+' | '-' | '*' | '/') Int
	| (indeg | outdeg) '('Node')'
	| length '('AtomVar | StringVar | ListVar')'
	| card'('(SetNodeVar | SetEdgeVar)')'
String	::= ' " ' Char ' " ' | CharVar | StringVar | String '.' String
Node	::= NodeVar | (s | t) '(' EdgeVar')'
EMark	::= none | red | green | blue | dashed | any | mₑ'('EdgeVar')'
VMark	::= none | red | blue | green | grey | any | mᵥ'('Node')
Equal	::= Node ('=' | '≠') Node | EdgeVar ('=' | '≠') EdgeVar
	| Lst ('=' | '≠') Lst | VMark ('=' | '≠') VMark
	| EMark ('=' | '≠') EMark

Fig. 2. Abstract syntax of monadic second-order formulas

For brevity, we write $c \Rightarrow d$ for $\neg c \vee d$, $c \Leftrightarrow d$ for $(c \Rightarrow d) \wedge (d \Rightarrow c)$, $\forall_\mathsf{V} \mathsf{x}(c)$ for $\neg \exists_\mathsf{V} \mathsf{x}(\neg c)$, and similarly with $\forall_\mathsf{e} \mathsf{x}(c), \forall_\mathsf{I} \mathsf{x}(c), \forall_\mathsf{V} \mathsf{X}(c)$, and $\forall_\mathsf{E} \mathsf{X}(c)$. We also sometimes write $\exists_\mathsf{V} \mathsf{x}_1, \ldots, \mathsf{x}_n(\mathsf{c})$ for $\exists_\mathsf{V} \mathsf{x}_1(\exists_\mathsf{V} \mathsf{x}_2(\ldots \exists_\mathsf{V} \mathsf{x}_n(\mathsf{c}) \ldots))$ (also for other quantifiers). Terms in MSO formulas are defined as usual and may contain function symbols, constants and variables.

Example 1 (Monadic second-order formulas).
1) $\exists_V X(\forall_v x(x \in X \Rightarrow m_V(x) = \text{none}) \wedge \text{card}(X) \geq 2)$ expresses "there exists at least two unmarked nodes".
2) $\exists_V X(\forall_V x(m_V(x) = \text{grey} \Leftrightarrow x \in X) \wedge \exists_I n(\text{card}(X) = 2 * n))$ expresses "the number of grey nodes is even".

Note that the first-order formula $\exists_v x, y(m_V(x) = \text{none} \wedge m_V(y) = \text{none} \wedge x \neq y)$ is equivalent to the first formula. But it is unlikely that the second formula can be expressed in the first-order fragment of our MSO logic because pure first-order logic on graphs (without built-in functions and relations) cannot specify that the number of nodes is even [6].

The truth value of an MSO formula over a graph is defined via assignments, which are functions mapping free variables to their domains.

Definition 1 (Assignment). Consider an MSO formula c. Let A, B, C, D, E be the set of free node, edge, list, node-set, and edge-set variables in c, respectively. Given a free variable x, we write $\text{dom}(x)$ for the domain of x as defined by Table 1. A *formula assignment* for c over a host graph G is a pair $\alpha = \langle \alpha_G, \alpha_L \rangle$ where $\alpha_G = \langle \alpha_V : A \rightarrow V_G, \alpha_E : B \rightarrow E_G, \alpha_{2V} : D \rightarrow 2^{V_G}, \alpha_{2E} : E \rightarrow 2^{E_G} \rangle$ and $\alpha_L : C \rightarrow \mathbb{L}$, such that for each free variable x, $\alpha(x) \in \text{dom}(x)$. We denote by c^α the (first-order) formula resulting from c after replacing each term y with y^α, where y^α is defined inductively as follows:

1. If y is a free variable, $y^\alpha = \alpha(y)$;
2. If y is a constant, $y^\alpha = y$;
3. If $y = \text{length}(x)$ for some list variable x, y^α equals to the number of characters in x^α if x is a string variable, 1 if x is an integer variable, or the number of atoms in x^α if x is a list variable;
4. If $y = \text{card}(X)$ for some node-set or edge-set variable X, y^α is the number of elements in X^α;
5. If y is the functions $s(x), t(x), l_E(x), m_E(x), l_V(x), m_V(x), \text{indeg}(x)$, or $\text{outdeg}(x)$, y^α is $s_G(x^\alpha), t_G(x^\alpha), \ell_G^E(x^\alpha), m_G^E(x^\alpha), \ell_G^V(x^\alpha), m_G^V(x^\alpha)$, indegree of x^α in G , or outdegree of x^α in G, respectively;
6. If $y = x_1 \oplus x_2$ for $\oplus \in \{+, -, *, /\}$ and integers $x_1{}^\alpha, x_2{}^\alpha$, $y^\alpha = x_1 \oplus_{\mathbb{Z}} x_2$;
7. If $y = x_1.x_2$ for some terms $x_1{}^\alpha, x_2{}^\alpha$, y^α is string concatenation x_1 and x_2;
8. If $y = x_1 : x_2$ for some lists $x_1{}^\alpha, x_2{}^\alpha$, y^α is list concatenation x_1 and x_2 □

A graph G satisfies a formula c, denoted by $G \models c$, if there exists an assignment α for c over G such that c^α is true. Table 2 shows how the truth value of c^α is determined.

Table 2. Truth value of c^α in graph G

c^α	true iff
true	true
false	false
int(x)	$x \in \mathbb{Z}$
char(x)	$x \in$ Char
string(x)	$x \in$ Char*
atom(x)	$x \in \mathbb{Z} \cup$ Char*
root(x)	$p_G(x) = 1$
$t_1 \otimes t_2$	$t_1 \otimes_{\mathbb{Z}} t_2$
$X \oslash Y$	$X \oslash_{\mathbb{Z}} Y$
$x \in X$	$x \in_{\mathbb{Z}} X$

c^α	true iff
edge(x, y, l, m)	$s_G(e) = x$ and $t_G(e) = y$ for some $e \in E_G$ where $l_G^E(e) = l$ and $m_G^E(e) = m$
path(x, y, E)	for some $e_1, \ldots, e_n \in E_G - E$, $s_G(e_1) = a$, $s_G(e_n) = b$, $t_G(e_i) = s_G(e_{i+1})$ for every $i = 1, \ldots, n-1$
$t_1 \ominus t_2$	if t_1 (or t_2) is any: t_2 (or t_1) $\ominus_{\mathbb{B}}$ blue, red, green, gray, or dashed; otherwise: $t_1 \ominus_{\mathbb{B}} t_2$

c^α	true iff
$\neg b$	b is false in G
$b_1 \vee b_2$	b_1 is true in G or b_2 is true in G
$b_1 \wedge b_2$	both b_1 and b_2 are true in G
$\exists_V x(b)$	$b^{[x \mapsto v]}$ is true in G for some $v \in V_G$
$\exists_e x(b)$	$b^{[x \mapsto e]}$ is true in G for some $e \in E_G$
$\exists_l x(b)$	$b^{[x \mapsto l]}$ is true in G for some $l \in \mathbb{L}$
$\exists_V X(b)$	$b^{[X \mapsto V]}$ is true in G for some $V \in 2^{V_G}$
$\exists_E X(b)$	$b^{[X \mapsto E]}$ is true in G for some $E \in 2^{E_G}$

In the table, $\otimes \in \{>, >=, <, <=\}$, $\ominus \in \{=, \neq\}$, $\oslash \in \{=, \neq, \subset, \subseteq\}$, $\otimes_{\mathbb{Z}}$ is the integer operation represented by $\otimes$, and $\ominus_{\mathbb{B}}$ (or $\oslash_{\mathbb{B}}$) is the Boolean operation represented by $\ominus$ (or $\oslash$). Also, given a Boolean expression b, a (set) variable x, and a constant i, we denote by $b^{[x \mapsto i]}$ the expression obtained from b by changing every occurrence of x to i.

4 Constructing a Strongest Liberal Postcondition

In this section, we present a construction that can be used to obtain a strongest liberal postcondition from a given precondition and a rule schema. Here, we limit the precondition to closed MSO formulas.

Definition 2 (Strongest liberal postcondition over a conditional rule schema). An assertion d is a *liberal postcondition* w.r.t. a conditional rule schema r and a precondition c, if for all host graphs G and H, $G \vDash c$ and $G \Rightarrow_r H$ implies $H \vDash d$. A *strongest liberal postcondition* w.r.t. c and r, denoted by $\mathrm{SLP}(c, r)$, is a liberal postcondition w.r.t. c and r that implies every liberal postcondition w.r.t. c and r. $\qquad\square$

In [21], we show how to construct a strongest liberal postcondition over FO formulas. Here, we use the same approach in the construction, that is, by obtaining a left-application condition, which then be used to obtain a right-application condition, so that finally we can obtain a strongest liberal postcondition.

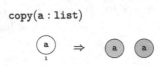

Fig. 3. GP 2 conditional rule schema copy

As a running example, let us consider the rule schema copy of Fig. 3 and the MSO formula e expressing "the number of grey nodes is even":

$$e \equiv \exists_V X (\neg \exists_v x ((m_V(x) = \text{grey} \land x \notin X) \lor (m_V(x) \neq \text{grey} \land x \in X)) \land \exists_I n(\text{card}(X) = 2 * n)).$$

Note that the interface of the rule copy is the empty graph. We intentionally do not preserve the node 1 and have two new nodes instead to see the effect of both removal and addition of an element in the construction of a strongest liberal postcondition.

Remark 1. In the following subsections we explain the transformations involved in the construction of a strongest liberal postcondition. For this purpose, we consider a generalised form of MSO formulas called *conditions*, which may contain node and edge constants. Also, we consider a generalised form of rule schemata which have both a left and a right application condition, where the conditions can be more expressive than the application conditions of GP 2 rule schemata.

4.1   From Precondition to Left-Application Condition

We start with the transformation of a precondition to a left-application condition with respect to a conditional rule schema $r = \langle L \leftarrow K \rightarrow R, \Gamma \rangle$. Intuitively, the transformation is done by:

1. Expressing the dangling condition as a condition over L, denoted by $\text{Dang}(r)$.
2. Finding all possibilities of variables in c representing nodes/edges in a match of L and of forming a disjunction from all possibilities, denoted by $\text{Split}(c, r)$.
3. Evaluating terms and Boolean expression we can evaluate in $\text{Split}(c, r)$, $\text{Dang}(r)$, and Γ with respect to the left-hand graph of the given rule, then form a conjunction from the result of evaluation, and simplify the conjunction.

4.1.1 Condition Dang

The dangling condition must be satisfied by an injective morphism g if $G \Rightarrow_{r,g} H$ for some rule schema $r = \langle L \leftarrow K \rightarrow R \rangle$ and host graphs G, H. A graph G with an injective morphism $g : L \rightarrow G$ satisfies the dangling condition if every node $v \in g(L - K)$ is not incident to any edge outside $g(L)$. That is, all edges incident to a deleted node must be in $g(L)$. This means that the indegree and outdegree of each deleted node $g^{-1}(v) \in L - K$ are the same as the indegree and outdegree of v in G.

Definition 3 (Condition Dang). Consider a rule schema $r = \langle L \leftarrow K \rightarrow R \rangle$ where $\{v_1, \cdots, v_n\}$ is the set of nodes in $L - K$. Let $indeg_L(v)$ and $outdeg_L(v)$ denote the indegree and outdegree of a node v in L. The condition $\mathrm{Dang}(r)$ is defined as follows:

1. if $V_L - V_K = \emptyset$ then $\mathrm{Dang}(r) = \mathsf{true}$
2. if $V_L - V_K \neq \emptyset$ then
   $\mathrm{Dang}(r) = \bigwedge_{i=1}^{n} \mathsf{indeg}(\mathsf{v_i}) = indeg_L(v_i) \wedge \mathsf{outdeg}(\mathsf{v_i}) = outdeg_L(v_i)$ □

Example 2.
For the rule $r = \mathsf{copy}$ (see Fig. 3): $\mathrm{Dang}(r) = \mathsf{indeg}(1) = 0 \wedge \mathsf{outdeg}(1) = 0$

4.1.2 Transformation Split

A node (or edge) variable x in c can represent any node (or edge) in an input graph, in the sense that we can substitute any node (or edge) in G to check the truth value of c in G (see point 5 and 6 of Definition 5). Also, a node (or edge) set variable X in c can represent any set of nodes (or edges) in the input graph, where each node (or edge) in the image of a match may or may not be an element of the set (see point 8 and 9 of Definition 5).

To express that a set of nodes/edges in L is a subset of a set of nodes/edges represented by a set variable, we define subset formulas.

Definition 4 (Subset Formula). Given a set of nodes $N = \{v_1, \ldots, v_n\}$, a *subset formula* for N with respect to a node set variable X has the form $c_1 \wedge c_2 \wedge \ldots \wedge c_n$ where for $i = 1, \ldots, n$, $c_i = v_i \in X$ or $v_i \notin X$. The formula true is the only subset formula for the empty set with respect to any set variable.   □

Definition 5 (Transformation Split). Let us consider a rule schema $r = \langle L \leftarrow K \rightarrow R, \Gamma \rangle$, where $V_L = \{v_1, \ldots, v_n\}$ and $E_L = \{e_1, \ldots, e_m\}$. Let $\{V_1, \ldots, V_{2^n}\}$ be the power set of V_L, and $d_1, \ldots, d_{2^n}$ be subset formulas of V_L w.r.t. X where for every $i = 1, \ldots, 2^n$, d_i represents V_i. Similarly, let $\{E_1, \ldots, E_{2^m}\}$ be the power set of E_L, and $a_1, \ldots, a_{2^m}$ be subset formulas of E_L w.r.t. X where for every $i = 1, \ldots, 2^m$, a_i represents E_i.

Let c be a condition over L sharing no variables with r (note that it is always possible to replace the label variables in c with new variables that are distinct from variables in r). We define the condition $\mathrm{Split}(c, r)$ over L inductively as follows, where c_1, c_2 are conditions over L:

1) If c is either true, false, a predicate $\mathsf{int}(t), \mathsf{char}(t), \mathsf{string}(t), \mathsf{atom}(t), \mathsf{root}(t)$ for some term t, in the form $t_1 \ominus t_2$ for $\ominus \in \{= . \neq . <, \leq, >, \geq\}$ and some terms t_1, t_2, or in the form $x \in X$ or $x \notin X$,

$$\mathrm{Split}(c, r) = c$$

2) $\mathrm{Split}(c_1 \vee c_2, r) = \mathrm{Split}(c_1, r) \vee \mathrm{Split}(c_2, r)$,
3) $\mathrm{Split}(c_1 \wedge c_2, r) = \mathrm{Split}(c_1, r) \wedge \mathrm{Split}(c_2, r)$,
4) $\mathrm{Split}(\neg c_1, r) = \neg \mathrm{Split}(c_1, r)$,
5) $\mathrm{Split}(\exists_\mathsf{v} x(c_1), r) = (\bigvee_{i=1}^{n} \mathrm{Split}(c_1^{[x \mapsto v_i]}, r)) \vee \exists_\mathsf{v} x(\bigwedge_{i=1}^{n} x {\neq} v_i \wedge \mathrm{Split}(c_1, r)$,
6) $\mathrm{Split}(\exists_\mathsf{e} x(c_1), r) = (\bigvee_{i=1}^{m} \mathrm{Split}(c_1^{[x \mapsto e_i]}, r)) \vee \exists_\mathsf{e} x(\bigwedge_{i=1}^{m} x {\neq} e_i \wedge \mathrm{inc}(c_1, r, x))$,
where

$$\mathrm{inc}(c_1, r, x) = \bigvee_{i=1}^{n} (\bigvee_{j=1}^{n} \mathsf{s}(x) = v_i \wedge \mathsf{t}(x) = v_j \wedge \mathrm{Split}(c_1^{[\mathsf{s}(x) \mapsto v_i, \mathsf{t}(x) \mapsto v_j]}, r))$$
$$\vee (\mathsf{s}(x) = v_i \wedge \bigwedge_{j=1}^{n} \mathsf{t}(x) \neq v_j \wedge \mathrm{Split}(c_1^{[\mathsf{s}(x) \mapsto v_i]}, r))$$
$$\vee (\bigwedge_{j=1}^{n} \mathsf{s}(x) \neq v_j \wedge \mathsf{t}(x) = v_i \wedge \mathrm{Split}(c_1^{[\mathsf{t}(x) \mapsto v_i]}, r))$$
$$\vee (\bigwedge_{i=1}^{n} \mathsf{s}(x) \neq v_i \wedge \bigwedge_{j=1}^{n} \mathsf{t}(x) \neq v_j \wedge \mathrm{Split}(c_1, r))$$

7) $\mathrm{Split}(\exists_\mathsf{l} x(c_1), r) = \exists_\mathsf{l} x(\mathrm{Split}(c_1, r))$
8) $\mathrm{Split}(\exists_\mathsf{v} X(c_1), r) = \exists_\mathsf{v} X(\bigwedge_{i=1}^{2^n} d_i \Rightarrow \mathrm{Split}(c_1, r))$
9) $\mathrm{Split}(\exists_\mathsf{E} X(c_1), r) = \exists_\mathsf{E} X(\bigwedge_{i=1}^{2^m} a_i \Rightarrow \mathrm{Split}(c_1, r))$

where $c^{[a \mapsto b]}$ for a variable or function a and constant b represents the condition c after the replacement of all occurrence of a with b.   □

Intuitively, we only need to consider substituting nodes in L for each term in c representing a node (a node variable or a source or target function), and similarly, edges in L for all edge variables in c. In addition, we need to consider all possible ways in which nodes/edges in L are elements of a set in c.

Example 3.
Consider again the precondition e from our running example:

$$\exists_\mathsf{v} X(\neg \exists_\mathsf{v} x((\mathsf{m}_\mathsf{V}(x) = \mathsf{grey} \wedge x \notin X) \vee (\mathsf{m}_\mathsf{V}(x) \neq \mathsf{grey} \wedge x \in X)) \wedge \exists_\mathsf{l} n(\mathsf{card}(X) = 2 * n))$$

has the form of $\exists_\mathsf{v} X(c_1)$. From point 8 and 3 of Definition 5, for $d = \exists_\mathsf{v} x((\mathsf{m}_\mathsf{V}(x) = \mathsf{grey} \wedge x \notin X) \vee (\mathsf{m}_\mathsf{V}(x) \neq \mathsf{grey} \wedge x \in X))$, we have

$$Split(e, r) = \exists_\mathsf{v} X((1 \in X \Rightarrow \mathrm{Split}(\neg d, r) \wedge \mathrm{Split}(\exists_\mathsf{l} n(\mathsf{card}(X) = 2 * n), r))$$
$$\wedge (1 \notin X \Rightarrow \mathrm{Split}(\neg d, r) \wedge \mathrm{Split}(\exists_\mathsf{l} n(\mathsf{card}(X) = 2 * n), r))).$$

We know that $\text{Split}(\exists_i n(\text{card}(X) = 2 * n), r)$ is equal to $\exists_i n(\text{card}(X) = 2 * n)$ (see point 7 of Definition 5), while $\text{Split}(\neg d, r) = \neg\text{Split}(d, r)$ (see point 4 of Definition 5). Then from point 5 of Definition 5, we have

$\text{Split}(d, r) = (m_V(1) = \text{grey} \wedge 1 \notin X) \vee (m_V(1) \neq \text{grey} \wedge 1 \in X)$
$\qquad\qquad \vee \exists_V x(x \neq 1 \wedge ((m_V(x) = \text{grey} \wedge x \notin X) \vee (m_V(x) \neq \text{grey} \wedge x \in X))$

so that

$$\text{Split}(e, r) = \exists_V X((1 \in X \Rightarrow \neg\text{Split}(d, r) \wedge \exists_i n(\text{card}(X) = 2^*n))$$
$$\wedge (1 \notin X \Rightarrow \neg\text{Split}(d, r) \wedge \exists_i n(\text{card}(X) = 2^*n)))$$

4.1.3 Transformation Val

The condition resulting from transformation Split, the condition Dang, and the rule schema condition Γ may contain node/edge identifiers of the given left-hand graph. To simplify the conditions, we can check if there is a disjuntion with a true disjunc or a conjunction with a false conjunct so that we can ruled out because of its value in the left-hand graph. For a simple example, a conjunct condition $m_V(1) = \text{grey}$ can be replaced with false if node 1 in the given left-hand graph is not grey.

Let us consider a rule schema $r = \langle L \leftarrow K \rightarrow R, \Gamma \rangle$, a condition c over L, a host graph G, and a premorphism $g : L \rightarrow G$. Let c share no variables with L. To simplify c w.r.t. L, we apply the transformation $\text{Val}(c, r)$ as follows:

1. Obtain c' from c by replacing terms involving s, t, l_V, l_E, m_V, m_E, indeg and outdeg, that do not have node/edge variables as arguments, with their values in L. In addition, we also replace integer, string, and list operations with their values if their arguments are only constants.
   Note that the values of indeg and outdeg depend on the host graph, while here we evaluate them in the left-hand graph. Hence, we use the terms $\text{incon}(v)$ and $\text{outcon}(v)$ as constants representing the indegree resp. outdegree of $g(v)$ minus indegree resp. outdegree of v in L.
2. Obtain c'' from c' by evaluating Boolean operations $=, \neq, \leq, \geq$, root, if their arguments only consists of constants, to their values in L.
3. Consider any implication of the form $a \Rightarrow d$ for some subset formula a and condition d to $a \Rightarrow d^T$. d^T is obtained from d by changing every subcondition of the form $i \in X$ for $i \in V_L$, $i \in E_L$ and set variable X to true if $i \subset X$ is implied by a or false otherwise.
4. Simplify c''' by simplifying conjunct disjunct involving true or false. Also, change the subconditions of the forms $\neg \text{true}, \neg(\neg a), \neg(a \vee b), \neg(a \wedge b)$, and $a \Rightarrow \text{false}$ for some conditions a, b to false, a, $\neg a \wedge \neg b, \neg a \vee \neg b, \neg a$ resp.   □

The formal definition of $\text{Val}(c, r)$ is rather long [22] because the expressions we have in a condition may be nested. Hence, we do not present it in this paper.

Example 4. Let $f = \text{Split}(e, r)$ from Example 3. That is,

$$f = \exists_V X((1 \in X \Rightarrow \neg\text{Split}(d, r) \wedge \exists_i n(\text{card}(X) = 2^*n))$$
$$\wedge (1 \notin X \Rightarrow \neg\text{Split}(d, r) \wedge \exists_i n(\text{card}(X) = 2^*n)))$$

where $\text{Split}(d, r) = (\text{m}_V(1) = \text{grey} \wedge 1 \notin X) \vee (\text{m}_V(1) \neq \text{grey} \wedge 1 \in X)$
$$\vee\ \exists_V x (x \neq 1 \wedge ((\text{m}_V(x) = \text{grey} \wedge x \notin X) \vee (\text{m}_V(x) \neq \text{grey} \wedge x \in X))$$
Since node 1 in the left-hand graph of r is unmarked, then we can replace $\text{m}_V(1) = \text{grey}$ with false, and $\text{m}_V(1) \neq \text{grey}$ with true.
We also replace $1 \in X$ and $1 \notin X$ with true or false, based on the premise in the conjunct of f. That is, replace $1 \in X$ and $1 \notin X$ with true and false (resp.) for the first conjunct of f, and with false and true (resp.) for the second conjunct. Hence, we obtain the following condition

$$\exists_V X((1 \in X \Rightarrow \neg((\text{false} \wedge \text{false}) \vee (\text{true} \wedge \text{true}) \vee b) \wedge \exists_I n(\text{card}(X) = 2^* n))$$
$$\wedge (1 \notin X \Rightarrow \neg((\text{false} \wedge \text{true}) \vee (\text{true} \wedge \text{false}) \vee b) \wedge \exists_I n(\text{card}(X) = 2^* n)))$$

where $b = \exists_V x(x \neq 1 \wedge ((\text{m}_V(x) = \text{grey} \wedge x \notin X) \vee (\text{m}_V(x) \neq \text{grey} \wedge x \in X)))$.
Finally, we simplify $\neg((\text{false} \wedge \text{false}) \vee (\text{true} \wedge \text{true}) \vee b) \wedge \exists_I n(\text{card}(X) = 2^* n)$ to false. Also, $\neg((\text{false} \wedge \text{true}) \vee (\text{true} \wedge \text{false}) \vee b)$ to $\neg b$. Hence, we finally obtain $\text{Val}(f, r) = \exists_V X(1 \notin X \Rightarrow \neg b \wedge \exists_I n(\text{card}(X) = 2^* n))$

4.1.4   Transformation Lift
Finally, we define the transformation Lift, which takes a precondition and a rule schema as an input and gives a left-application condition as an output.

Definition 6 (Transformation Lift). Let $r = \langle L \leftarrow K \rightarrow, \Gamma \rangle$ be a rule schema, c be a precondition, and $\text{Lift}(c, r)$ is a left application condition w.r.t. c and r. Then, $\text{Lift}(c, r) = \text{Val}(\text{Split}(c \wedge \Gamma, r) \wedge \text{Dang}(r), r)$.

Example 5.
For the rule schema $r = \text{copy}$, $\Gamma = \text{true}$ and $\text{Dang}(r) = \text{indeg}(1) = 0 \wedge \text{outdeg}$
$(1) = 0$ such that $\text{Val}(\text{Dang}(r), r) = \text{true}$ and $\text{Split}(e \wedge \Gamma, r) = \text{Split}(e, r)$. Hence, $\text{Lift}(e, r^\vee) = \text{Val}(\text{Split}(e, r), r)$.

In [22], we show that by using the described construction, we can obtain a left-application condition that is satisfied by every possible match of the given rule schema.

Let us consider the transformation Split. From point 8 and 9 of Definition 5, we know that Split may gives us conjunction of implications in specific form (i.e. implications with subset formula as premise), and such form will still be exist in the resulting condition of the transformation Lift. From now on, we say that the obtained application condition (from Lift) is in 'lifted form'.

4.2   From Left- to Right-Application Condition

To obtain a right-application condition from a left-application condition, we need to consider what properties could be different in the initial and the result graphs. Recall that in constructing a left-application condition, we evaluate all functions with a node/edge constant argument and change them with constant.

4.2.1   Transformation Adj

Due to the deletion of nodes/edges by a rule schema, properties that hold in the initial graph may not hold anymore in the output graph. Hence, we need to adjust the obtained left application condition so that we can have a condition that can be satisfied by a comatch.

For example, the Boolean value for $x = i$ for any node/edge variable x and node/edge constant i that gets deleted must be false in the resulting graph. Analogously, $x \neq i$ is always true. Also, all variables in the left-application condition should not represent any new nodes and edges in the right-hand side. In addition, we also need to consider the case where we have set variables.

In a lifted form, we may have subformulas of the form $\exists_V X(\bigwedge_{i=1}^{2^n} d_i \Rightarrow \text{Split}(c_1, r))$ (or similar for edges), where each d_i represent the condition where a subset of V_L is a subset of the set represented by X. A node in V_L may or may not exist in the output graph. Hence, we need to do adjustment by use a property in standard logic.

Definition 7 (Transformation Adj). Given a rule schema $r = \langle L \leftarrow K \rightarrow R, \Gamma \rangle$ where $V_L = \{v_1, \ldots, v_n\}$, $E_L = \{e_1, \ldots, e_m\}$, $V_K = \{u_1, \ldots, u_k\}$, $V_R = \{w_1, \ldots, w_p\}$, and $E_R = \{z_1, \ldots, z_q\}$, where $v_i \neq w_j$ (or $e_i \neq z_j$) for all v_i and w_j (or e_i and x_j) not in K. Let $\{V_1, \ldots, V_{2^n}\}$ be the power set of V_L, and $d_1, \ldots, d_{2^n}$ be subset formulas of V_L w.r.t. X where for every $i = 1, \ldots, 2^n$, d_i represents V_i. Similarly, let $\{U_1, \ldots, U_{2^k}\}$ be the power set of V_K, and $b_1, \ldots, b_{2^k}$ be subset formulas of V_K w.r.t. X where for every $i = 1, \ldots, 2^k$, b_i represents U_i. Also, let $\{E_1, \ldots, E_{2^m}\}$ be the power set of E_L, and $a_1, \ldots, a_{2^m}$ be subset formulas of E_L w.r.t. X where for every $i = 1, \ldots, 2^m$, a_i represents E_i.

For a condition c over L in lifted form, the *adjusted* condition of c w.r.t. r is defined inductively as below, where $c_1, \ldots, c_s$ are conditions over L, for $s \geq 2^m$ and $s \geq 2^n$:

1. If c is the formulas true or false,
   $\text{Adj}(c, r) = c$
2. If c is predicate $\text{int}(x), \text{char}(x), \text{string}(x)$, or $\text{atom}(x)$ for some list variable x,
   $\text{Adj}(c, r) = c$
3. If c is a Boolean operation $f_1 = f_2$ or $f_1 \neq f_1$ where each f_1 and f_2 are terms representing a list and neither contains free node/edge variable,
   $\text{Adj}(c, r) = c$
4. If c is a Boolean operation $f_1 = f_2$ or $f_1 \neq f_1$ where each f_1 and f_2 are terms representing a node (or edge) and neither contains free node/edge variable or node/edge constant,
   $\text{Adj}(c, r) = c$
5. If c is a Boolean operation $f_1 \diamond f_2$ for $\diamond \in \{=, \neq, <, \leq, >, \geq\}$ and some terms f_1 and f_2 representing integers and neither contains free node/edge variable or any set variables,

$$\text{Adj}(c, r) = \begin{cases} \text{false}, & \text{if } \ominus \in \{=\} \text{ and } x_1 \in V_L - V_K \cup E_L \text{ or } x_2 \in V_L - V_K \cup E_L, \\ \text{true}, & \text{if } \ominus \in \{\neq\} \text{ and } x_1 \in V_L - V_K \cup E_L \text{ or } x_2 \in V_L - V_K \cup E_L, \\ c', & \text{otherwise} \end{cases}$$

6. If c is a Boolean operation $x \in X$ for a bounded set variable X and bounded edge variable x, or a bounded set variable X and a bounded node variable x, $x = s(y)$ or $x = t(y)$ for some bounded edge variable y,
$\mathrm{Adj}(c, r) = c$

7. If $c = \exists_I x(c_1$ for some condition c_1 over L in lifted form,
$\mathrm{Adj}(c, r) = \exists_I x(\mathrm{Adj}(c_1, r))$

8. If $c = \exists_V x \left(\bigwedge_{i=1}^{n}, x \neq v_i \wedge c_1 \right)$ for some condition c_1 over L in lifted form,
$\mathrm{Adj}(c, r) = \exists_V x(\bigwedge_{i=1}^{p}, x \neq w_i \wedge \mathrm{Adj}(c_1, r))$

9. If $c = \exists_e x \left(\bigwedge_{i=1}^{m}, x \neq e_i \wedge c_1 \right)$ for some condition c_1 over L in lifted form,
$\mathrm{Adj}(c, r) = \exists_e x(\bigwedge_{i=1}^{q}, x \neq z_i \wedge \mathrm{Adj}(c_1, r))$

10. If $c = \exists_V X(\bigwedge_{i=1}^{2^n} d_i \Rightarrow c_i)$ where each c_i is a condition over L in lifted form or contains $\mathrm{card}(X)$
$$\mathrm{Adj}(c, r) = \exists_V X(\bigwedge_{v \in V_R - V_K} v \notin X \bigwedge_{i=1}^{2^k}(b_i \Rightarrow \bigvee_{j \in W_i} c_j'))$$
where $c_j' = \mathrm{Adj}(c_j, r)^{[\mathrm{card}(X) \mapsto \mathrm{card}(X) + |(V_L - V_K) \cap V_j|]}$ and for $i = 1, \ldots, 2^k$, W_i is a subset of $\{1, \ldots, 2^n\}$ such that for all $j \in \{1, \ldots, 2^n\}$, $j \in W_i$ iff d_j implies b_i

11. If $c = \exists_E X(\bigwedge_{i=1}^{2^m} a_i \Rightarrow c_i)$ where each c_i is a condition over L in lifted form, construction of $\mathrm{Adj}(c, r)$ is analogous to point 10

12. If $c = c_1 \vee c_2$ for some conditions c_1, c_2 over L in lifted form,
$\mathrm{Adj}(c, r) = \mathrm{Adj}(c_1, r) \vee \mathrm{Adj}(c_2, r)$

13. If $c = c_1 \wedge c_2$ for some conditions c_1, c_2 over L in lifted form,
$\mathrm{Adj}(c, r) = \mathrm{Adj}(c_1, r) \wedge \mathrm{Adj}(c_2, r)$

14. If $c = \neg c_1$ for some condition c_1 over L in lifted form,
$\mathrm{Adj}(c, r) = \neg \mathrm{Adj}(c_1, r)$

Example 6. Let us consider $\mathrm{Lift}(e, r), r)$ from Example 5. That is, the condition $\exists_V X(1 \notin X \Rightarrow \neg b \wedge \exists_I n(\mathrm{card}(X) = 2^* n))$ where $b = \exists_V x(x \neq 1 \wedge ((m_V(x) = \mathsf{grey} \wedge x \notin X) \vee (m_V(x) \neq \mathsf{grey} \wedge x \in X)))$. From point 10 of Definition 7, we get $\mathrm{Adj}(\mathrm{Lift}(e, r))$ is $\exists_V X(2 \notin X \wedge 3 \notin X \wedge (\mathsf{true} \Rightarrow \neg \mathrm{Adj}(b, r) \wedge \exists_I n(\mathrm{card}(X) = 2^* n)))$ where $\mathrm{Adj}(b, r)$ is $\exists_V x(x \neq 2x \neq 3 \wedge ((m_V(x) = \mathsf{grey} \wedge x \notin X) \vee (m_V(x) \neq \mathsf{grey} \wedge x \in X)))$ (see point 5 and 8 of Definition 7. Hence,

$$\mathrm{Adj}(b, r) = \exists_V X(2 \notin X \wedge 3 \notin X \wedge \exists_I n(\mathrm{card}(X) = 2^* n)$$
$$\wedge \neg \exists_V x(x \neq 2x \neq 3 \wedge ((m_V(x) = \mathsf{grey} \wedge x \notin X) \vee (m_V(x) \neq \mathsf{grey} \wedge x \in X))))$$

4.2.2   Condition Spec and Transformation Shift

To have a right application condition that yield to strongest liberal postcondition, we need to have a condition that express properties of right-hand graph, in addition to the condition that derived from the given precondition. Hence, we need a condition that explicitly express the structure, labels, marks of the right-hand graph. Also, the right-application condition should express the dangling condition for any co-match.

To express the structure and properties of R, we use the condition $\mathrm{Spec}(R)$, which specify the right-hand graph uniquely up to the node/edge IDs and name of variables. $\mathrm{Spec}(R)$ is defined as the condition

$$\bigwedge_{i=1}^{k}\mathrm{Type}(x_i) \wedge \bigwedge_{i=1}^{n} \mathsf{l}_\mathsf{V}(\mathsf{v}_i) = \ell_R^V(v_i) \wedge \mathsf{m}_\mathsf{V}(\mathsf{v}_i) = m_R^V(v_i) \wedge \mathrm{Root}_R(v_i)$$
$$\wedge \bigwedge_{i=1}^{m} \mathsf{s}(\mathsf{e}_i) = s_R(e_i) \wedge \mathsf{t}(\mathsf{e}_i) = t_R(e_i) \wedge \mathsf{l}_\mathsf{E}(\mathsf{e}_i) = \ell_R^E(e_i) \wedge \mathsf{m}_\mathsf{E}(\mathsf{e}_i) = m_R^E(e_i)$$

where $\mathrm{Type}(x)$ for $x \in X$ is $\mathsf{int}(\mathsf{x})$, $\mathsf{char}(\mathsf{x})$, $\mathsf{string}(\mathsf{x})$, $\mathsf{atom}(\mathsf{x})$, or true if x is an integer, char, string, atom, or list variable respectively, and $\mathrm{Root}_L(v)$ for $v \in V_L$ is a function such that $\mathrm{Root}_L(v) = \mathsf{root}(\mathsf{v})$ if $p_L(v) = 1$, and $\mathrm{Root}_L(v) = \neg\mathsf{root}(\mathsf{v})$ otherwise.

Definition 8 (Shifting). Consider a rule schema $r = \langle L \leftarrow K \rightarrow R, \Gamma \rangle$, and a precondition c. The right-application condition w.r.t. c and r, denoted by $\mathrm{Shift}(c, r)$, is defined as:
$$\mathrm{Shift}(c, r) = \mathrm{Adj}(\mathrm{Lift}(c, r), r) \wedge \mathrm{Spec}(R) \wedge \mathrm{Dang}(r^{-1}) \qquad \square$$

Example 7.
$\mathrm{Adj}(\mathrm{Lift}(c, r), r)$ has been obtained from Example 6, where $\mathrm{Spec}(R)$ is the condition $\mathsf{m}_\mathsf{V}(2) = \mathsf{grey} \wedge \mathsf{m}_\mathsf{V}(3) = \mathsf{grey} \wedge \mathsf{l}_\mathsf{V}(2) = \mathsf{a} \wedge \mathsf{l}_\mathsf{V}(3) = \mathsf{a}$.
Also, $\mathrm{Dang}(r^{-1}) = \mathsf{indeg}(2) = 0 \wedge \mathsf{indeg}(3) = 0 \wedge \mathsf{outdeg}(2) = 0 \wedge \mathsf{outdeg}(3) = 0$ (see Definition 3). Hence, $\mathrm{Shift}(e, r)$ is
$$\exists_\mathsf{V}\mathsf{X}(\exists_\mathsf{I}\mathsf{n}(\mathsf{card}(\mathsf{X}) = 2^*\mathsf{n}) \wedge 2 \notin \mathsf{X} \wedge 3 \notin \mathsf{X}$$
$$\wedge \neg\exists_\mathsf{V}\mathsf{x}(\mathsf{x} \neq 2 \wedge \mathsf{x} \neq 3 \wedge ((\mathsf{x} \notin \mathsf{X} \wedge \mathsf{m}_\mathsf{V}(\mathsf{x}) = \mathsf{grey}) \vee (\mathsf{m}_\mathsf{V}(\mathsf{x}) \neq \mathsf{grey} \wedge \mathsf{x} \in \mathsf{X}))))$$
$$\wedge \mathsf{m}_\mathsf{V}(2) = \mathsf{grey} \wedge \mathsf{m}_\mathsf{V}(3) = \mathsf{grey} \wedge \mathsf{l}_\mathsf{V}(2) = \mathsf{a} \wedge \mathsf{l}_\mathsf{V}(3) = \mathsf{a}$$
$$\wedge \mathsf{indeg}(2) = 0 \wedge \mathsf{indeg}(3) = 0 \wedge \mathsf{outdeg}(2) = 0 \wedge \mathsf{outdeg}(3) = 0$$

4.3 From Right-Application Condition to Postcondition

The right-application condition we obtain from transformation Shift is strong enough to express properties of the result graph, w.r.t the comatch. To turn the condition c obtained from Shift to a postcondition, we only need to generalised the condition by the transformation $\mathrm{Var}(c)$, which is obtained from c by substituting fresh variables to node/edge identifiers and adding a constraint that different fresh variables represent different nodes/edges that there is no two new variables express the same node/edge. Finally, we need to bind all free variables to obtain a closed MSO formula.

Definition 9 (Slp). Given a rule $r = \langle r, \Gamma \rangle$ for a rule schema $r = \langle L \leftarrow K \rightarrow R \rangle$ and a precondition c. A postcondition w.r.t. c and r, denoted by $\mathrm{Slp}(c, r)$, is the MSO formula $\exists_\mathsf{V}x_1, \ldots, x_n(\exists_\mathsf{e}y_1, \ldots, y_m(\exists_\mathsf{l}z_1, \ldots, z_k(\mathrm{Var}(\mathrm{Shift}(c, r)))))$, where $\{x_1, \ldots, x_n\}$, $\{y_1, \ldots, y_m\}$, and $\{z_1, \cdots, z_k\}$ denote the set of free node, edge, and label (resp.) variables in $\mathrm{Var}(\mathrm{Shift}(c, r))$.

Example 8. First, we need to obtain $\mathrm{Var}(\mathrm{Shift}(e, r))$ by substituting fresh variables to node/edge identifiers in $\mathrm{Shift}(e, r)$ of Example 7. The condition $\mathrm{Shift}(e, r)$ has two node variables, that are 2 and 3. We can then to y and z respectively because we do not both variables in $\text{Shift}(e,r)$. In addition, we also need to add a constraint that $y \neq z$. Hence, we have

$$Var(Shift(e,r)) = y \neq z$$
$$\wedge \exists_V X (\exists_I n (\text{card}(X) = 2^* n) \wedge y \notin X \wedge z \notin X$$
$$\wedge \neg \exists_V x (x \neq y \wedge x \neq z \wedge ((x \notin X \wedge m_V(x) = \text{grey})$$
$$\vee (m_V(x) \neq \text{grey} \wedge x \in X))))$$
$$\wedge m_V(y) = \text{grey} \wedge m_V(z) = \text{grey} \wedge l_V(y) = a \wedge l_V(z) = a$$
$$\wedge \text{indeg}(y) = 0 \wedge \text{indeg}(z) = 0 \wedge \text{outdeg}(y) = 0 \wedge \text{outdeg}(z) = 0$$

so that
$$\text{Slp}(e,r) = \exists_V y, z (\exists_I a (Var(Shift(e,r))))$$

Theorem 1 (Strongest liberal postconditions). Given a precondition c and a conditional rule schema $r = \langle\langle L \leftarrow K \rightarrow R\rangle, \Gamma\rangle$. Then, $\text{Slp}(c,r)$ is a strongest liberal postcondition w.r.t. c and r.

In [22], we prove Theorem 1 by showing that $\text{Lift}(c,r)$ and $\text{Shift}(c,r)$ must be satisfied by every match and comatch (resp.).

5   Proof Calculus

In this section, we define a syntactic proof calculus in the sense of total correctness, called SYN.

5.1   The Calculus

Our calculus is a total correctness calculus, which means that a Hoare triple $\{c\}\ P\ \{d\}$ is totally correct if the execution of P on G satisfying c either yields a proper graph or fails (divergence is excluded).

Definition 10 (Partial and total correctness [17]). Consider a precondition c and a postcondition d. A graph program P is *partially correct* with respect to c and d, denoted by $\vDash_{\text{par}} \{c\}\ P\ \{d\}$, if for every host graph G and every graph H in $[\![P]\!]G$, $G \vDash c$ implies $H \vDash d$. The triple $\{c\}\ P\ \{d\}$ is *totally correct*, denoted by $\vDash_{\text{tot}} \{c\}\ P\ \{d\}$, if it is partially correct and if for every host graph G satisfying c, P does not diverge or get stuck.

A program can get stuck if it contains a command if/try C then P else Q where C can diverge from a graph G, or it contains a loop $B!$ whose body B can diverge from a graph G. Hence, getting stuck is always a signal of divergence. To prove that a program does not diverge, we use a termination function # which assigns a natural number to every host graph. The proof rule for loops will require that loop bodies decrease the #-value of graphs satisfying the loop invariant. This concept was introduced in [18], but only for loop bodies that are rule set calls.

Definition 11 (Termination function; #-decreasing). A *termination function* is a mapping $\#\colon \mathcal{G}(\mathcal{L}) \to \mathbb{N}$ from host graphs to natural numbers. Given an assertion c and a graph program P, we say that P is *#-decreasing* (under c) if for all graphs $G, H \in \mathcal{G}(\mathcal{L})$ such that $G \vDash c$,

$$\langle P, G\rangle \to^* H \text{ implies } \#G > \#H.$$

To define a proof calculus, we need assertions that can express preconditions of failing or successful executions. For this, we also use the assertion Success and Fail as defined in [21] which can be defined if we consider the classes loop-free programs and iteration commands. A loop-free program simply is a program that has no loop, while an iteration command is inductively defined as: 1) every loop-free program and non-failing command is an iteration command, and 2) a command in the form $C; P$ is an iteration command if C is a loop-free program and P is an iteration command.

Theorem 2. For any loop-free program P and precondition c, there exists MSO formula Success(P) and Slp(c, P) such that a graph $G \vDash$ Success(P) if and only if there exists a host graph $H \in \llbracket P \rrbracket G$ and $G \vDash$ Slp(c, P) if and only if G is a strongest liberal postcondition w.r.t c and P. Also, for any iteration command S, there exists MSO formula Fail(P) such that $G \vDash$ Fail(S) if and only if fail $\in \llbracket P \rrbracket G$.

Intuitively, MSO formulas Success(P) and Fail(P) are preconditions that assert the existence of successful and failing (resp.) execution of P. In addition, we consider the predicate Break(c, P, d) for graph command P and assertions c, d as a predicate that is true if and only if for all derivations $\langle P, G\rangle \to^*$ $\langle \texttt{break}, H\rangle, G \vDash c$ implies $H \vDash d$.

From [21], we know that we have constructions for Slp, Success, and Fail as mentioned in Theorem 2 if we have the construction of a strongest liberal postcondition over a rule schema. Since we have it, we can can define the constructions of Slp(c, P), Success(P), and Fail(P) to prove the theorem. As an example, for Slp(c, P), we can define it inductively as: (i) if P is a rule set call $\mathcal{R} = \{r_1, \dots, r_n\}$ then Slp(c, P)=Slp(c, S) = Slp(c, r_1) $\vee \dots \vee$ Post(c, r_n), (ii) if $P = Q$ **or** S for some programs Q, S then Slp(c, P)=Slp(c, Q) $\vee$ Slp(c, S), (iii) if $P = Q; S$ then Slp(c, p)=Slp(Slp(c, Q), S), (iv) if $P = $ **if** C **then** Q **else** S for some program C then Slp(c, P)=Slp($c \wedge$ Success(C), Q) $\vee$ Slp($c \wedge$ Fail(C), S), and (v) if $P = $ **try** C **then** Q **else** S then Slp(c, P)=Slp($c \wedge$ Success(C) , $C; Q$) $\vee$ Slp($c \wedge$ Fail(C), S). The construction for Success and Fail can be seen in Appendix.

Definition 12 (Proof rules). The total correctness proof rules is defined in Fig. 4, where c, d, and d' are any conditions, r is any conditional rule schema, $\mathcal{R}$ is any set of rule schemata, C is any loop-free program, P and Q are any control commands, and S is any iteration command.

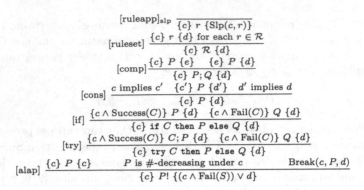

Fig. 4. Total correctness proof rules of calculus SYN

The proof rules are used to construct proof trees.

Definition 13 (Provability; proof tree[16]). A triple $\{c\}\ P\ \{d\}$ is provable in the calculus, denoted by $\vdash \{c\}\ P\ \{d\}$, if one can construct a *proof tree* from the axioms and inference rules of the calculus with that triple as the root. If $\{c\}\ P\ \{d\}$ is an instance of an axiom X then $\left(X\ \dfrac{}{\{c\}\ P\ \{d\}}\right)$ is a proof tree, and $\vdash \{c\}\ P\ \{d\}$. If $\{c\}\ P\ \{d\}$ can be instantiated from the conclusion of an inference rule Y, and there are proof trees $T_1, \ldots, T_n$ with conclusions that are instances of the n premises of Y, then $\left(Y\ \dfrac{T_1\ \ \cdots\ \ T_n}{\{c\}\ P\ \{d\}}\right)$ is a proof tree, and $\vdash \{c\}\ P\ \{d\}$.

5.2 Soundness

In [21], we show that our partial correctness calculus is sound. Now, we extend it to total correctness calculus, which is also proven to be sound in [23]. We prove the soundness by considering the induction on proof trees.

Theorem 3 (Soundness of the calculus). Given graph program P and MSO formulas c, d. Then, $\vdash \{c\}\ P\ \{d\}$ implies $\vDash_{\text{tot}} \{c\}\ P\ \{d\}$.

In the calculus, we use [ruleapp]$_{\text{slp}}$ as an axiom. Alternatively, we can change the axiom to [ruleapp]$_{\text{wlp}}$ $\dfrac{}{\{\neg \text{Slp}(\neg d, r^{-1})\}\ r\ \{d\}}$ and we still have a sound proof calculus [23].

However, relative completeness of the calculus is still an open problem. If we consider FO Hoare-triples, there is a strong evidence that we may have a correct FO Hoare-triple but we can not prove it by our FO proof calculus (see [21]) while we can prove it if by MSO proof calculus, which shows that the expressiveness of assertions play important role in relative completeness.

Courcelle [4,5] has proven that the following properties are not expressible in MSO logic without counting (either with set of node or set of edges quantifier):

1. The graph has even number of nodes
2. The number of nodes in a graph is a prime number
3. The graph has the same number of red nodes and grey nodes

However, we can express the three properties by the following MSO formulas, respectively:

1. $\exists_V X(\forall_v x(x \in X) \wedge \exists_I n(card(x) = 2 * n))$
2. $\exists_V X(\forall_v x(x \in X) \wedge \neg\exists_I n, m(n \neq 1 \wedge m \neq 1 \wedge card(x) = n * m))$
3. $\exists_V X, Y(\forall_v x(m_V(x) = red \Leftrightarrow x \in X) \wedge \forall_v x(m_V(x) = grey \Leftrightarrow x \in Y) \wedge card(X) = card(Y))$

With the existence of function card, our formula can express more properties if we compare it with counting MSO logic in [5] because we can compare cardinality between two sets with ours. However, what kind of properties can not be expressed by our formulas is still an open problem in this paper. Hence, the relative completeness of our MSO Hoare-triple is still unknown.

6 Case Study

In this section, we present the graph programs is-connected [3] and we verify the graph program with respect to the given specifications. Due to page limitation, we do not show the proof of implications in this paper. The proof can be found in [22] and other examples can be found in [22].

```
Main = try init then (DFS!; Check)
DFS = forward!; try back else break
Check = if match then fail
```

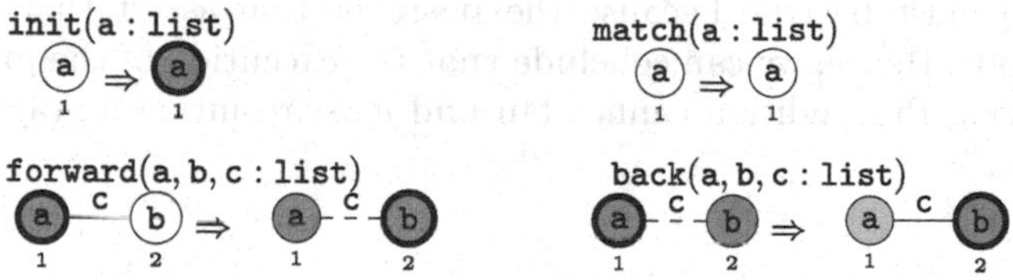

Fig. 5. Graph program is-connected (Color figure online)

Here we consider the graph program is-connected as seen in Fig. 5. The program is executed by checking the existence of an unrooted node with no marks and change it to a red rooted node. The program then execute depth first-search procedure by finding unrooted node that is adjacent with the red rooted node and change the node to red, swap the rootedness, and mark the edge between them by dashed and repeat it as long as possible. The procedure continue by searching a red node that adjacent to red unrooted node by dashed edge and change the mark of the rooted node to grey while unmarking it, and move the root to the other node, then reply the procedure. Finally, the program checks if there still exists an unmarked node. If so, then the program yields fail.

For the specification, here we consider the case where the input graph is connected. For the case with disconnected graph, please see [22].

Precondition:
All nodes and edges are unmarked, and all nodes are unrooted. Also, the graph is connected, that is, for every nodes x, y, there exists an undirect path from x to y)
Postcondition:
Either the graph is empty, or there is a node that is marked with red and is rooted while other nodes are grey and unrooted. All edges are unmarked, and the graph is connected.

Now let us consider loops we have in the program **is-connected**. There are two loops: **forward!** and DFS!. For the former, we can consider #-function that count the number of unmarked nodes. By the application of the rule schema **forward**, the number of unmarked nodes obviously decreasing. Hence **forward** is #-decreasing. For DFS!, we can consider a #-function that count unmarked nodes and red nodes. From the initial graph, the application of **forward!** will not change the value of #, while **try** either will decrease the value of # by 1 or make us reach **break**. Hence, DFS is #-decreasing as well.

The total correctness proof for this case study is given by the proof tree of Fig. 6. We refer to [22] for the assertions in the proof tree, which we omit here because of the lack of space. For the same reason, we omit #-decreasing requirement in the premise of proof rule [alap].

From the proof tree we know the triple $\{pre\}$ **init** $\{c\}$ and $\{c\}$ DFS! $\{post\}$ are totally correct so that by the proof rule [comp] we can conclude that $\{pre\}$ **init**; DFS! $\{post\}$ is totally correct as well. Implication $post \Rightarrow \neg$Fail(**match**) must be true because the postcondition assert that there is no unmarked node. Hence, we can conclude that the execution of the program on a graph satisfying Precondition cannot fail and must resulting a graph satisfying Postcondition.

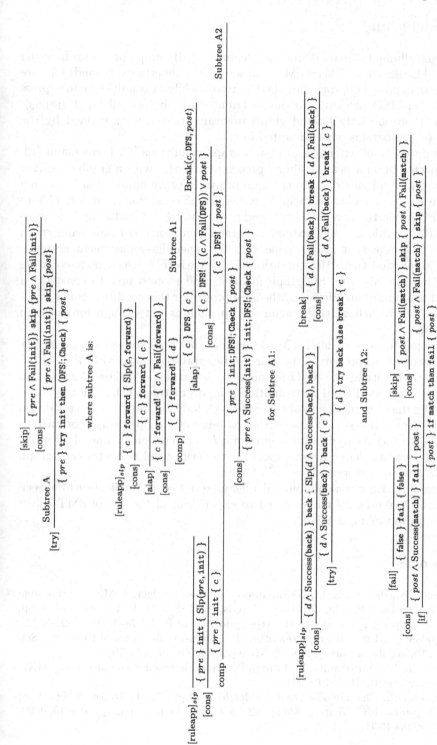

Fig. 6. Proof tree for is-connected

7   Conclusion

Poskitt and Plump [17] have defined a calculus to verify graph programs by using a so-called E-conditions [16] and M-conditions [19] as assertions. E-conditions are only able to express FO properties of GP 2 graph, while M-conditions can express properties of MSO properties of non-attributed graph (not all GP 2 graphs). However, there are only limited graph programs that can be verified by the calculus (e.g. programs with no nested loop).

E-condition is an extension of nested graph conditions [7]. Pennemann [13] shows how to obtain a weakest liberal precondition (wlp) w.r.t a graph condition and a program and introduced a theorem prover to prove implication between a precondition and the obtained wlp. However, graph conditions also only able to express FO properties of a non-attributed graph. Habel and Radke [9] then introduced HR* conditions, which extend the graph conditions by introducing graph variables that represent graphs generated by hyperedge-replacement systems. Radke [20] showed that HR* conditions is somewhere between node-counting MSO graph formulas and SO graph formulas and showed how to construct a wlp w.r.t the conditions. However, theorem prover for this condition is not available yet, and we believe that having a wlp alone is not enough for program verifications.

In this paper, we have defined MSO formulas that can express local properties of GP 2 graphs, even properties that can not be expressed in counting MSO graph formulas [6]. By using the MSO formulas as assertions, we show that we can construct a strongest liberal postcondition (Slp) over a rule schema. Moreover, we also can use the construction to obtain Slp over a loop-free program, precondition $Success(P)$ (or $Fail(P)$) that asserts the existence of a proper graph (or path to failure) in the execution of loop-free program P (or iteration command S). With this result, we can define a proof calculus to verify total correctness of graph programs with nested loops in certain forms.

As usual for Hoare calculi, our calculus does not cover implications between assertions. Currently, we have started to experiment of the use of SMT solver Z3 [1] to prove the implication.

References

1. Bjørner, N., de Moura, L., Nachmanson, L., Wintersteiger, C.M.: Programming Z3. In: Bowen, J.P., Liu, Z., Zhang, Z. (eds.) SETSS 2018. LNCS, vol. 11430, pp. 148–201. Springer, Cham (2019). https://doi.org/10.1007/978-3-030-17601-3_4
2. Campbell, G.: Efficient graph rewriting. BSc thesis, Department of Computer Science, University of York (2019). ArXiv e-print arXiv:1906.05170
3. Campbell, X., Courtehoute, B., Plump, D.: Fast rule-based graph programs. ArXiv e-print arXiv:2012.11394 (2020)
4. Courcelle, B.: The monadic second-order logic of graphs. I. Recognizable sets of finite graphs. Inf. Comput. **85**(1), 12–75 (1990). https://doi.org/10.1016/0890-5401(90)90043-H

5. Courcelle, B.: Monadic second-order definable graph transductions. In: Raoult, J.-C. (ed.) CAAP 1992. LNCS, vol. 581, pp. 124–144. Springer, Heidelberg (1992). https://doi.org/10.1007/3-540-55251-0_7
6. Courcelle, B., Engelfriet, J.: Graph Structure and Monadic Second-Order Logic: A Language-Theoretic Approach. Encyclopedia of Mathematics and Its Applications, vol. 138. Cambridge University Press (2012). https://doi.org/10.1017/CBO9780511977619
7. Habel, A., Pennemann, K.-H.: Correctness of high-level transformation systems relative to nested conditions. Math. Struct. Comput. Sci. **19**, 245–296 (2009). https://doi.org/10.1017/S0960129508007202
8. Habel, A., Plump, D.: Relabelling in graph transformation. In: Corradini, A., Ehrig, H., Kreowski, H.-J., Rozenberg, G. (eds.) ICGT 2002. LNCS, vol. 2505, pp. 135–147. Springer, Heidelberg (2002). https://doi.org/10.1007/3-540-45832-8_12
9. Habel, A., Radke, H.: Expressiveness of graph conditions with variables. Electron. Commun. EASST **30** (2010). https://doi.org/10.14279/tuj.eceasst.30.404
10. Nipkow, T.: Hoare logics in Isabelle/HOL. In: Schwichtenberg, H., Steinbrüggen, R., (eds.), Proof and System-Reliability, pp. 341–367. Kluwer Academic Publishers (2002). https://doi.org/10.1007/978-94-010-0413-8_11
11. Nipkow, T., Klein, G.: Concrete Semantics - With Isabelle/HOL. Springer, Heidelberg (2014). https://doi.org/10.1007/978-3-319-10542-0
12. Paulin-Mohring, C.: Introduction to the Coq proof-assistant for practical software verification. In: Meyer, B., Nordio, M. (eds.) LASER 2011. LNCS, vol. 7682, pp. 45–95. Springer, Heidelberg (2012). https://doi.org/10.1007/978-3-642-35746-6_3
13. Pennemann, K.-H.: Development of Correct Graph Transformation Systems. PhD thesis, Universität Oldenburg (2009)
14. Plump, D.: The graph programming language GP. In: Bozapalidis, S., Rahonis, G. (eds.) CAI 2009. LNCS, vol. 5725, pp. 99–122. Springer, Heidelberg (2009). https://doi.org/10.1007/978-3-642-03564-7_6
15. Plump, D.: From imperative to rule-based graph programs. J. Logic Algebraic Meth. Program. **88**, 154–173 (2017). https://doi.org/10.1016/j.jlamp.2016.12.001
16. Poskitt, C.M.: Verification of Graph Programs. PhD thesis, The University of York (2013)
17. Poskitt, C.M., Plump, D.: Hoare-style verification of graph programs. Fundam. Informaticae **118**(1–2), 135–175 (2012). https://doi.org/10.3233/FI-2012-708
18. Poskitt, C.M., Plump, D.: Verifying total correctness of graph programs. In: Graph Computation Models (GCM 2012), Revised Selected Papers, volume 61 of Electronic Communications of the EASST (2013). https://doi.org/10.14279/tuj.eceasst.61.827
19. Poskitt, C.M., Plump, D.: Verifying monadic second-order properties of graph programs. In: Giese, H., König, B. (eds.) ICGT 2014. LNCS, vol. 8571, pp. 33–48. Springer, Cham (2014). https://doi.org/10.1007/978-3-319-09108-2_3
20. Radke, H.: A Theory of HR* Graph Conditions and their Application to Meta-Modeling. PhD thesis, University of Oldenburg, Germany (2016)
21. Wulandari, G., Plump, D.: Verifying graph programs with first-order logic. In: Proceedings of GCM 2020, volume 330 of EPTCS, pp. 181–200 (2020). https://doi.org/10.4204/EPTCS.330.11
22. Wulandari, G.S.: Verification of graph programs with monadic second-order logic. Submitted PhD thesis (2021)
23. Wulandari, G.S., Plump, D.: Verifying graph programs with monadic second-order logic (extended version). Technical report, University of York (2021). https://uoycs-plasma.github.io/GP2/publications

On the Complexity of Simulating Probabilistic Timed Graph Transformation Systems

Christian Zöllner$^{(\boxtimes)}$, Matthias Barkowsky , Maria Maximova ,
and Holger Giese

Hasso Plattner Institute at the University of Potsdam, Potsdam, Germany
{christian.zoellner,matthias.barkowsky,maria.maximova,holger.giese}@hpi.de

Abstract. To develop future cyber-physical systems, like networks of autonomous vehicles, the modeling and simulation of huge networks of collaborating systems acting together on large-scale topologies is required. Probabilistic Timed Graph Transformation Systems (PTGTSs) have been introduced as a means of modeling a high-level view of these systems of systems. In our previous work, we proposed a simulation scheme based on local search incremental graph matching that can handle large-scale real-world topologies. However, the prohibitive complexity of the graph matching problem underlying the simulation of any GTS variant makes this setup potentially problematic.

In this paper, we present an improved simulation algorithm and identify restrictions that hold for PTGTS high-level models of cyber-physical systems and real-world topologies, for which we can establish favorable worst-case complexity bounds. We show that the worst-case amortized complexity per simulation step is only logarithmic in the number of active collaborating systems (like vehicles) and constant concerning the size of the topology. The theoretical results are confirmed by experiments.

Keywords: Graph Transformation Systems · Probabilistic behavior · Timed behavior · Simulator · Complexity · Algorithm

1 Introduction

One of the dominant trends in current technological development are autonomous systems, such as autonomous vehicles. These autonomous systems are on the one hand expected to locally interact with their environment and with each other in a safe and reliable way, while on the other hand being part of a large-scale system of systems that must function globally as a whole. With the number of interconnected cyber-physical systems being expected to dramatically increase in the future, developing these systems with the global system of systems perspective in mind becomes increasingly complex.

Funded by the Deutsche Forschungsgemeinschaft (DFG, German Research Foundation) - 241885098.

F. Gadducci and T. Kehrer (Eds.): ICGT 2021, LNCS 12741, pp. 262–279, 2021.
https://doi.org/10.1007/978-3-030-78946-6_14

A modeling technique for any kind of cyber-physical system of systems must include timed behavior, to reflect the real-time nature of the individual system, probabilistic behavior, to capture probabilistic phenomena like failures, and structure dynamics, to allow changing interconnections between the autonomous subsystems at runtime. Given the potential criticality of cyber-physical systems for human lives and safety, verification and simulation are important use-cases for any modeling technique. Therefore, the increasing size of cyber-physical systems of systems additionally requires a high scalability for simulation.

In [13], we introduced Probabilistic Timed Graph Transformation Systems (PTGTSs) as a modeling technique for a high-level view of these systems of systems based on the formal foundations of graph transformations. We further presented a simulator for PTGTS models [21] that can import and efficiently simulate complex large-scale real-world topologies and automatically detect violations of state properties in them. We argued and empirically demonstrated this by simulating the movement of vehicles on representations of the actual tram track network of several Germany cities, the largest of which was modeled with a graph of more than 10.000 nodes.

However, no theoretical bounds on the complexity of simulating PTGTS models have been established. The expected growth in the scale of future cyber-physical systems of systems makes it worthwhile to study the complexity of the simulation problem for PTGTS models in light of typical restrictions on the rules and topologies. Note that in addition to the graph matching problem, the question of how the time is efficiently advanced globally is also relevant here.

Therefore, in this paper, we present improvements to our simulation algorithm that address the remaining potential scalability bottlenecks. Furthermore, we identify restrictions that hold for the PTGTS rules of high-level models of typical cyber-physical systems of systems and for the real-world topologies on which these systems of systems operate. We formally show that given the typical restrictions and using our improved algorithm, PTGTS can be simulated efficiently with an amortized runtime effort per simulation step that is, in the worst-case, logarithmically dependent on the number of (actively) collaborating systems and independent from the size of the topology after a certain number of simulation steps. Finally, we support the findings of our analysis by a series of experiments where we also compare our improved and former algorithm.

Employing graph transformation systems (GTSs) and incremental graph pattern matching techniques for the simulation of complex systems has been proposed in [17]. A link between GTSs and discrete event simulation has been considered in [18]. Also, an extension of GTSs with stochastic behavior has been considered in [1] and related simulators like GraSS [20] and SimSG [4] have been developed. Finally, PTGTS [13] supporting timed and probabilistic behavior and related support for scalable simulation [21] have been presented. However, so far only empirical studies but no theoretical complexity analysis as approached in this paper have been done.

This paper is structured as follows: As preliminaries, the PTGTS formalism is introduced in Sect. 2. The simulator implementations, the simulation algorithm, and in particular the novel improvements are described in Sect. 3. The restrictions and the amortized complexity of the simulation algorithm are then formally analyzed in Sect. 4 and empirically evaluated in Sect. 5. The paper is closed with a conclusion and an outlook on future work.

2   Preliminaries

In this section, we briefly recall typed graphs, graph transformation, and the formalism of PTGTSs [13]. Moreover, we introduce our running example, where we model autonomous shuttles driving on a large track topology as a PTGTS.

In PTGTSs, we employ typed graphs [5] to capture the states of a system. A *graph* $G = (G_N, G_E, s_G, t_G)$ is given by a set G_N of nodes, a set G_E of edges, and source and target functions $s_G, t_G : G_E{\rightarrow}G_N$. For graphs $G = (G_N, G_E, s_G, t_G)$ and $H = (H_N, H_E, s_H, t_H)$, a *graph morphism* $f : G{\rightarrow}H$ is defined as a pair of mappings $f_N : G_N{\rightarrow}H_N$, $f_E : G_E{\rightarrow}H_E$ that are compatible with the source and target functions, i.e., $f_N \circ s_G = s_H \circ f_E$ and $f_N \circ t_G = t_H \circ f_E$.

For a distinguished graph TG, called a *type graph*, a *typed graph* $(G, type)$ consists of a graph G and a graph morphism $type : G{\rightarrow}TG$. For two given typed graphs $G'_1 = (G_1, type_1)$ and $G'_2 = (G_2, type_2)$, a *typed graph morphism* $f : G'_1{\rightarrow}G'_2$ is a graph morphism $f : G_1{\rightarrow}G_2$ that is compatible with the typing functions, i.e., $type_2 \circ f = type_1$.

In PTGTSs, we also use *attributes* but, for brevity, we omit a technical introduction of the attribution concept for graphs here, which can be straightforwardly added to the presented formalization of typed graphs.

As a running example, we model a scenario inspired by the RailCab project [16] where autonomous shuttles drive on a fixed large topology given by interconnected tracks. The type graph for this scenario is given in Fig. 1. It contains *Track* nodes connected by *next* edges that form a topology and *Shuttle* nodes representing shuttles that are located *at Track* nodes, can move across *next* edges from one track to another, can stop or brake to avoid collisions, and can establish connections with other nearby shuttles (represented by *Connection* nodes).

For PTGTSs, we require that the type graph contains a distinguished type node *Clock*. Furthermore, for every graph G, we use the function $CN(G) = \{n \mid n \in G_N \wedge type_N(n) = Clock\}$ to identify in every graph G the nodes of type *Clock*, which are used for time measurement only. In the following, we call these

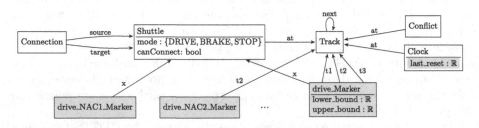

Fig. 1. Type graph of shuttle scenario with the generated extensions (gray, see Sect. 3).

nodes in $CN(G)$ simply *clocks*. For our running example, each *Track* node is equipped with a single attached clock (cf. Fig. 1). For brevity, we omit these clocks in our later visualizations and refer to a clock c of an element e as $e.c$.

For PTGTSs, we use graph transformation according to the Double Pushout Approach [5] to perform rule-based modifications of graphs. A *graph transformation rule* (short *rule*) $\rho = L \xleftarrow{l} K \xrightarrow{r} R$ is a span of injective graph morphisms where the graphs L and R are the left-hand side (LHS) and the right-hand side (RHS) of the rule, respectively. A *match* m from L into the graph G under transformation identifies the substructure $m(L)$ in G where the rule is to be applied. Transformation of the graph G is then realized by applying a rule ρ for a match m on G. Intuitively, the rule ρ specifies the removal of all elements in $L - \ell(K)$ and the addition of all elements in $R - r(K)$ (see [5] for more technical details).

PTGTSs have been introduced in [13], as a combination of Probabilistic Graph Transformation Systems (PGTSs) [12] and Timed Graph Transformation Systems (TGTSs) [2,7,14]. Similarly to PGTSs, *probabilistic timed graph transformation (PTGT) rules* have multiple right-hand sides, each equipped with a probability. While the choice for a match of a PTGT rule remains nondeterministic, the effect of rule application is probabilistic. Similarly to TGTSs, each PTGT rule is annotated with a *guard* over the clocks contained in its left-hand side graph L. The usage of guards restricts the applicability of PTGT rules w.r.t. the current clock values. Moreover, each PTGT rule is equipped with the information about clocks to be *reset* when that PTGT rule is applied.

Definition 1 (Probabilistic Timed Graph Transformation Rule [13]). *A probabilistic timed graph transformation (PTGT) rule R is a tuple (L, P, μ, ϕ, r_C) where L is a common left-hand side graph, P is a finite set of rules with the common left-hand side graph L, $\mu \in Dist(P)$ is a probability distribution on the rules from P, $\phi \in \Phi(CN(L))$ is a guard over clocks contained in L, and $r_C \subseteq CN(L)$ is the set of clocks from L to be reset.*

In PTGT rules, we also employ *negative application conditions (NACs)* [9] over the common left-hand side graph L. They increase the descriptive expressiveness of rules specifying that the match used for PTGT rule application cannot be extended to match further graph elements forbidden by the NAC. The use of NACs can be added straightforwardly to the presented formalization of rules.

We model the behavior of our running example using 14 PTGT rules in HEN-SHIN [10], which describe the driving, stopping, and braking of shuttles as well as their attempts for connection and disconnection. With connection attempts potentially failing, the PTGT rules for establishing a connection between two shuttles are probabilistic. In the following, we discuss only one exemplary PTGT rule (for more details see [15,21]). The PTGT rule *drive* (see Fig. 2a) allows a shuttle to move forward if there are no shuttles located too close in front of it, which is ensured using the NACs #1 to #5.

We also make use of *priorities*, which allow to state urgency of PTGT rule applications by disabling all PTGT rules with lower priority if a rule with higher priority is applicable. Note that all rules with higher priority are not allowed to state constraints on clocks to ensure that they can always be applied first.

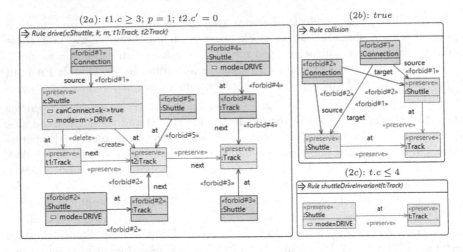

Fig. 2. PTGT rule *drive (a)*, atomic proposition *collision (b)*, and invariant *shuttleDriveInvariant (c)* of the shuttle scenario PTGTS in HENSHIN notation.

To reflect real-time behavior of driving shuttles, we require that each shuttle spends 3 to 4 time units for moving on a single track. This temporal constraint is specified in the PTGT rules for driving using the corresponding guards and invariants formulated over the track clocks. For the PTGT rule *drive* from Fig. 2a, the lower bound of 3 time units is encoded using the guard $t1.c \geq 3$. To measure the time spent on a track, we reset the clock of the track to which a shuttle is moving, which is annotated for the PTGT rule *drive* in Fig. 2a by $t2.c' = 0$.

Invariants and atomic propositions of PTGTSs are state properties defined as non-changing PTGT rules. *Invariants* are used to express upper bounds on clocks. In our shuttle scenario, the invariant *shuttleDriveInvariant* from Fig. 2c is used to state that a driving shuttle cannot stay on a track longer than 4 time units annotated by $t.c \leq 4$. *Atomic propositions* are used to identify graphs satisfying certain structural conditions. Note that atomic propositions are not allowed to state constraints on clocks and have the highest priority to ensure that they are evaluated for all graphs. For our shuttle scenario, the atomic proposition *collision* from Fig. 2b identifies graphs where two shuttles collide by being on the same track without being connected (ensured using the NAC #1).

We now define $PTGTSs^1$ comprising the notions discussed above.

Definition 2 (Probabilistic Timed Graph Transformation System [13]).
A probabilistic timed graph transformation system (PTGTS) S is a tuple (TG, $G_0, v_0, \Pi, I, AP, prio$) where TG is a finite type graph with the node type Clock,

---

[1] Note that the additional restrictions in the following definition, compared to [13], do not restrict the expressiveness of PTGTSs as (a) lower bounds for invariants can be replaced by additional conditions with clock constraints for all PTGT rules and (b) higher priority PTGT rules with constraints for the clocks can be emulated by additional pre-conditions for all lower-priority PTGT rules.

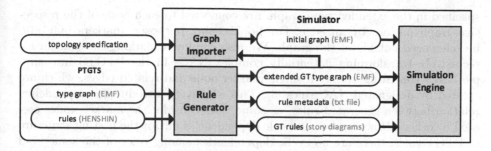

Fig. 3. Architecture of the PTGTS simulator. Active component are marked in gray.

G_0 is a finite initial graph typed over TG, $v_0 : CN(G_0) \rightarrow \mathbb{R}$ is the initial clock valuation assigning the value 0 to every clock of G_0, Π is a finite set of PTGT rules, I is a finite set of invariants, AP is a finite set of atomic propositions, and prio : $\Pi \rightarrow \mathbb{N}$ is a priority function assigning a priority to each PTGT rule.

The semantics of a PTGTS is given in terms of its induced *state space* where (a) *states* are given by pairs (G, v) of graphs G and valuations v assigning the clocks of G to real-valued time points and (b) *steps* between such states are either *timed steps* advancing all clocks by a common duration or *discrete steps* adapting the graph and valuation of a state according to one of the PTGT rules. For further technical details on the semantics of PTGTSs see [13].

3   Simulation Algorithm

In this section, we present the fundamental concepts behind our approach for simulating PTGTS models [15,21]. The idea is that, after an initial matching step, graph transformations are only applied locally so that the runtime impact of these possibly costly operations is kept independent from the model size. At the same time, the local clocks and relative time constraints are translated to a global simulation time. Thus, for each simulation step, the global considerations are limited to comparing lower and upper bound time values against a global simulation time, which are efficient numeric operations.

3.1   PTGTS Simulator Architecture

Our simulator consists of three active components highlighted in Fig. 3.

The first two components prepare suitable inputs for the simulation. The graph importer constructs input graphs and supports both real-world public transport network topologies and synthetically generated examples. The rule generator creates an extended GT type graph (see Fig. 1) and GT rules from a PTGTS. The generation process is described with specific examples in [21].

In order to enable incremental, local updates, these GT rules mark all pattern occurrences with marker nodes. Theses marker nodes, whose types are also specified in the extended type graph, are connected to each node of the respective graph pattern by an edge. For bookkeeping purposes, markers may also be referenced outside of the graph. We also employ a marking mechanism for recursively transforming potentially complex NACs in the PTGTS into simple NACs consisting of only a single marker node and related edges, which are created by designated NAC rules [3]. This effectively introduces a dependency relation between generated rules, as some rules consider the marker nodes created by others. However, because rule dependencies mirror the nesting structure of PTGT rules, there are no cyclic dependencies among rules of the same kind (FIND, UPDATE, and CHECK). For representing the timed behavior, the local clocks of the original PTGTS are translated to clock nodes with a last reset time attribute, while guards and invariants are translated to lower and upper bound attributes in the marker nodes.

Finally, based on the extended GT type graph with additional marker node types and a last reset attribute for the clock nodes, the rule generator creates the following GT rules for each PTGT rule, invariant and atomic proposition:

1. **FIND rules** detect the original rule's and the NAC's left-hand side patterns. They add a marker node to mark pattern occurrences and calculate upper and lower bound attributes based on adjacent clocks' last reset times.
2. **APPLY rules** that require a marker on the left-hand side and that apply the PTGT rule's right-hand sides and perform clock resets. Multiple APPLY rules represent the multiple right-hand sides of a PTGT rule.
3. **CHECK rules** that use a negative application condition to delete a marker if the pattern associated with it is not complete anymore.
4. **UPDATE rules** that recompute lower and upper bounds of markers after adjacent clock nodes' last reset attributes have been altered.

Note that the rule generator creates rules with Single Pushout semantics. In order to retain the Double Pushout behavior and avoid the accidental deletion of dangling edges, for each LHS node we create a NAC corresponding to each edge type that in the type graph is connected to the node's type. The final component from Fig. 3 is the simulation engine, which selects and applies the generated GT rules, following the algorithm described in Subsect. 3.2. The engine is implemented in Java. It uses the Eclipse Modeling Framework (EMF) [6] and an interpreter for Story Diagrams [8] that allows graph pattern matching starting with a fixed partial match.

3.2 Algorithm

The simulation engine's three-step algorithm is sketched in Fig. 4.

1. **Step 1: Add Initial Markers.** Patterns occurring in the input graph are marked by generated FIND rules. The employed rule ordering respects dependencies so that created markers can be used in dependent rules [3]. This step requires graph pattern matching on the complete graph, but only once.

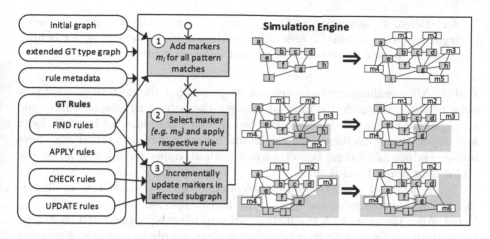

Fig. 4. PTGTS simulation algorithm based on marking pattern matches.

2. **Step 2: Apply a Rule and Advance Time.** Out of the created markers, the engine randomly selects one that (a) represents an enabled rule application, (b) has the highest available rule priority, (c) fulfills the invariants and its time bounds if present regarding the global simulation time. If a suitable rule marker is found, the engine computes a new global time so that no invariants are violated and uses the APPLY rule to apply the actual PTCT rule at the marked pattern as well as to reset the clocks.
3. **Step 3: Update Affected Subgraph.** After a rule application, the subgraph affected by the application (incl. all markers) is determined and FIND and CHECK rules are used to locally add or remove markers in places where the graph has changed. UPDATE rules are used to update marker time bounds. As in step 1, the rule ordering respects the dependencies.

The simulation stops when no applicable rule is found in step 2 (due to a lack of markers or due to violated invariants or time constraints) or when an atomic proposition (e.g. *collision* from Fig. 2b) is matched in step 1 or 3.

3.3 Optimized Data Structures

Due to the idea of global time management, in each simulation step, *all* invariant and rule markers must be considered to select a suitable rule. Since the contents of the input graph govern the potential number of markers, the runtime of this part of the algorithm's step 2 scales with the input graph size. In order to ensure scalability, data structures outside the graph are used to keep track of all relevant rule and invariant markers.

Invariants. As introduced in Sect. 2, invariants only have upper bounds. Since in each simulation step, *all* invariants must be fulfilled with respect to the global simulation time, it suffices to only find the lowest invariant upper bound t_{min_inv}.

An efficient identification of t_{min_inv} can be implemented using a balanced binary tree that maintains an ordering of all invariant markers by upper bounds and has logarithmic insertion, deletion and selection runtime.

Rules. After evaluating the invariants, we know that the next simulation step must be executed inside the interval between the current simulation time t and t_{min_inv}, and that a rule marker must be selected whose interval between lower and upper bound overlaps with the former interval. Either interval may have infinite endpoints (if there is no invariant with an upper bound or if a rule has no lower or upper bound).

For a balanced nondeterministic selection of an applicable rule r, all rule markers where $r.lower\_bound \leq t_{min\_inv}$ and $r.upper\_bound \geq t$ must be collected. Also, the rules with the highest priority must be applied first, but since high-priority rules have no time bounds, they can simply be stored in a separate data structure per priority that is checked before proceeding to the primary, timed data structure. Since any one-dimensional indexing structure can only be sorted either by the lower or by the upper bound, finding *all* applicable rules efficiently would require maintaining two structures and merging the results from both.

To resolve this, we propose that rules are selected from a balanced binary tree where markers are sorted only by the lower bounds, which are compared only against t_{min_inv}. While the selection can now be done in logarithmic time, it requires that when t advances, all rules where the upper bound is smaller than the new t are removed from the data structure. We use a second balanced binary tree to sort all rule markers by their upper bounds and to identify any markers that might be disabled by the advancing t. For higher-priority rules, a single balanced binary tree suffices to allow insertions and random retrieval in logarithmic time.

Removing a single marker from a balanced binary tree is possible in logarithmic time. However, in the worst case *all* markers are disabled in the same step. Since each removal incurs logarithmic cost, the complete maintenance operation would require $\mathcal{O}(n \cdot log(n))$ runtime. We analyze how the impact of this wost-case runtime can be amortized in Subsect. 4.3.

Implementation. For our simulator, the balanced binary tree data structures were implemented using AVL trees.

4   Efficiency

4.1   Input Restrictions

Host Graph. We assume an input graph G, typed over a type graph TG, with n nodes in total, represented as bidirectional adjacency lists. We define the *target cardinality* of an edge type $e_{TG} \in TG_E$ in G as $c_T(e_{TG}, G) = max_{v \in G_N} |\{e \in G_E \mid type(e_G) = e_{TG} \wedge s_G(e) = v\}|$. We define *source cardinality* analogously. We assume that the number of edge types has a constant upper bound T and that the source and target cardinalities of all edge types have a common constant upper bound C, which is typical for real-world physical topologies. Thus, the number of edges in G is linear in the number of nodes.

Since our use case are cyber-physical systems that interact inside a topology, a sizable part of n usually represents the static topology, while a number of active nodes n_{active} of certain types represents interacting subsystems, with $n_{active} \ll n$. In our example, the set of active nodes is the set of *Shuttle* nodes.

Rule Properties. We assume the satisfaction of the following assumptions regarding the PTGTS and thus generated GT rules:

- The number of nodes in each PTGT rule's LHS and RHS has a constant upper bound. Therefore, the number of nodes in the LHS and RHS of each generated GT rule has a constant upper bound Q.
- The number of edges, attribute constraints, attribute assignments, and NACs of each PTGT rule and thus in each GT rule have a constant upper bound.
- Checking of attribute constraints and computation of attribute assignments takes $\mathcal{O}(1)$ computational steps.
- The LHS of each PTGT rule as well as each NAC, and thus the LHS of each GT rule, are weakly connected.
- The LHS of all PTGT rules and NACs, and thus the LHS of all generated GT rule, contain at least one node of a type representing an active node.
- Successive application of PTGT and thus APPLY rules never increases an edge type's source or target cardinality beyond C and never increases n_{active}.
- The number of NACs of a PTGT rule and the nesting depth of such NACs both have a constant upper bound. Therefore, the number of dependencies of each rule has a constant upper bound B and the length of the longest path in the rule dependency graph has a constant upper bound D.
- The number of PTGT rules has a constant upper bound. Therefore, considering the upper bounds regarding nesting in the PTGTS, the number of FIND, UPDATE, and CHECK rules has a constant upper bound N.

All introduced assumptions are satisfied for our running example.

4.2   Pattern Matching Efficiency

We now analyze the worst-case runtime complexity of the graph transformation tasks involved in the simulation algorithm based on assumptions in Sect. 4.1.

Step 1: Add Initial Markers. Because the LHS L of any FIND rule contains at most Q nodes and (when considering all edges to be undirected) has a spanning tree where both source and target cardinality of each involved edge type have a constant upper bound C, the number of matches in a host graph G involving a fixed mapping for any one node in L is at most C^{Q-1}. By construction, the creation of redundant marking nodes is avoided by appropriate NACs, each type of marking node is only created by a single FIND rule, and each node in G can take the role of a fixed mapping for at most Q nodes in L. Therefore, each node in G can be associated with at most C^{Q-1} marking nodes via marking edges of a certain type and with at most $Q \cdot C^{Q-1}$ marking nodes of each type overall.

Given a fixed mapping for one node in L, we enumerate a complete set of match candidates containing at most C^{Q-1} elements in $\mathcal{O}(Q \cdot C^{Q-1})$ using an efficient algorithm [11]. The time required to check a candidate is dominated by the time required for checking NACs, which is in $\mathcal{O}(T \cdot Q + B \cdot Q \cdot C^{Q-1})$.

Considering the upper bounds induced by our assumptions, finding all matches for L with a fixed mapping for one node takes $\mathcal{O}(1)$ computational steps. Since the application of the rule creates only a single marking node and marking edges to Q nodes for each match, the effort for search and application remains in $\mathcal{O}(1)$.

Since the dependency relation among FIND rules is acyclic, we can create the initial set of marking nodes in an initial host graph G by executing all FIND rules in a topological order. To execute an individual FIND rule, we select one node with a type representing an active node in its LHS as a starting point for the search. We then successively fix the mapping for the starting point to each node with a matching type in G, find all corresponding matches and apply the transformation part of the rule. Note that consequently, the total number of created marking nodes is in $O(n_{active})$. Overall the effort for executing a single FIND rule is thus in $\mathcal{O}(n_{active})$, which is also the runtime complexity of executing all FIND rules since the number of such rules has a constant upper bound N.

Step 2: Apply a Rule. Since the application of an APPLY rule to G in a simulation step starts with a fixed mapping for the associated marking node, all marking edges have a target cardinality of 1, and apply rules do not perform any checks, finding a match for the LHS takes only $\mathcal{O}(Q)$ computational steps. By deleting nodes of the match, the application to G may require the deletion of at most $Q \cdot C^{Q-1}$ dangling marking edges of each marking edge type. Because of the introduced NACs, deletion of dangling original edges is not allowed. Due to the assumptions on cardinality, explicit deletion of original edges removes at most $T \cdot Q \cdot C$ edges. Because the upper bound on original edge cardinality has to remain unchanged, at most $2 \cdot T \cdot Q \cdot C$ edges are created.

By our assumptions, all attribute assignments performed as part of the rule application take $\mathcal{O}(1)$ computational steps. The execution time for modifying the adjacency lists is in $O(1)$ per created edge. Considering the effort for deleting elements from adjacency lists with up to C elements in the case of original edge types and C^{Q-1} elements in the case of marking edge types, the effort for search and application combined is thus in $\mathcal{O}(T \cdot Q \cdot C^2 + N \cdot Q^2 \cdot C^{2 \cdot (Q-1)}) = \mathcal{O}(1)$. The application modifies at most $Q + N \cdot Q^2 \cdot C^{Q-1}$ nodes in G, that is, it creates or deletes the node itself or an adjacent edge or modifies an attribute of the node.

Step 3: Update Affected Subgraph. The incremental update of the marking after the application of an apply rule a consists of executing all FIND, UPDATE, and CHECK rules r in the context of changed elements in an order $r_1, r_2, ..., r_m$ that respects the dependencies. The search for each rule r_i is executed with fixed input mappings, which are derived based on the *relevant* nodes modified by a and all rules r_j with $j < i$. A modified node is relevant to a rule if its type matches the type of a node in the rule's LHS and it was (i) created or deleted, (ii) an adjacent edge with a type that appears in the rule's LHS or its NACs was created or deleted, or (iii) a considered attribute was modified.

In the case of FIND rules, relevant nodes are directly used as fixed input mappings for each node with a matching type in the LHS. Thus, at most $Q \cdot S$ such local searches are executed, where S is the number of relevant modified nodes. Analogously to the initial execution, each individual local search takes $\mathcal{O}(1)$ computational steps and finds at most C^{Q-1} matches. The execution of a FIND rule may therefore create up to C^{Q-1} marking nodes and associated edges for the matches found in a single local search, thus overall modifying at most $S \cdot Q \cdot C^{Q-1}$ additional elements in G and causing effort in $\mathcal{O}(S \cdot Q \cdot C^{Q-1}) = \mathcal{O}(S)$. However, since no redundant markings are created, this cannot increase the number of marking nodes of each type connected to an original node in G beyond C^{Q-1}.

In the case of UPDATE and CHECK rules, all marking nodes of the corresponding type that are associated with a relevant node are used as starting points for the search. Since the LHS of each rule only contains one marking node and each node in the host graph can only be associated with at most $Q \cdot C^{Q-1}$ marking nodes of each type, the number of starting points is at most $S \cdot Q \cdot C^{Q-1}$. Since all elements in the LHS of the rule are connected to the marking node via a marking edge of an edge type with target cardinality 1 and, analogously to FIND rules, all required checks require $\mathcal{O}(1)$ computational steps, the effort for a single local search is in $\mathcal{O}(1)$ for both UPDATE and CHECK rules. Note that at most one match for UPDATE and CHECK rules can be found by a local search starting with a fixed mapping for the marking node.

The application of an UPDATE rule only updates an attribute of a single marking node per match, thus taking $\mathcal{O}(1)$ computational steps. Therefore, at most $S \cdot Q \cdot C^{Q-1}$ additional elements are modified and the combined overall effort for search and application is in $\mathcal{O}(S \cdot Q \cdot C^{Q-1}) = \mathcal{O}(S)$.

The application of a CHECK rule deletes a single marking node and all adjacent marking edges. Since marking nodes cannot have any additional adjacent edges, no dangling edges have to be considered. Thus, at most Q elements are modified for each match, with a computational effort for deleting the corresponding entries from the adjacency lists in $\mathcal{O}(Q \cdot C^{Q-1})$. The combined overall effort for search and application is thus in $\mathcal{O}(S \cdot Q^2 \cdot C^{2 \cdot (Q-1)}) = \mathcal{O}(S)$ and at most $S \cdot Q^2 \cdot C^{Q-1}$ nodes are modified in total.

All nodes modified by a simulation step's apply rule may be relevant to each FIND, UPDATE, and CHECK rule. However, FIND, UPDATE, and CHECK rules only create and delete marking nodes and marking edges of certain types that are only considered by dependent rules. Thus, the number of relevant modified elements for a rule r_i in the sequence is upper bounded by $s(r_i) \le S_0 + \sum_{r_d \in dep(r_i)} s(r_d) \cdot X$, where $dep(r_i)$ is the set of dependencies of r_i,

$S_0 = Q + N \cdot Q^2 \cdot C^{Q-1}$ and $X = Q^2 \cdot C^{Q-1}$. Unraveling the recursion yields an upper bound $s(r_i) \leq \sum_{j=0}^{D} S_0 \cdot (B \cdot X)^j \leq (D+1) \cdot S_0 \cdot (B \cdot X)^D = S_i$. Consequently, the overall complexity of the execution of any rule r_i is in $\mathcal{O}(Q^2 \cdot C^{2 \cdot (Q-1)} \cdot S_i)$. The execution of the entire rule sequence associated with a simulation step $a, r_1, r_2, ..., r_m$ thus takes $\mathcal{O}(Y + N \cdot Q^2 \cdot C^{2 \cdot (Q-1)} \cdot S_i)$ computational steps, where $Y = T \cdot Q \cdot C^2 + N \cdot Q^2 \cdot C^{2 \cdot (Q-1)}$ is the complexity of applying the initial apply rule. Finally, because of the constant upper bound induced on P and S_i by our assumptions, this complexity is still in $\mathcal{O}(1)$.

Considerations. While the established upper bounds involving C^{Q-1} imply a potentially very large constant factor in the runtime complexity of the tasks related to pattern matching and rule application, we note that these bounds are highly pessimistic, as they essentially assume that matches for the LHS of any rule are only constrained by the spanning tree of the LHS.

In our example, most of the involved constants are also rather small: $Q = 15$, $N = 158$, $B = 6$, and $D = 1$. Regarding the upper bound on edge cardinality C, the only edge type with source or target cardinality greater than 1 is the *next* edge type. A *Track* can have up to two incoming or outgoing *next* edges if it functions as a switch, thus $C = 2$. However, our example topologies are restricted to never contain successive switches, so that the worst case scenario for pattern matching with $C^{Q-1} = 2^{14}$ candidate matches for a fixed starting point never occurs. Moreover, as realistic topologies only have few switches, it is unlikely that multiple switch nodes have to be considered during a single search with a fixed starting point, which reduces the average-case effort for pattern matching.

The assumption regarding n_{active} not growing through rule application may seem rather restrictive. However, in most application scenarios, unbounded growth of the simulated system has to be avoided anyway. In such scenarios, instead of performing actual node creation, the introduction of new active nodes can be modeled by picking and enabling a node from a preallocated, limited pool of currently disabled nodes. Since the size of this pool then determines n_{active}, the assumption is satisfied for any system that does not exhibit unbounded growth given appropriate modeling decisions.

4.3   Marker Data Structures Efficiency

The locality that is exploited for the graph pattern matching efficiency comes at the price of handling simulation time globally, so that for each simulation step, the balanced binary trees for invariant upper bound, rule lower bound and rule upper bounds as introduced in Subsect. 3.3 must be considered. Following the arguments in the previous subsection, we know that the amount of invariant and rule markers created by the graph transformations and stored in the data structures is always in $\mathcal{O}(n_{active})$.

Step 1: Add Initial Markers. When executing the FIND rules, rule and invariant markers must be inserted into the data structures. Since the number of markers to be inserted is in $\mathcal{O}(n_{active})$ and each insert is logarithmic to the maximum number of inserted elements, the total data-structure-related runtime for this step is $\mathcal{O}(n_{active} \cdot log(n_{active}))$.

Step 2: Apply a Rule and Advance Time. As mentioned in Subsect. 3.3, we must first traverse the invariant marker tree to reach the invariant marker with the smallest upper bound, which takes $\mathcal{O}(log(n_{active}))$ time. Next, an enabled rule marker must be nondeterministically selected. For the markers with normal priority and with time bounds, the random traversal of the binary tree that sorts the markers by lower bound and the random selection of one that is below the threshold of t_{min_inv} is possible in $\mathcal{O}(log(n_{active}))$. If there is a marker with elevated priority (and without time bounds) in the respective lists, they can also be selected randomly in $\mathcal{O}(log(n_{active}))$.

Executing the apply rule includes deleting the original rule marker, which needs $\mathcal{O}(log(n_{active}))$. Now, when the simulation time is advanced, the rule data structures must be purged of the markers that are disabled by the new time. The respective markers with an upper bound smaller than the simulation time can be identified using the second binary tree that is sorted by rule upper bounds. The resulting k rule markers then must be removed from both binary trees, costing $\mathcal{O}(log(n_{active}))$ per marker. In the worst case, k is in $\mathcal{O}(n_{active})$.

In order to determine if this cost can be amortized during a simulation run of s steps, we have to consider the amortized runtime cost during each step. Amortized analysis [19] is a technique to determine a runtime upper bound r for a sequence of s steps, despite some individual steps potentially having a higher cost than $\frac{r}{s}$. It uses the argument that these higher costs are *amortized* by the cheaper costs of the remaining steps.

In our case, we start with the fact that in each simulation step we might add a number of markers in $\mathcal{O}(1)$, and that the insertion requires $\mathcal{O}(log(n_{active}))$ runtime per marker. When we double this cost, we stay in the same complexity class but we have an additional *time budget* that already amortizes the future cost of the markers' removal. However, the data structure also contains $\mathcal{O}(n_{active})$ marker that were added during step 1 of the algorithm that must be removed over time. For their successive removal, we can add an average $\frac{\mathcal{O}(n_{active} \cdot log(n_{active}))}{s}$ runtime cost to each of the s steps. Now, if $s \geq n_{active}$, we achieve a total amortized runtime in $\mathcal{O}(log(n_{active}))$ per simulation step for s steps.

Step 3: Update Affected Subgraph. Finally, FIND, UPDATE and CHECK rules are executed. As argued previously, the number of applied graph transformation rules as well as the number of markers that are potentially added by one rule in this step are both in $\mathcal{O}(1)$. This leads to a number of delete and insert operations in $\mathcal{O}(1)$ and thus to a runtime of $\mathcal{O}(log(n_{active}))$ for this step.

4.4   Overall Worst-Case Complexity

Following the arguments of this section, we can summarize the computational complexity of the individual steps as follows:

Algorithm		Pattern matching	Data structures	Overall
Initialization	Step 1	$\mathcal{O}(n_{active})$	$\mathcal{O}(n_{active} \cdot log(n_{active}))$	$\mathcal{O}(n_{active} \cdot log(n_{active}))$
Simulation loop	Step 2	$\mathcal{O}(1)$	$\mathcal{O}(log(n_{active}))$ *	$\mathcal{O}(log(n_{active}))$ *
	Step 3	$\mathcal{O}(1)$	$\mathcal{O}(log(n_{active}))$	

* Amortized runtime for at least n_{active} steps.

Note that all complexities depend on n_{active} rather than n, making them independent from the topology size. For a simulation of s steps and simulation time t, with s larger than n_{active}, we get a total amortized runtime in $\mathcal{O}((n_{active} + s) \cdot log(n_{active}))$. For a typical simulation, s is in $\mathcal{O}(t \cdot n_{active})$ with $t > 1$, so that the overall simulation runs in $\mathcal{O}(t \cdot n_{active} \cdot log(n_{active}))$.

5   Evaluation

Runtime in Real-World Examples. We already showed in [21] that for example graphs generated from real-world tram networks, the average runtime for a simulation step is a) stationary (i.e. the average value does not change according to a trend after some time) after an initial non-stationary interval and b) seemingly independent of the model size. However, with $n_{active} = 918$, even the largest of these examples was comparatively small, so that the linear growth imposed by the list data structures in the old algorithm was not noticeable besides the cost for executing the graph transformations. In order to analyze the data structure access cost in particular, we re-ran these experiments[2] and explicitly measured the average cost of the respective individual operations. For each topology, we conducted three different experiments with different random initial shuttle placements, which among other things led to different numbers of simulation steps until an invariant violation was reached. Each experiment was run three times for up to 10,000 steps each, with one step representing one PTGT rule application and being executed through several hundred applications of various of the generated rules. In the following table we show the highest and lowest average data structure access times observed in the different experiments with the same n_{active}. While the linearly growing cost of accessing the list data structure is obvious, the binary tree operations are far more efficient, but have a high variance and thus no observable growth.

Experiment	List operations		Tree operations	
	Modify	Access	Modify	Access
Potsdam City ($n_{active} = 9$)	0.5–0.5 ns	16.0–17.0 ns	3.3–3.5 ns	2.4–2.7 ns
Potsdam ($n_{active} = 142$)	0.7–1.3 ns	99.2–112.5 ns	5.5–7.7 ns	4.5–6.3 ns
Frankfurt ($n_{active} = 316$)	0.5–1.4 ns	155.9–167.7 ns	5.7–8.0 ns	4.8–7.0 ns
Leipzig ($n_{active} = 593$)	0.7–0.9 ns	275.5–291.5 ns	5.8–6.6 ns	4.8–5.9 ns
Berlin ($n_{active} = 918$)	0.6–0.8 ns	437.2–455.7 ns	5.1–5.2 ns	4.1–4.3 ns

---

[2] The experiments were run on a server with 256 GB RAM and two Intel Xeon E5-2643 CPUs (4 cores/3.4 GHz). Our single-threaded implementation runs using Java 1.8.

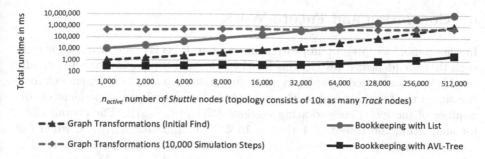

Fig. 5. Runtimes for the simulation (10,000 steps) of large-scale synthetic examples. Note that both axes are scaled logarithmically.

Runtime Scalability for Very Large Host Graphs. In order to demonstrate the scalability of our approach, we constructed artificial input graphs for several thousand shuttles driving around in a circle, assuming a density of one shuttle per 10 tracks. We ran each of these experiments five times for 10,000 simulation steps. As can be seen in Fig. 5, the empirical results seem to confirm our arguments in Sect. 4: The cost of accessing the list data structures grows linearly by several orders of magnitude and for very large examples even outgrows the (significant) cost for graph transformations, while the cost for accessing the binary tree data structure is much smaller in the first place and only grows logarithmically. At the same time, the cost for graph transformations during the initial find step grows linearly with n_{active}, while the cost for graph transformations during the simulation steps hardly grows at all.

Note however, that there is a very slight increase in graph transformation time for the simulation steps that is related to model size. With a runtime growth by a factor of less than 1.4 for a model that grew by a factor of 512 at the same time, this effect is comparatively very small. Also, more than 80% of the growth can be linked to initialization tasks of the graph transformation tool rather than to the iterative pattern matching process itself. While we were not completely able to isolate the effect, we assume that imperfect, hash-based indexing structures used by the graph transformation tool are responsible.

Threats to Validity. We remark that we only considered one example scenario and our empirical results thus are not necessarily generalizable to other PTGTSs. Furthermore, our experiments on very large graphs were only conducted on synthetic example graphs, which do not necessarily have all characteristics of realistic scenarios and thus might behave differently.

Finally, it must be considered that the main contribution of this paper is the identification of a worst-case time complexity. Repeated experiments on realistic examples, however, naturally tend to have average-case runtimes, which could have disproved our worst-case assumptions, but cannot actually support them.

6   Conclusion and Future Work

In this paper, we present improvements to our PTGTS simulation algorithm such that for identified restrictions for the rules and topologies we can formally establish an amortized complexity per simulation step that is independent from the size of the topology and, in the worst-case, logarithmically dependent on the number of (actively) collaborating systems ($\mathcal{O}(log(n_{active}))$). The overall effort for simulating the time $t > 1$ then is in $\mathcal{O}(t \cdot n_{active} \cdot log(n_{active}))$, when t is simulated in $\mathcal{O}(t \cdot n_{active})$ steps. The theoretical results are confirmed by a series of experiments, which also compare our improved and former algorithm.

As future work, we plan to improve the efficiency of our tool so that the remaining weaknesses compared to the theoretical results are further reduced, to support checking for more than simple state properties and to extend the complexity analysis accordingly, so that complexity bounds can also be established for checking more complex properties.

References

1. Bapodra, M., Heckel, R.: Abstraction and training of stochastic graph transformation systems. In: Cortellessa, V., Varró, D. (eds.) FASE 2013. LNCS, vol. 7793, pp. 312–326. Springer, Heidelberg (2013). https://doi.org/10.1007/978-3-642-37057-1_23
2. Becker, B., Giese, H.: On safe service-oriented real-time coordination for autonomous vehicles. In: Proceedings of ISORC 2008. IEEE (2008). https://doi.org/10.1109/ISORC.2008.13
3. Beyhl, T., Blouin, D., Giese, H., Lambers, L.: On the operationalization of graph queries with generalized discrimination networks. In: Echahed, R., Minas, M. (eds.) ICGT 2016. LNCS, vol. 9761, pp. 170–186. Springer, Cham (2016). https://doi.org/10.1007/978-3-319-40530-8_11
4. Ehmes, S., Fritsche, L., Schürr, A.: SimSG: rule-based simulation using stochastic graph transformation. J. Object Technol. 18(3), 1–17 (2019). https://doi.org/10.5381/jot.2019.18.3.a1
5. Ehrig, H., Ehrig, K., Prange, U., Taentzer, G.: Fundamentals of Algebraic Graph Transformation. MTCSAES. Springer, Heidelberg (2006). https://doi.org/10.1007/3-540-31188-2
6. Eclipse modeling framework. https://www.eclipse.org/modeling/emf
7. Giese, H.: Modeling and verification of cooperative self-adaptive mechatronic systems. In: Kordon, F., Sztipanovits, J. (eds.) Monterey Workshop 2005. LNCS, vol. 4322, pp. 258–280. Springer, Heidelberg (2007). https://doi.org/10.1007/978-3-540-71156-8_14
8. Giese, H., Hildebrandt, S., Seibel, A.: Improved flexibility and scalability by interpreting story diagrams. Electron. Commun. Eur. Assoc. Softw. Sci. Technol. 18 (2009). https://doi.org/10.14279/tuj.eceasst.18.268
9. Habel, A., Heckel, R., Taentzer, G.: Graph grammars with negative application conditions. Fundam. Inf. 26(3/4), 287–313 (1996). https://doi.org/10.3233/FI-1996-263404
10. EMF Henshin. https://www.eclipse.org/modeling/emft/henshin

11. Hildebrandt, S.: On the performance and conformance of triple graph grammar implementations. Ph.D. thesis, University of Potsdam (2014)
12. Krause, C., Giese, H.: Probabilistic graph transformation systems. In: Ehrig, H., Engels, G., Kreowski, H.-J., Rozenberg, G. (eds.) ICGT 2012. LNCS, vol. 7562, pp. 311–325. Springer, Heidelberg (2012). https://doi.org/10.1007/978-3-642-33654-6_21
13. Maximova, M., Giese, H., Krause, C.: Probabilistic timed graph transformation systems. J. Log. Algebraic Methods Program. **101**, 110–131 (2018). https://doi.org/10.1016/j.jlamp.2018.09.003
14. Neumann, S.: Modellierung und Verifikation zeitbehafteter Graphtransformationssysteme mittels GROOVE. Master's thesis, University of Paderborn (2007)
15. PTGTS simulator project website. https://mdelab.de/ptgts-simulator
16. RailCab project. https://www.hni.uni-paderborn.de/cim/projekte/railcab
17. Ráth, I., Vago, D., Varró, D.: Design-time simulation of domain-specific models by incremental pattern matching. In: IEEE Symposium on Visual Languages and Human-Centric Computing, VL/HCC 2008, Herrsching am Ammersee, Germany, 15–19 September 2008, Proceedings, pp. 219–222. IEEE Computer Society (2008). https://doi.org/10.1109/VLHCC.2008.4639089
18. Syriani, E., Vangheluwe, H.: A modular timed graph transformation language for simulation-based design. Softw. Syst. Model. **12**(2), 387–414 (2013). https://doi.org/10.1007/s10270-011-0205-0
19. Tarjan, R.E.: Amortized computational complexity. SIAM J. Algebraic Discrete Methods **6**(2), 306–318 (1985). https://doi.org/10.1137/0606031
20. Torrini, P., Heckel, R., Ráth, I.: Stochastic simulation of graph transformation systems. In: Rosenblum, D.S., Taentzer, G. (eds.) FASE 2010. LNCS, vol. 6013, pp. 154–157. Springer, Heidelberg (2010). https://doi.org/10.1007/978-3-642-12029-9_11
21. Zöllner, C., Barkowsky, M., Maximova, M., Schneider, M., Giese, H.: A simulator for probabilistic timed graph transformation systems with complex large-scale topologies. In: Gadducci, F., Kehrer, T. (eds.) ICGT 2020. LNCS, vol. 12150, pp. 325–334. Springer, Cham (2020). https://doi.org/10.1007/978-3-030-51372-6_20

Tool Presentations

Automated Checking and Completion of Backward Confluence for Hyperedge Replacement Grammars

Ira Fesefeldt[1]([⊠]) [iD], Christoph Matheja[2] [iD], Thomas Noll[1] [iD],
and Johannes Schulte[3]

[1] Software Modeling and Verification Group, RWTH Aachen University,
Aachen, Germany
{fesefeldt,noll}@cs.rwth-aachen.de
[2] Programming Methodology Group, ETH Zürich, Zurich, Switzerland
cmatheja@inf.ethz.ch
[3] Aachen, Germany

Abstract. We present a tool that checks for a given context-free graph grammar whether the corresponding graph reduction system in which all rules are applied backward, is *confluent*—a question that arises when using graph grammars to guide state space abstractions for analyzing heap-manipulating programs; confluence of the graph reduction system then guarantees the abstraction's uniqueness. If a graph reduction system is *not* confluent, our tool provides symbolic representations of counterexamples to confluence, i.e., non-joinable critical pairs, for manual inspection. Furthermore, it features a heuristics-based completion procedure that attempts to turn a graph reduction system into a confluent one without invalidating the properties mandated by the abstraction framework. We evaluate our implementation on various graph grammars for verifying data structure traversal algorithms from the literature.

Keywords: Graph grammars · Confluence · Critical pairs · Completion

1 Introduction

Confluence is a central property of many rewriting formalisms, including term rewriting and graph transformation systems: Confluent systems require no backtracking since all terminating sequences of rule applications produce the same result. In this paper, we present a tool that checks confluence for certain *graph reduction systems*—more precisely: hyperedge replacement grammars (HRG) [9] in which all rules are reversed—based on the algorithm in [14].

Our work is motivated by the usage of HRGs as an abstraction mechanism for verifying pointer programs. This approach is at the core of ATTESTOR[1]—a graph-based model-checking tool for analyzing Java programs operating on

---

[1] https://github.com/moves-rwth/attestor.

© Springer Nature Switzerland AG 2021
F. Gadducci and T. Kehrer (Eds.): ICGT 2021, LNCS 12741, pp. 283–293, 2021.
https://doi.org/10.1007/978-3-030-78946-6_15

dynamic data structures [2,12]. To cope with large or even unbounded state spaces arising in this context, ATTESTOR performs a symbolic shape analysis based on a user-supplied HRG that characterizes the data structures handled by the program. Here, a suitable HRG generates graphs modeling concrete heaps where hyperedges labeled with nonterminal symbols act as placeholders for the (partial) data structures under consideration, e.g., doubly-linked lists or binary trees with a fixed root. Abstracting of a concrete heap then corresponds to applying HRG rules backward until a normal form is reached. While termination of this procedure is guaranteed for the HRGs admitted in our setting, *confluence*, i.e., uniqueness of normal forms, is not. Confluence is vital for the performance of verification tools, such as ATTESTOR, because—rather than abstracting a heap in all possible ways—it suffices to apply abstraction rules exhaustively and in arbitrary order. Furthermore, confluence can be exploited to decide whether an abstract state is subsumed by an already computed one—a particular instance of the graph language inclusion problem that is crucial for ensuring termination of the overall analysis in the presence of loops or recursive procedures (cf. [21]).

Apart from checking whether the graph reduction system induced by an HRG is confluent, our tool supports a heuristics-based completion procedure to transform it into a confluent one. In particular, the heuristics can be chosen such that properties of the HRG required by ATTESTOR, e.g., those ensuring termination of the abstraction, are preserved during completion. We evaluate our implementation on various heap abstractions that have been proposed in the literature (as HRGs or equivalent inductive predicates in separation logic).

Related Tools. While algorithms for deciding confluence have been extensively studied in the context of graph transformations (cf. [6,11,14,20,23,24]), there are, to the best of our knowledge, only few tools that support computing critical pairs—a key component for confluence checking.

However, we are not aware that any of the tools below support proving backward confluence for HRGs, which additionally requires checking whether the computed critical pairs are joinable. Moreover, they do not support completion.

AGG[2] is a development environment for attributed graph transformation systems supporting an algebraic approach to graph transformation [25,27]. For analyzing critical pairs, it implements the algorithm developed in [11].

VeriGraph [7] is a tool for simulation and analysis of transformation systems given by graph grammars, which appears not to be developed further anymore. It implements the critical-pair analysis described in [19]. A performance comparison with AGG is given in [4], analyzing both critical pairs and sequences to capture conflicts and dependencies between rules. The evaluation shows that Verigraph outperforms AGG in realistic test cases, which indicates that AGG is more sensitive to the size of the graphs contained in rewriting rules.

Henshin [1] is a model transformation environment that is based on the Eclipse Modeling Framework. It first integrated a critical-pair analysis as presented in [5], which has later been superseded by a more efficient and flexible conflict and dependency analysis [18].

---

[2] https://www.user.tu-berlin.de/o.runge/agg/.

SyGraV [8] is a graph analysis tool that supports checking local confluence of attributed graph transformation systems. It was the first to fill the gap between theoretical results and practical usability of symbolic graph analysis.

2   The Tool

We implemented a confluence checking and a completion component within the software model checker ATTESTOR; our confluence checker can also be used as a standalone tool. The tool, its source code, and our benchmarks are available online.[3] In this article, we do not distinguish between ATTESTOR and its confluence checker. This section briefly outlines our tool's *input*, its main steps for *proving confluence*, the *feedback* it provides, and the extent to which it supports automatic *completion* of graph grammars. A detailed discussion of the underlying algorithm (which is based on [14]), its implementation, and the heuristics applied for guiding completion is found in [26].

2.1   Input

Our confluence checker targets a subset of HRGs suitable for modeling dynamic data structures (cf. [2,12] for details). While HRGs are context-free graph grammars—and thus always confluent—ATTESTOR checks whether the corresponding *graph reduction system* (GRS) [13] in which all rules are applied in reverse direction, is confluent as well. In our setting, graph rewriting through reverse rule applications always terminates because we require all HRG rules to be increasing, i.e., every hyperedge connected to $n \geq 0$ nodes is mapped to a hypergraph consisting of at least $n + 1$ nodes and edges. HRGs are specified as a set of (forward) rules in a JSON-style format. Both nodes and hyperedges are equipped with attributes indicating the type of nodes and edges; these types can be consulted to differentiate elements when checking for graph isomorphism.

2.2   Proving Confluence

ATTESTOR implements the confluence checking algorithm in [14] for the set of GRSs from above. That is, it systematically computes all *critical pairs*—overlappings of two hypergraphs appearing on the left-hand side of graph transformation rules—and, for each critical pair, checks whether it is *strongly joinable*, i.e., exhaustive rewriting after applying either of the two possible rules leads to isomorphic normal forms. Here, "strongly" refers to an additional requirement while searching for graph isomorphisms: we distinguish between nodes in the original overlapping that are not deleted by rule applications. The above condition is necessary because a GSR is confluent iff all critical pairs are strongly joinable. ATTESTOR reports if a critical pair is joinable but not strongly joinable since this case seems to be a frequent error when manually designing supposedly confluent grammars for heap abstraction (see evaluation).

---

[3] https://github.com/moves-rwth/attestor-confluence.

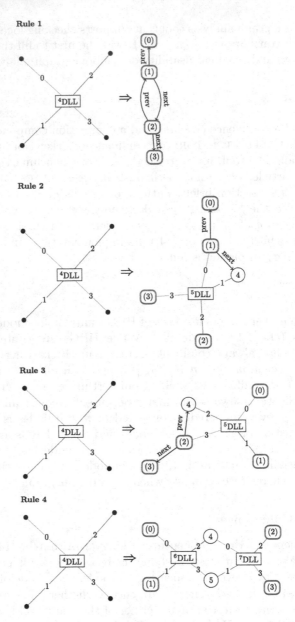

Fig. 1. Rules of an HRG modelling doubly-linked list segments; the graphical representation was automatically generated by ATTESTOR. In our graphical notation, circles indicate nodes; double circles capture the sequence of external nodes (ordered by their label). Hyperedges are drawn as rectangles; the numbered connections indicate the sequence of nodes attached to a hyperedge, i.e., all hyperedges labeled with DLL are attached to four (not necessarily different) nodes. For simplicity, hyperedges connected to exactly two nodes are drawn as (labeled) arrows. In all hypergraphs, nodes and edges are additionally equipped with an integer, e.g., 4 for the single hyperedge on the LHS of rules), to identify them across rule applications.

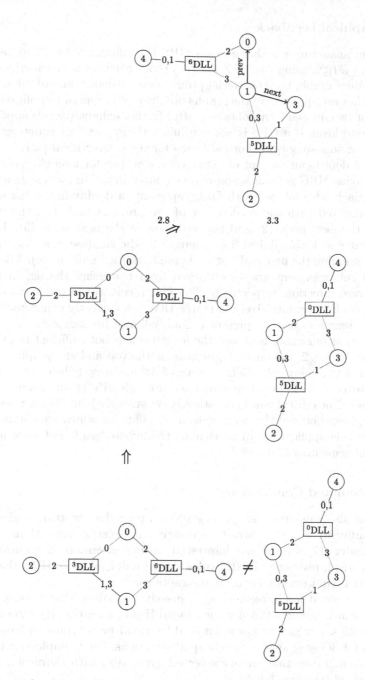

Fig. 2. Example of the feedback generated for a critical pair. It displays a non-joinable pair at the bottom, a trivial derivation to the graphs in the middle and a derivation using the second and third rule from Fig. 1 to obtain the context graph of the critical pair at the top.

2.3   Graphical Feedback

Apart from answering whether a given GRS is confluent, ATTESTOR generates a report (in LaTeX, using the TikZ library) that visualizes for each critical pair: (a) the context graph, i.e., the overlapping of two left-hand sides of rules, (b) the two rules that are applied, (c) the graphs obtained after one rule application, and (d) at most two normal forms obtained after further exhaustive rule applications up to isomorphism. If a GRS is not confluent, the report lists counterexamples consisting of non-isomorphic normal forms for the same critical pair.

Figure 2 depicts an excerpt of ATTESTOR's output for a single critical pair. The underlying HRG generates non-empty doubly-linked list segments as shown in Fig. 1. Each edge labeled with DLL represents a doubly-linked list segment that is connected with the predecessor of the previous node (0), the previous node (1), the next node (2) and the successor of the next node (3). The first rule generates a doubly-linked list of length 2 (the smallest this grammar can generate assuming the first and last node are both the "null" node). The second and third rule represent graphs obtained from traversing the list in forward and backward direction, respectively. The fourth rule was introduced to achieve backward confluence: intuitively, it states that two correctly connected doubly-linked lists segments again represent a doubly-linked list segment.

However, as our analysis shows, the fourth rule is not sufficient to guarantee confluence. In Fig. 2, the context graph is at the top and the graphs resulting from the first two rule applications (2.8 and 3.3)[4] are directly below; the numbers assigned to each node and hyperedge serve to identify them throughout rule applications. The critical pair in question is not strongly joinable because there is no isomorphism between the two graphs at the bottom, which are obtained after exhaustive rule application. In particular, the hyperedges 6 and 0 are attached to different sequences of nodes.

2.4   Automated Completion

ATTESTOR also supports a simple completion procedure to turn a given GRS into a confluent one. In contrast to existing completion algorithms, such as Knuth-Bendix [17], we are not interested in *any* extension that ensures confluence. Instead, reversing all GRS rules should still lead to an HRG that meets ATTESTOR's requirements for heap abstractions.

We thus opted for implementing a greedy procedure that applies various heuristics which preserve the aforementioned HRG properties. By choosing suitable heuristics, our greedy approach enables rapid prototyping of handcrafted strategies for devising appropriate heap abstractions. For example, one heuristic attempts to add rules that group connected hyperedges with identical labels into a single one of the same label.

---

[4] That is, rules 2 and 3 were applied. To improve performance, ATTESTOR generates specialized rules in which two or more external nodes are identical; the number after the dot indicates which case of a rule has been applied.

Table 1. The experimental results for our confluence checker; confluent grammars are marked with ✓. All runtimes are in milliseconds.

Grammar	Critical pairs				Runtime			
	Not	Weak	Strong	Total	Node	Edge	Validity	Total
✗ InTree	22	1	18	41	9.6	19.6	32.8	63.5
✓ InTreeLinked	0	0	83	83	11.1	7.0	32.8	51.7
✗ LinkedTree1	5	0	10	15	6.0	9.3	4.0	19.7
✗ LinkedTree2	217	4	20	241	61.1	111.5	95.7	270.1
✓ BT	0	0	33	33	2.1	4.8	4.9	12.2
✓ SLList	0	0	15	15	0.3	0.4	0.8	1.6
✗ SimpleDLL	1	0	2	3	0.1	0.4	0.2	0.7
✗ DLList	63	12	137	212	89.8	22.5	78.7	192.6

3 Evaluation

We evaluated our implementation on both confluent and non-confluent graph grammars proposed for modeling dynamic data structures. For the non-confluent grammars, we also experimented with feeding our greedy completion procedure with various heuristics to turn them into confluent ones.

Setup. All experiments were performed on a Thinkpad X1 Carbon 2019 with an Intel Core i7-8565U, 1.8 GHz and 16 GB Ram, which runs Ubuntu 20.04.1.

Confluence Checking. Table 1 shows our experimental results for checking whether a given graph reduction system induced by an HRG is confluent. As noted in the previous section, ATTESTOR checks for each critical pair whether it is *strongly-joinable*, *weakly joinable* or *not joinable* at all. To conclude that a graph grammar is confluent (and thus mark it with ✓), all critical pairs must be strongly joinable. Furthermore, we measured the time for computing *overlappings of nodes*, *overlappings of edges*, *a validity check* for possibly spurious pairs (we discard graphs that do not model heaps and thus cannot appear in our setting) as well as the total runtime. Starting the java virtual machine and parsing the grammar took 0.9 s CPU time and was thus dominant for small examples.

   The HRGs InTree [15] and InTreeLinked generate "in-trees", i.e., binary trees in which the direction of edges is inverted such that child nodes point to their parent. InTreeLinked additionally connects all leaves of the in-tree from left to right via a singly-linked list. It is noteworthy that—despite having more than twice as many rules—ATTESTOR managed to prove confluence for InTreeLinked faster than determining all critical pairs that are not strongly joinable for InTree. One possibe explanation is that the different edge labels (for the left and right child as well as the successor in the list) used by InTreeLinked reduce the number of edge overlappings that need to be computed.

Table 2. Completion results (✓ if successful, ✗ if unsuccessful) for different combinations of completion heuristics. All runtimes are in seconds.

	InTree		LinkedTree1		LinkedTree2		SimpleDLL		DLList	
CA	✗,	0.003	✗,	0.003	✗,	0.337	✗,	<0.001	✗,	0.127
RNN	✗,	1.538	✗,	0.048	✗,	55762.111	✗,	0.002	✗,	533.773
JGN	✗,	2.415	✗,	0.051	✗,	53327.031	✗,	0.003	✗,	540.818
SNR	✗,	1.576	✗,	0.048	✗,	55341.839	✓,	0.006	✗,	526.315
OR	✗,	1.640	✗,	0.046	✗,	55018.761	✓,	0.004	✗,	527.410
ORL	✗,	0.847	✗,	0.213	✗,	31075.274	✗,	0.004	✗,	362.976
A1	✗,	1.570	✓,	0.049	✓,	55042.314	✓,	0.004	✗,	392.728
A1L	✗,	1.175	✓,	0.203	✗,	27297.678	✗,	0.004	✗,	258.643
A2	✓,	0.333	✓,	0.024	✓,	12.405	✓,	0.004	✗,	26.421
A2L	✗,	0.838	✓,	0.038	✗,	27053.035	✗,	0.004	✗,	261.090

The HRGs `LinkedTree1` and `LinkedTree2` generate binary trees with a given root, where each node has a back pointer to its parent and all leaves are connected from left to right. While `LinkedTree2`, which is taken from [22], consists of only two rules with up to 7 nodes and 7 hyperedges (of rank at most 4), it turned out to be quite complex, leading to a large number of possible non-joinable critical pairs. `LinkedTree1` is an early attempt to turn a simplified version of `LinkedTree2` into an HRG that induces a confluent GRS.

The HRG `BT` [3] generates binary trees with a given root that has been handcrafted for verifying tree traversal algorithms. Ensuring confluence required two different nonterminal symbols and fourteen rules in total. Similarly, `SLList` is a handcrafted grammar modeling singly-linked lists.

`SimpleDLL` and `DLL` generate doubly-linked lists. The former version only admits list traversals from left to right whereas the latter version admits traversals in both directions. Both HRGs do *not* induce confluent GRSs. This surprised us as DLL has been successfully applied for analyzing pointer programs [2]. While confluence is not required for the soundness of such program analyses, non-confluence typically leads to performance penalties. Upon closer inspection, we discovered that the list-manipulating programs analyzed in [2] with these grammars did not lead to states containing a non-joinable critical pair.

Completion. We applied our confluence completion procedure to the non-confluent grammars shown in Table 1, i.e., `InTree`, `LinkedTree1`, `LinkedTree2`, `SimpleDLL` and `DLList`. The results of our tests are given in Table 2.

`CA` adds application conditions to rule out some non-joinable critical pairs; `RNN` introduces new nonterminals to join critical pairs. Moreover, `JGN` and `SNR` extend `RNN` by joining or using existing nonterminals. The remaining heuristics combine subsets of the above heuristics in different order (details are found in [10, 26]). We require that heuristics ending with 'L' preserve local concretizability—a property ensuring that those parts of the heap that can be manipulated by one program instruction are obtainable using a single rule application (cf. [16]).

We observe that combining heuristics increases the chance to successfully complete a grammar, as is the case for InTree, LinkedTree1 and LinkedTree2. Moreover, enforcing local concretizability can both increase runtime (see A2 vs. A2L), but also affect the completion result (see InTree, LinkedTree2, and SimpleDLL). For complex grammars, such LinkedTree2, completion is expensive, which prohibits its usage with every invocation of ATTESTOR. In such cases, it is preferable to perform completion as a preprocessing step and store the completed grammar for re-use.

4   Conclusion

We presented a tool for checking backward confluence of (certain) attributed hyperedge replacement grammars. The tool is implemented as a component of the graph-based software model checker ATTESTOR but can also be used standalone. Within ATTESTOR, checking for backward confluence of user-supplied grammars did not lead to substantial performance penalties. However, it did improve the overall verification pipeline in at least two aspects: First, incompleteness issues, i.e., failed verification attempts of correct programs; our tool enables detecting such issues before running an expensive state space generation. Second, rather than manually inspecting thousands of graphs in a generated state space, our tool creates a counterexample to backward confluence that allows fixing the supplied grammar.

Furthermore, we implemented a heuristics-driven algorithm that attempts to turn a given HRG into a backward confluent one. Although this algorithm is expensive, it can be run independently of the actual verification pipeline; if completion succeeds, the resulting grammar can be exported for future use.

A possible future improvement of our tool's performance is to detect and omit critical pairs that are not relevant for ATTESTOR's state space generation. Such an approach would amount to proving confluence up to garbage [6].

References

1. Arendt, T., Biermann, E., Jurack, S., Krause, C., Taentzer, G.: Henshin: advanced concepts and tools for in-place EMF model transformations. In: Petriu, D.C., Rouquette, N., Haugen, Ø. (eds.) MODELS 2010. LNCS, vol. 6394, pp. 121–135. Springer, Heidelberg (2010). https://doi.org/10.1007/978-3-642-16145-2_9
2. Arndt, H., Jansen, C., Katoen, J.-P., Matheja, C., Noll, T.: Let this graph be your witness!. In: Chockler, H., Weissenbacher, G. (eds.) CAV 2018. LNCS, vol. 10982, pp. 3–11. Springer, Cham (2018). https://doi.org/10.1007/978-3-319-96142-2_1
3. Arndt, H., Jansen, C., Matheja, C., Noll, T.: Graph-based shape analysis beyond context-freeness. In: Johnsen, E.B., Schaefer, I. (eds.) SEFM 2018. LNCS, vol. 10886, pp. 271–286. Springer, Cham (2018). https://doi.org/10.1007/978-3-319-92970-5_17

4. Azzi, G.G., Bezerra, J.S., Ribeiro, L., Costa, A., Rodrigues, L.M., Machado, R.: The verigraph system for graph transformation. In: Heckel, R., Taentzer, G. (eds.) Graph Transformation, Specifications, and Nets. LNCS, vol. 10800, pp. 160–178. Springer, Cham (2018). https://doi.org/10.1007/978-3-319-75396-6_9
5. Born, K., Arendt, T., Heß, F., Taentzer, G.: Analyzing conflicts and dependencies of rule-based transformations in Henshin. In: Egyed, A., Schaefer, I. (eds.) FASE 2015. LNCS, vol. 9033, pp. 165–168. Springer, Heidelberg (2015). https://doi.org/10.1007/978-3-662-46675-9_11
6. Campbell, G., Plump, D.: Confluence up to garbage. In: Gadducci, F., Kehrer, T. (eds.) ICGT 2020. LNCS, vol. 12150, pp. 20–37. Springer, Cham (2020). https://doi.org/10.1007/978-3-030-51372-6_2
7. Costa, A., et al.: Verigraph: a system for specification and analysis of graph grammars. In: Ribeiro, L., Lecomte, T. (eds.) SBMF 2016. LNCS, vol. 10090, pp. 78–94. Springer, Cham (2016). https://doi.org/10.1007/978-3-319-49815-7_5
8. Deckwerth, F.: Static Verification Techniques for Attributed Graph Transformations. Ph.D. thesis, Technische Universität, Darmstadt (2017)
9. Drewes, F., Kreowski, H.J., Habel, A.: Hyperedge replacement graph grammars. In: Handbook of Graph Grammars and Computing by Graph Transformation, vol. I: Foundations, pp. 95–162. World Scientific (1997)
10. Fesefeldt, I.: moves-rwth/attestor-confluence. https://github.com/moves-rwth/attestor-confluence/blob/master/readme/HEURISTICS.md
11. Heckel, R., Küster, J.M., Taentzer, G.: Confluence of typed attributed graph transformation systems. In: Corradini, A., Ehrig, H., Kreowski, H.-J., Rozenberg, G. (eds.) ICGT 2002. LNCS, vol. 2505, pp. 161–176. Springer, Heidelberg (2002). https://doi.org/10.1007/3-540-45832-8_14
12. Heinen, J., Jansen, C., Katoen, J.P., Noll, T.: Verifying pointer programs using graph grammars. Sci. Comput. Programm. 97(1), 157–162 (2015)
13. Hoffmann, B., Plump, D.: Implementing term rewriting by jungle evaluation. RAIRO Theor. Inform. Appl. 25, 445–472 (1991)
14. Hristakiev, I., Plump, D.: Checking graph programs for confluence. In: Seidl, M., Zschaler, S. (eds.) STAF 2017. LNCS, vol. 10748, pp. 92–108. Springer, Cham (2018). https://doi.org/10.1007/978-3-319-74730-9_8
15. Jansen, C.: Static Analysis of Pointer Programs - Linking Graph Grammars and Separation Logic. Ph.D. thesis, RWTH Aachen University, Germany (2017)
16. Jansen, C., Heinen, J., Katoen, J.-P., Noll, T.: A local greibach normal form for hyperedge replacement grammars. In: Dediu, A.-H., Inenaga, S., Martín-Vide, C. (eds.) LATA 2011. LNCS, vol. 6638, pp. 323–335. Springer, Heidelberg (2011). https://doi.org/10.1007/978-3-642-21254-3_25
17. Knuth, D., Bendix, P.: Simple word problems in universal algebra. Computational Problems in Abstract Algebra, pp. 263–297 (1970)
18. Lambers, L., Born, K., Kosiol, J., Strüber, D., Taentzer, G.: Granularity of conflicts and dependencies in graph transformation systems: a two-dimensional approach. J. Logic. Algebraic Meth. Program. 103, 105–129 (2019)
19. Lambers, L., Ehrig, H., Orejas, F.: Conflict detection for graph transformation with negative application conditions. In: Corradini, A., Ehrig, H., Montanari, U., Ribeiro, L., Rozenberg, G. (eds.) ICGT 2006. LNCS, vol. 4178, pp. 61–76. Springer, Heidelberg (2006). https://doi.org/10.1007/11841883_6
20. Lambers, L., Ehrig, H., Orejas, F.: Efficient conflict detection in graph transformation systems by essential critical pairs. Electron. Notes Theor. Comput. Sci. 211, 17–26 (2008)

21. Matheja, C.: Automated reasoning and randomization in separation logic. Ph.D. thesis, RWTH Aachen University, Germany (2020)

22. Matheja, C., Jansen, C., Noll, T.: Tree-like grammars and separation logic. In: Feng, X., Park, S. (eds.) APLAS 2015. LNCS, vol. 9458, pp. 90–108. Springer, Cham (2015). https://doi.org/10.1007/978-3-319-26529-2_6

23. Plump, D.: Confluence of graph transformation revisited. In: Middeldorp, A., van Oostrom, V., van Raamsdonk, F., de Vrijer, R. (eds.) Processes, Terms and Cycles: Steps on the Road to Infinity. LNCS, vol. 3838, pp. 280–308. Springer, Heidelberg (2005). https://doi.org/10.1007/11601548_16

24. Plump, D.: Checking graph-transformation systems for confluence. ECEASST **26** (2010)

25. Runge, O., Ermel, C., Taentzer, G.: AGG 2.0 – new features for specifying and analyzing algebraic graph transformations. In: Schürr, A., Varró, D., Varró, G. (eds.) AGTIVE 2011. LNCS, vol. 7233, pp. 81–88. Springer, Heidelberg (2012). https://doi.org/10.1007/978-3-642-34176-2_8

26. Schulte, J.: Automated Detection and Completion of Confluence for Graph Grammars. Master thesis, RWTH Aachen University, Germany (2019). http://www-i2.informatik.rwth-aachen.de/pub/index.php?type=download&pub_id=1765

27. Taentzer, G.: AGG: a graph transformation environment for modeling and validation of software. In: Pfaltz, J.L., Nagl, M., Böhlen, B. (eds.) AGTIVE 2003. LNCS, vol. 3062, pp. 446–453. Springer, Heidelberg (2004). https://doi.org/10.1007/978-3-540-25959-6_35

GrapePress - A Computational Notebook for Graph Transformations

Jens H. Weber$^{(\boxtimes)}$ (iD)

University of Victoria, Victoria, BC, Canada
`jens@acm.org`

Abstract. Computational notebooks (CNs) have gained popularity in data science, artificial intelligence, and engineering. CNs are used to document experiments *and* make them repeatable by incorporating executable segments and renderings of computational results. Graphs and computations by graph transformations have applications in many problem domains but they are not supported by current CNs. Existing graph transformation tools require a steep learning curve and are not integrated with CNs. In order to close this gap, we have developed *GrapePress* a CN that incorporates graphs and computations by graph transformations. We present the fundamental concepts for *GrapePress* and describe its use in documenting executable experiments involving graph transformations to approach real world problems in science and engineering.

Keywords: Graph transformations · Computational notebook · Tools

1 Introduction

Computational notebooks (CN) like *Colab*[1], *Jupyter*[2], and *nteract*[3] blend technical writing with interactive computation and visualization [17]. CNs have gained particular popularity in data science and artificial intelligence, as they facilitate collaborative exploration, experimentation and repeatable documentation of complex analyses [21,26]. While graphs are sometimes used for visualizing the results of computation, abstract graph-manipulations, e.g. graph rewriting by means of graph transformation (GT) rules, are not supported by current CNs. This is unfortunate, since the level of abstraction afforded by the definition of computations with GTs has shown to facilitate the analysis and solution of many complex problems in engineering and computer science [18,22,23].

There is indeed a sizable set of tools that have been developed in support of computing with GTs. These tools range from tightly integrated, visual environments (e.g., [15,16,19,24]) over extensions of integrated software development environments (IDEs) with interoperable meta models (e.g., [2,9,10]) to domain-specific languages (DSLs) that are embedded in general purpose programming

---

[1] https://colab.research.google.com/.
[2] https://jupyter.org/.
[3] https://nteract.io.

© Springer Nature Switzerland AG 2021
F. Gadducci and T. Kehrer (Eds.): ICGT 2021, LNCS 12741, pp. 294–302, 2021.
https://doi.org/10.1007/978-3-030-78946-6_16

languages (e.g., [12,27,28]). However, none of these tools combine technical writing and graph computation in a notebook paradigm [17].

This paper introduces *GrapePress*, a tool that seeks to close this gap. *GrapePress* utilizes and extends the *Graph Replacement And Persistence Engine* (*Grape*) [27] and combines it with *Gorilla*, a notebook-style REPL for Clojure.[4] Following the CN paradigm [17], *GrapePress* allows users to document experimental applications of graph transformation, while treating the notebook pages as executable artifacts, that can be used to repeat or modify experiments. Since its release into open source, *GrapePress* has been used in multiple projects in industry and academia. The rest of this paper is structured as follows: We discuss related work in the following section. Section 3 provides an overview of *GrapePress* and its fundamental concepts. Section 4 offers concluding remarks.

2   Related Work

Oakes et al. define the concept of *computational notebooks* (CN) as tools that blend the traditional concept of laboratory notebooks as a means to record designs, experiments and results with capabilities to (re-)produce computational workflows [17]. Today, CNs have particularly become popular in AI and data science [26]. They often combine textual notes with other media types like interactive charts, graphs, images and audio.

The development of CNs was inspired by earlier ideas of *literate programming* as pioneered by Knuth [13,14]. A major difference to early notions of *literate programming* lies with the *incremental interactivity* afforded by modern CNs, which allow authors to execute computational code snippets and display results "*on the fly*". This type of functionality is commonly enabled by a *Read-eval-print-loop (REPL)*, a mechanism provided by many modern programming language environments. A REPL allows programmers to run code fragments and display results of their computation [5]. The integration of REPLs in modern integrated development environments (IDEs) has made programming an interactive and exploratory activity. Indeed, the previously recommended IDE for *Grape* (the GT engine for *GrapePress*) used *InstaREPL*, a REPL that would instantly evaluate code fragments entered in the *LightTable* IDE [27].

Despite these similarities, there are significant difference between our earlier work and the work described in this paper: In contrast to programming with *Grape* within the *LightTable* IDE, *GrapePress* incorporates mechanisms for querying and visualizing graphs. *GrapePress* is geared to be used as a notebook by writers rather than to be used by software engineers. It supports markdown word processing and multimedia content. It is lightweight in the sense that it requires only a Web browser on the client side. This facilitates sharing and collaboration on documents that include graph computations by graph transformations. The goal to easily share and reproduce experiments related to research papers is related to work by Van Grorp et al., who propose the use of virtual machines (VMs) in a Web-based archive called SHARE [25]. However, in

---

[4] http://gorilla-repl.org.

contrast to CNs that truly blend documentation and computation into a form on an *executable paper*, the SHARE VMs have primarily been used as appendices to classical research papers.

3   GrapePress

3.1   Overview

GrapePress is a Web browser based CN tool using the *Grape* GT engine [27] and the notebook-style Gorilla REPL for Clojure.[5] Notebook pages in *GrapePress* (Gorilla) are called *worksheets*. Worksheets consist of *static* and *dynamic segments*. Static segments are authored in an extended *Markdown* markup language which can include text, images, video and other media. Writing a worksheet with only static segments is analogous to using other digital notebook tools. Dynamic segments contain executable code and possibly a rendering of the result of executing that code. *Grape* is well suited to be integrated into a CN paradigm, since it uses a hybrid GT definition language, i.e., rules are defined textually but visualized graphically [27]. Moreover, since *Grape* defines an internal DSL in Clojure, *GrapePress* authors can utilize the full capabilities of Clojure and Gorilla, including libraries of computational and visualization functions, beyond graphs and graph transformations. Code (re)execution is on demand by the user who can select which segments to (re)execute. The renderings of the results of code executions (e.g., graphs, charts, rule visualizations) are persisted when worksheets are saved, i.e., they are available upon re-opening the worksheet without the need to rerun the code execution. Worksheets can also be published in view-only mode to prevent unwanted changes.

The *GrapePress* language tutorial[6] itself has been written with *GrapePress* and can serve as a first example worksheet. Figure 1 shows the start of this 42 page long worksheet. Dynamic segments are framed by a grey border, with the computational code shown with grey background and the rendering of its computational result shown with a white background.

3.2   Foundations

Graph Model. We use *directed, attributed, node- and edge-labeled (danel) graphs*. Formally, a (danel) *graph* is a tuple $G : (N, E, s, t, l, a)$ where

- N and E are sets of *nodes* and *edges*, respectively;
- $s, t : E \rightarrow N$ are total functions mapping each edge to its respective *source* and *target* node, respectively;
- $l : N \cup E \rightarrow STRING$ is a partial *labeling* function mapping nodes and edges to string labels;
- $a : N \times STRING \rightarrow VAL$ is a partial *attribution* function mapping nodes and strings (attribute names) to values VAL.

---

[5] http://gorilla-repl.org.
[6] https://github.com/jenshweber/grape.

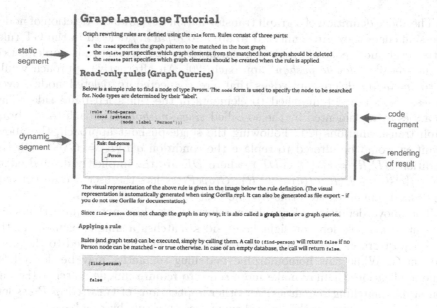

Fig. 1. A sample *GrapePress* worksheet

To facilitate rapid exploration and experimentation, *GrapePress* follows a *schema-less* approach, i.e., graphs are by default untyped and there is no need to declare a graph schema. *GrapePress* does, however, allow the user to define *graph constraints*, which can be used to ensure schema conformance as needed. The tool provides a range of built-in constraint types, e.g., for typing the source and target nodes of edges, limiting the edge cardinality, and enforcing attribute key constraints. Moreover, *Grape* provides a flexible mechanisms for users to add more complex constraints by defining graph tests to check for violations.

Graph Transformations. A *graph transformation rule* $p : (L, R)$ is a pair of graphs, commonly referred to as the *left-hand side* L and the *right-hand side* R with a defined union $L \cup R$.

A *graph transformation* is an application of a graph transformation rule to a given graph G (pre-graph) to derive a graph G' (post-graph). Formally, a graph transformation is denoted by $G \xoverset{p(o)}{\Longrightarrow} G'$, where o is a graph homomorphism $o : L \cup R \to G \cup G'$ called occurrence, such that [4]

- $o(L) \subseteq G$ and $o(R) \subseteq G'$, i.e., the left-hand side is fully contained in the pre-graph and the right-hand side is fully contained in the post-graph, and
- $o(L \backslash R) = G \backslash G'$ and $o(R \backslash L) = G' \backslash G$, i.e., only those graph elements are deleted that are matched to elements in L that do not also appear R, and only those elements are created that are matched to elements in R that do not also appear in L.

The above definition of a graph transformation disallows the deletion of nodes unless all edges they are connected to are also *explicitly* deleted in the GT rule. In the algebraic theory of graph transformations, this semantics is implemented by the so-called *double-pushout* approach [20]. An alternative approach would be to *implicitly* delete any edges that are connected to deleted nodes, even if these edges are not matched to elements in the rule's left-hand side. That semantics is implemented by the so-called *single-pushout* approach to algebraic graph transformations [20]. Following the single-pushout approach, the above definition would be altered to replace the condition $o(L \backslash R) = G \backslash G'$ with the condition $o(L \backslash R) = G \backslash (G' \cup DE)$, where DE are the implicitly deleted edges $DE = \{e | s(e) \in L \backslash R \vee t(e) \in L \backslash R\}$. *GrapePress* lets the user choose between the desired semantics on a rule-by-rule basis.

The above definition of graph transformations does not require that each element in a rule's left- or right-hand sides matches a unique element in the pre- or post-graph. For example, two nodes in L may be matched to the same node in G. While this homomorphic matching semantics may be desired for some applications, it may make more sense to require unique matches (i.e., an isomorphic matching semantics) for other applications of GTs. *GrapePress* lets the user choose between the desired semantics on a rule-by-rule basis.

(Negative) Application Conditions. The application of GTs can be further restricted by adding *application conditions* (ACs) to a rule definition. ACs need to hold when a rule is applied [8]. *GrapePress* provides two kinds of ACs: Firstly, the tool supports *attribute conditions*, which are Boolean conditions on the values of attributes of graph elements matched to a rule's left-hand side. Secondly, *GrapePress* supports *path expressions*, which are queries on the sub-graph connecting two nodes on a rule's left-hand side [1]. *GrapePress* also supports *Negative Application Conditions* (NACs), which are conditions that would prohibit the application of a graph transformation rule [11].

Operationalization of Graph Transformations. The operationalization of applying a graph transformation rule $p : (L, R)$ with a set of application conditions $AC(p)$ and a set of negative application conditions $NAC(p)$ to a pre-graph G with a set of graph constraints C is performed in the following steps:

1. **CHOOSE** a (possibly injective) occurrence o for L in G
2. **CHECK** that all application conditions in $AC(p)$ hold for o and that none of the negative application conditions in $NAC(p)$ hold for o
3. **DELETE** $o(L \backslash R)$ from G; also delete dangling edges if *single-pushout* semantics was chosen, otherwise *abort* if dangling edges exist.
4. **ADD** $o(R \backslash L)$ to the graph resulting from step 3 and **GLUE** the added elements to the context graph $o(R \cup L)$, yielding the post-graph G';
5. **VERIFY** that all graph constraints C hold on G'; *abort* if this is not the case.

Extension: Optional Elements. *GrapePress* offers extensions to the definition of graph transformation rules that have been introduced for convenience
based on practical experiences with using the language in real-world applications. *GrapePress* transformation rules can have *optional* graph elements. The
semantics of applying rules with optional elements is derived from the above
definition:

Rules with optional elements can be defined as a pair of regular GT rules
$(p : (L, R), \bar{p} : (\bar{L}, \bar{R}))$ where $L \subset \bar{L}$ and $R \subseteq \bar{R}$. The application of such a rule to
a graph G is defined by first attempting to find an occurrence for all the graph
elements (including the optional ones) $G \overset{\bar{p}(o)}{\Longrightarrow} G'$ and, if that fails, attempting to
find an occurrence for the graph elements that are not optional, i.e., $G \overset{p(o)}{\Longrightarrow} G'$.

Extension: Merged Elements. Another extension that allows for a more
concise notation in some applications is that of *merged* grape elements. Merged
elements can be used in situations where the writer wants to *create* graph elements only if they do not already exist (in context of a rule application). That
behaviour can, of course, be achieved by using negative application conditions
(at the cost of a less concise rule definition language). However, there is a notable
difference in the semantics of executing a rule with *merged* elements when compared to a NAC-controlled rule that *creates* elements: the application of the
merge rule succeeds even of the elements to be merged (created) already exist.
Formally, a graph transformation rule with merged elements can be defined as a
triple $\hat{p} : (L, R, M)$ with $M \subseteq R$ and $M \cap L = \emptyset$ representing the elements to be
merged. We can derive two regular graph transformation rules from $\hat{p}$, namely

- $\overset{+}{p} : (L, R)$ with a negative application condition *"the chosen occurrence $o(L)$
  cannot be extended to $o(L \cup M)$"* , and
- $\overset{0}{p} : (L \cup M, R)$

The semantics of applying $\hat{p}$ to a graph G is then defined as first attempting
to apply the derived regular transformation rule $\overset{0}{p}$ to G and, if unsuccessful,
attempting to apply $\overset{+}{p}$.

Control Structures. Real world applications of graph transformations often
require some form of *control structure* to govern the execution of transformation
rules [6]. Since *GrapePress* is an internal DSL in Clojure, programmers can use
regular Clojure control structures, e.g., loops, conditionals, etc. GT rules can be
parameterized with input and output parameters. However, one disadvantage
of using the regular Clojure control structures is that they do not guarantee
atomicity of composite graph operations, e.g., operations that consist of multiple
graph transformations. *GrapePress* therefore provides the concept of *transactions*
and a set of dedicated control structures that can be used to define operations
that should be executed "all or nothing". In fact, since *GrapePress* uses a graph
database (Neo4J), its transactions provide not only atomicity but all of the usual

ACID properties [3]. *GrapePress* control structures implement a backtracking-based search mechanism when executing complex transactions.

3.3 Interacting with GrapePress

As previously mentioned, *GrapePress* uses a notebook-style interaction model. Users can simply add pages (called *worksheet*) and start to write. While all user input is textual, text can be used to define non-textual content, for example by using plugins like *Mermaid*[7] to author diagrams. All worksheets created for a given project share a graph as a common data structure. *GrapePress* provides commands to render the graph's state (or a part of it). Worksheets can import other worksheets. This allows writers to utilize definitions they have made elsewhere. Indeed, worksheets are saved as valid Clojure files, which means that worksheets can also be imported by regular Clojure programs. In that case, the dynamic code segments (e.g., any defined graph transformation rules) are available to the Clojure program, while the static segments of these worksheets would simply be ignored. *GrapePress* has been used in multiple applications with industrial relevance within and outside our lab. An example application available on Github is the development of an approach to detect temporal conflicts between a set of medication prescriptions by means of temporal constraint network graphs.[7] The project including its *GrapePress* worksheet has been published at https://github.com/simbioses/chaos.

4 Conclusions

The value proposition of computational notebooks (CN) is that they provide a lightweight means to document experiments and make them repeatable and executable at the same time. Graphs and computations by GTs are widely applicable to many industrial problems. However, current GT tools and environments often require a steep learning curve, do not readily integrate with notebooks, and generally do not provide a lightweight experience akin to that offered by browser-based applications. *GrapePress* has been designed to close this gap. Users only require a Web browser to use the *GrapePress* notebook. While there is currently no publicly available server that hosts *GrapePress* notebooks, such servers can easily be set up using the instructions posted at the *GrapePress* github site.

References

1. Abiteboul, S., Vianu, V.: Regular path queries with constraints. J. Comput. Syst. Sci. **58**(3), 428–452 (1999)
2. Arendt, T., Biermann, E., Jurack, S., Krause, C., Taentzer, G.: Henshin: advanced concepts and tools for in-place EMF model transformations. In: Petriu, D.C., Rouquette, N., Haugen, Ø. (eds.) MODELS 2010. LNCS, vol. 6394, pp. 121–135. Springer, Heidelberg (2010). https://doi.org/10.1007/978-3-642-16145-2_9

---

[7] https://mermaid-js.github.io.

3. Baldan, P., Corradini, A., Foss, L., Gadducci, F.: Graph transactions as processes. In: Corradini, A., Ehrig, H., Montanari, U., Ribeiro, L., Rozenberg, G. (eds.) ICGT 2006. LNCS, vol. 4178, pp. 199–214. Springer, Heidelberg (2006). https://doi.org/10.1007/11841883_15

4. Baresi, L., Heckel, R.: Tutorial introduction to graph transformation: a software engineering perspective. In: Corradini, A., Ehrig, H., Kreowski, H.-J., Rozenberg, G. (eds.) ICGT 2002. LNCS, vol. 2505, pp. 402–429. Springer, Heidelberg (2002). https://doi.org/10.1007/3-540-45832-8_30

5. van Binsbergen, L.T., Verano Merino, M., Jeanjean, P., et al.: A principled approach to REPL interpreters. In: ACM SIGPLAN, pp. 84–100 (2020)

6. Bunke, H.: Programmed graph grammars. In: Claus, V., Ehrig, H., Rozenberg, G. (eds.) Graph Grammars 1978. LNCS, vol. 73, pp. 155–166. Springer, Heidelberg (1979). https://doi.org/10.1007/BFb0025718

7. Dechter, R., Meiri, I., Pearl, J.: Temporal constraint networks. AI **49**(1–3), 61–95 (1991)

8. Ehrig, H., Habel, A.: Graph grammars with application conditions. In: The Book of L, pp. 87–100. Springer, Berlin (1986). https://doi.org/10.1007/978-3-642-95486-3_7

9. Fritsche, L., Kulcsár, G.: eMoflon: a tool for tools and transformations. Modellierung **2018** (2018)

10. Giese, H., Hildebrandt, S., Lambers, L.: Toward bridging the gap between formal semantics and implementation of triple graph grammars. In: Workshop on Model-Driven Engineering, Verification, and Validation, pp. 19–24. IEEE (2010)

11. Habel, A., Heckel, R., Taentzer, G.: Graph grammars with negative application conditions. Fundamenta Informaticae **26**(3, 4), 287–313 (1996)

12. Hinkel, G., Goldschmidt, T.: Tool support for model transformations: on solutions using internal languages. Modellierung **2016** (2016)

13. Kery, M.B., Radensky, M., Arya, M., et al.: The story in the notebook: exploratory data science using a literate programming tool, pp. 1–11. ACM (2018)

14. Knuth, D.E.: Literate programming. Comput. J. **27**(2), 97–111 (1984)

15. de Lara, J., Vangheluwe, H.: Defining visual notations and their manipulation through meta-modelling and graph transformation. J. Vis. Lang. Comput. **15**(3–4), 309–330 (2004)

16. Nickel, U., Niere, J., Zündorf, A.: The FUJABA environment. In: ICSE, pp. 742–745 (2000)

17. Oakes, B.J., Franceschini, R., Van Mierlo, S., et al.: The computational notebook paradigm for multi-paradigm modeling. In: MODELS, pp. 449–454. IEEE (2019)

18. Westfechtel, B.: AGTIVE'03: summary from a tool builder's viewpoint. In: Pfaltz, J.L., Nagl, M., Böhlen, B. (eds.) AGTIVE 2003. LNCS, vol. 3062, pp. 493–495. Springer, Heidelberg (2004). https://doi.org/10.1007/978-3-540-25959-6_44

19. Rensink, A.: The GROOVE simulator: a tool for state space generation. In: Pfaltz, J.L., Nagl, M., Böhlen, B. (eds.) AGTIVE 2003. LNCS, vol. 3062, pp. 479–485. Springer, Heidelberg (2004). https://doi.org/10.1007/978-3-540-25959-6_40

20. Rozenberg, G.: Handbook of Graph Grammars and Computing by Graph Transformation, vol. 1. World Scientific, Singapore (1997)

21. Rule, A., Tabard, A., Hollan, J.D.: Exploration and explanation in computational notebooks, pp. 1–12. ACM, New York (2018)

22. Rensink, A., Taentzer, G.: AGTIVE 2007 graph transformation tool contest. In: Schürr, A., Nagl, M., Zündorf, A. (eds.) AGTIVE 2007. LNCS, vol. 5088, pp. 487–492. Springer, Heidelberg (2008). https://doi.org/10.1007/978-3-540-89020-1_33

23. Schürr, A., Rensink, A.: Software and systems modeling with graph transformations theme issue. Softw. Syst. Model. **13**(1), 171–172 (2014)
24. Schürr, A., Winter, A.J., Zündorf, A.: Graph grammar engineering with PROGRES. In: Schäfer, W., Botella, P. (eds.) ESEC 1995. LNCS, vol. 989, pp. 219–234. Springer, Heidelberg (1995). https://doi.org/10.1007/3-540-60406-5_17
25. Van Gorp, P., Mazanek, S.: Share: a web portal for creating and sharing executable research papers. Proc. Comput. Sci. **4**, 589–597 (2011)
26. Wang, A.Y., Mittal, A., et al.: How data scientists use computational notebooks for real-time collaboration. In: Proceedings of ACM HCI 3(CSCW), November 2019
27. Weber, J.H.: GRAPE – a graph rewriting and persistence engine. In: de Lara, J., Plump, D. (eds.) ICGT 2017. LNCS, vol. 10373, pp. 209–220. Springer, Cham (2017). https://doi.org/10.1007/978-3-319-61470-0_13
28. Zündorf, A., George, T., Lindel, S., et al.: Story driven modeling libary (SDMLib). In: 6th Transformation Tool Contest (TTC 2013), ser. EPTCS (2013)

Author Index

Printed in the United States
by Baker & Taylor Publisher Services